Lecture Notes in Physics

Springer-Verlag Berlin Heidelberg GmbH

The Editorial Policy for Proceedings

The series Lecture Notes in Physics reports new developments in physical research and teaching – quickly, informally, and at a high level. The proceedings to be considered for publication in this series should be limited to only a few areas of research, and these should be closely related to each other. The contributions should be of a high standard and should avoid lengthy redraftings of papers already published or about to be published elsewhere. As a whole, the proceedings should aim for a balanced presentation of the theme of the conference including a description of the techniques used and enough motivation for a broad readership. It should not be assumed that the published proceedings must reflect the conference in its entirety. (A listing or abstracts of papers presented at the meeting but not included in the proceedings could be added as an appendix.)

When applying for publication in the series Lecture Notes in Physics the volume's editor(s) should submit sufficient material to enable the series editors and their referees to make a fairly accurate evaluation (e.g. a complete list of speakers and titles of papers to be presented and abstracts). If, based on this information, the proceedings are (tentatively) accepted, the volume's editor(s), whose name(s) will appear on the title pages, should select the papers suitable for publication and have them refereed (as for a journal) when appropriate. As a rule discussions will not be accepted. The series editors and Springer-Verlag will normally not interfere with the detailed editing except in fairly obvious cases or on technical matters.

Final acceptance is expressed by the series editor in charge, in consultation with Springer-Verlag only after receiving the complete manuscript. It might help to send a copy of the authors' manuscripts in advance to the editor in charge to discuss possible revisions with him. As a general rule, the series editor will confirm his tentative acceptance if the final manuscript corresponds to the original concept discussed, if the quality of the contribution meets the requirements of the series, and if the final size of the manuscript does not greatly exceed the number of pages originally agreed upon. The manuscript should be forwarded to Springer-Verlag shortly after the meeting. In cases of extreme delay (more than six months after the conference) the series editors will check once more the timeliness of the papers. Therefore, the volume's editor(s) should establish strict deadlines, or collect the articles during the conference and have them revised on the spot. If a delay is unavoidable, one should encourage the authors to update their contributions if appropriate. The editors of proceedings are strongly advised to inform contributors about these points at an early stage.

The final manuscript should contain a table of contents and an informative introduction accessible also to readers not particularly familiar with the topic of the conference. The contributions should be in English. The volume's editor(s) should check the contributions for the correct use of language. At Springer-Verlag only the prefaces will be checked by a copy-editor for language and style. Grave linguistic or technical shortcomings may lead to the rejection of contributions by the series editors. A conference report should not exceed a total of 500 pages. Keeping the size within this bound should be achieved by a stricter selection of articles and not by imposing an upper limit to the length of the individual papers. Editors receive jointly 30 complimentary copies of their book. They are entitled to purchase further copies of their book at a reduced rate. As a rule no reprints of individual contributions can be supplied. No royalty is paid on Lecture Notes in Physics volumes. Commitment to publish is made by letter of interest rather than by signing a formal contract. Springer-Verlag secures the copyright for each volume.

The Production Process

The books are hardbound, and the publisher will select quality paper appropriate to the needs of the author(s). Publication time is about ten weeks. More than twenty years of experience guarantee authors the best possible service. To reach the goal of rapid publication at a low price the technique of photographic reproduction from a camera-ready manuscript was chosen. This process shifts the main responsibility for the technical quality considerably from the publisher to the authors. We therefore urge all authors and editors of proceedings to observe very carefully the essentials for the preparation of camera-ready manuscripts, which we will supply on request. This applies especially to the quality of figures and halftones submitted for publication. In addition, it might be useful to look at some of the volumes already published. As a special service, we offer free of charge LATEX and TEX macro packages to format the text according to Springer-Verlag's quality requirements. We strongly recommend that you make use of this offer, since the result will be a book of considerably improved technical quality. To avoid mistakes and time-consuming correspondence during the production period the conference editors should request special instructions from the publisher well before the beginning of the conference. Manuscripts not meeting the technical standard of the series will have to be returned for improvement.

For further information please contact Springer-Verlag, Physics Editorial Department II, Tiergartenstrasse 17, D-69121 Heidelberg, Germany

E. Martínez -González   J. L. Sanz (Eds.)

The Universe at High-z, Large-Scale Structure and the Cosmic Microwave Background

Proceedings of an Advanced Summer School
Held at Laredo, Cantabria, Spain, 4–8 September 1995

 Springer

Editors

Enrique Martínez-González
Jose Luis Sanz
Instituto de Física de Cantabria
Facultad de Ciencias
Av. de Los Castros
E-39005 Santander, Spain

Cataloging-in-Publication Data applied for.

Die Deutsche Bibliothek - CIP-Einheitsaufnahme

The universe at high-z, large scale structure and the cosmic microwave background : proceedings of a advanced summer school, held at Laredo, Cantabria, Spain, 4 - 8 September 1995 / E. Martínez-González ; J. L. Sanz (ed.).

(Lecture notes in physics ; Vol. 470)
  ISBN 978-3-662-14084-0   ISBN 978-3-540-68386-5 (eBook)
  DOI 10.1007/978-3-540-68386-5
NE: Martínez-González, Enrique [Hrsg.]; GT

ISBN 978-3-662-14084-0

Typesetting: Camera-ready by the authors
Cover design: Design & Production, Heidelberg
SPIN: 10520060   55/3142-543210 - Printed on acid-free paper

Preface

The advanced summer school "The Universe at High-z, Large-Scale Structure and the Cosmic Microwave Background" was held in Laredo (Cantabria, Spain), September 4 - 8 1995, at the XI Cursos de Verano de Laredo. The aim was to review and discuss, in a pedagogical way, the most recent developments in the field and the future perspectives. The topics covered in the 26 one-hour lectures included: QSO absorption systems, identification of objects at high redshift, radiogalaxies, galaxy formation and evolution, galaxy number counts, clustering, theories of structure formation, large-scale structure and streaming motions, gravitational lensing, and spectrum and anisotropies of the cosmic microwave background radiation. During the school, observational developments, data analysis, and theoretical aspects were covered. The lectures were delivered by the following speakers: T. Broadhurst (Univ. of California, Berkeley), L. Cayón (LBL & Univ. of California, Berkeley), S. Charlot (IAP, Paris), R. Giovanelli (Cornell Univ., Ithaca), W. Hu (Princeton Univ., Princeton), E. Martínez-González (IFCA, Santander), S. Matarrese (Univ. of Padova, Padova), G. Miley (Leiden Univ., Leiden), R. Rebolo (IAC, Tenerife), P. Schneider (MPIfA, Munich), J. L. Sanz (IFCA, Santander), J. Silk (Univ. of California, Berkeley), M. Dickinson (STScI, Baltimore), and D. Tytler (Univ. of California, San Diego).

Cosmology has dramatically evolved during the last decade. On the one hand, data coming from 4-m telescopes and more recently the 8-m Keck telescope and the Hubble Space Telescope (HST), have changed our ideas about the universe considerably (e.g. the large-scale structure and, more recently, imaging and spectroscopy of galaxies up to redshift $z \simeq 1$). On the other hand, there has been a certain theoretical development (e.g., theories of galaxy formation in connection with the early universe, gravitational lensing) dealing with different topics. These new developments motivated us to promote a school covering all of these new ideas and observations in a pedagogical way, with the lectures given by seniors and young researchers, to account for different views.

These proceedings contain the review talks given in the advanced summer school. The organizers express their cordial thanks to all participants and specially to our speakers who kindly accepted our invitation. We are also indebted to the sponsoring institutions: Universidad de Cantabria, XI Cursos de Verano de Laredo, and also M.E.C. and Facultad de Ciencias (Univ. de Cantabria).

Santander, March 1996

E. Martínez-González
J.L. Sanz

Contents

I

The Universe at High-z

Galaxy Evolution from Cluster and Absorption-Selected Samples

Mark Dickinson

Space Telescope Science Institute, 3700 San Martin Dr., Baltimore MD 21218 USA

Abstract. Rich clusters of galaxies and QSO absorption line systems provide two alternative venues for studying the evolution of galaxies at high redshift. Here I review some recent observational results in both areas of investigation, and consider their implications for the history of galaxy evolution since $z \sim 1$.

1 Introduction: Approaches to Distant Galaxies

The past few years have seen an explosion of observational data on the properties of ordinary galaxies at large redshifts. Regimes of cosmic lookback time previously known only through active objects like radio galaxies and QSOs are now being rapidly populated with ordinary galaxies selected by a variety of techniques.

The most general approach to the study of distant field galaxies is via deep magnitude–limited redshift surveys. The largest such surveys (e.g. the Canada–France Redshift Survey of Lilly *et al.* 1995 and the Autofib survey described by Ellis *et al.* 1996) are now sufficient to provide a reasonable characterization of the galaxy luminosity function and its evolution out to redshifts ~ 1.

Magnitude–limited field galaxy surveys have the advantage of conceptual simplicity – one simply selects galaxies down to some brightness limit and then goes out and measures redshifts. Two aspects complicate the interpretation of such data. Firstly, with a fixed apparent magnitude limit, any survey will sample different ranges of *absolute* magnitudes at different redshifts. Therefore some care must be taken when inter–comparing galaxy samples across a wide redshift range – at the most distant redshift limits probed by a magnitude limited survey, only the very brightest galaxies will be represented. Secondly, galaxies are typically selected in a single photometric passband. As redshift increases, the selection passband is in fact measuring rest–frame emission from shorter and shorter wavelengths. Because star formation produces hot blue stars while old stellar populations are red, magnitude limited surveys may be increasingly subject to redshift dependent biases toward more actively star forming objects. Recent surveys have tried to minimize this by selecting galaxies at longer wavelengths such as the I or K–bands.

Here I will consider other means of selecting high redshift galaxies: through the study of rich clusters of galaxies, and from QSO absorption line systems. Each has its own advantages and disadvantages when compared to blind field surveys, and provides complementary information.

2   Galaxy Evolution in Distant Clusters

Before the advent of efficient multiplexing spectrographs made large field galaxy surveys possible, rich galaxy clusters provided the best–studied samples of high redshift galaxies. Initially, this was largely because they provided an easier observational target. Very rich clusters can be recognized in deep images out to $z \sim 1$. Once redshifts have been measured for a few galaxies in the cluster, one can (optimistically) assume that many other galaxies visible in the image are also located at the same distance. This is only a statistical assumption – one must still correct for the projected foreground/background population when quantifying properties (colors, luminosity functions, etc.) of cluster galaxies from imaging data alone. A cluster then presents a volume–limited sample of galaxies collected together in one convenient place for investigation.

However, rich clusters do not offer not typical samples of galaxies. In the local universe there is a strong morphology–density relation: the core regions of clusters are heavily dominated by early–type galaxies (ellipticals and S0s), and exhibit a morphological mix quite different from that of the field. As rare and highly over–dense regions of space, rich clusters are far from representative of the general galaxy population, and it would be risky to draw conclusions too broadly from studying the evolution of cluster galaxies. The processes at work in rich clusters which led to todays morphology–density relation may be quite different from those which drive galaxy evolution in the field. On the other hand, elliptical galaxies probably represent the oldest galactic stellar populations at any epoch and are therefore interesting in their own right. Any measure of spectral evolution in the elliptical population may thus point back to the earliest epoch of galaxy formation.

Here I will briefly consider two aspects of galaxy evolution in distant clusters of galaxies: the fate of the disappearing "Butcher–Oemler" blue galaxy population, and the history of the red cluster ellipticals.

2.1   The Butcher–Oemler Effect

It was in galaxy clusters that the first direct evidence for galaxy evolution was observed by Butcher and Oemler (1978, 1984 and subsequent papers), who found an increasingly numerous population of blue cluster galaxies at higher redshifts. This "Butcher–Oemler effect" was initially established through imaging photometry alone (using clusters whose redshifts had been measured with only a few galaxies each), but was later confirmed by more extensive multiobject spectroscopy. The blue cluster galaxies were shown to exhibit a range of spectral properties, ranging from emission line spectra characteristic of ongoing star formation, to galaxies dominated by strong Balmer absorption lines. These latter have been termed "post–starburst" or "E+A" galaxies, and their spectra suggest that they are being observed some time after the termination of a strong burst of star formation.

Most recently, *HST* imaging of clusters at $z \approx 0.4$ has shed new light on the nature of the blue galaxies in high redshift clusters by revealing their kpc–scale morphologies (Dressler *et al.* 1994*a*, *b*; Couch *et al.* 1994). In general, the blue galaxies are disk systems – many are readily classifiable as Hubble Sequence spirals. However, Oemler *et al.* (1996) have noted that the star formation patterns seen in the spiral arms of these spirals often follows unusual patterns (e.g. rings) which are uncommon in nearby field spirals, as if they are suffering some sort of disturbance. Spatially, the blue galaxies are widely distributed throughout the cluster, while the red E/S0s are more strongly concentrated toward the cluster center. In this regard, the high–z clusters are similar to those nearby – there is a morphology–density relation, or perhaps a morphology–radius relation, at work which segregates galaxy types according to the nature of their local environment. However, Oemler *et al.* describe the difference between low– and high–z clusters as being an offset in the "normalization" of that relation, such that there are simply more blue, disk galaxies present in high–redshift clusters at all radii.

At present, the physical processes which removed disk galaxies from rich clusters by the present day, or which transformed them morphologically, are unknown. A tempting suggestion (e.g. Lavery & Henry 1988) is that galaxy–galaxy interactions and mergers may have transformed the spirals into the S0s and ellipticals which dominate today. While indications of interactions are indeed found in the *HST* images, Oemler *et al.* argue that this is unlikely to be solely responsible for driving the observed evolution. The majority of the disk galaxies do not show direct morphological evidence for interaction. There is no clear indication that blue cluster galaxies are any more likely to have close companions than do red ones. Moreover, on theoretical grounds, ordinary interactions and mergers should be inefficient in the high–velocity environment of a rich cluster. Finally, the remarkable homogeneity observed for the spectrophotometric properties of cluster ellipticals (see below) seems to allow little room for them to be the end–product of whatever evolution eliminates the star–forming disk galaxies.

Recently, Moore *et al.* (1996) have proposed that cluster spirals are eliminated through a process which they refer to as "galaxy harassment." Although genuine mergers may be uncommon in the high–velocity environment of rich clusters, repeated impulsive encounters with other galaxies can cause serious damage to galaxy disks. The global tidal field of a cluster can drive bar instabilities in spirals, and high speed interactions with other galaxies strip away substantial fractions of the dark matter halos. Moore *et al.* simulate this process with numerical N–body + hydrodynamic models, and find that during the process of harassment, angular momentum in the stellar and gaseous content of disk galaxies is lost to the dark matter halo and to other cluster galaxies. The spirals cease to be strongly rotationally supported, the stars in the galaxy are dynamically heated, and gas is driven to the center of the galaxy where it may fuel starburst activity. The end–product is something which resembles a dwarf spheroidal, a galaxy type which dominates the faint end of the luminosity function in rich clusters today.

The harassment scenario must face further tests, both observational and theoretical. The models have only been tested with static cluster potentials, whereas the actual environment of clusters may have been considerably less dynamically relaxed in their high–redshift youth. The faint end of the cluster galaxy population at high redshift has yet to be characterized – under this scenario, one would expect to see a *decrease* in the population of dwarf spheroidal galaxies at higher redshift. Moreover, the stellar populations of cluster dwarf spheroidals today should reflect their relatively recent origins (although the stars in these remnants should be comprised of both an older population left over from the progenitor plus younger stars created during any starburst phase which may accompany the disruption process). Overall, however, harassment provides an attractive alternative to traditional merging or stripping scenarios, and one which has the virtue (see below) of leaving the bright E/S0 galaxies relatively unscathed.

2.2 Distant Cluster Ellipticals

The traditional view of giant elliptical (gE) galaxies has them forming at high redshift with a single, major episode of star formation, followed thereafter by a long period of simple, passive evolution. Ellipticals obey a tidy color–magnitude (c–m) sequence, with smaller, fainter galaxies exhibiting bluer colors. This is generally interpreted as a mass–metallicity relation. The small dispersion in gE colors around the mean c–m relation has been cited as evidence that the galaxies formed quickly and within a time interval that is short relative to their present ages. Bower *et al.* (1992) have compared the color–magnitude sequences of E/S0 galaxies in the Coma and Virgo clusters, concluding that the galaxies in those two clusters at least are uniformly old and coeval, with little residual star formation at later times.

This simple, traditional picture of ellipticals has been challenged. Detailed spectral analyses of age and metallicity indicators suggest that ellipticals may have formed over a broad interval of cosmic time, with many galaxies displaying signs of substantial star formation as recently as a few Gyr ago (cf. Worthey *et al.* 1995). In addition, careful scrutiny of ellipticals often reveals shells and other non–uniformities in their light profiles which probably result from the ingestion of other galaxies. This has led to the extreme alternative viewpoint that many ellipticals formed at late times as the byproducts of major mergers between disk galaxies. Numerical simulations show that the products of such mergers resemble elliptical galaxies, with spheroidal stellar distributions and $r^{1/4}$–law light profiles.

We may hope to make progress in understanding the evolutionary history of elliptical galaxies by identifying and studying them at higher redshifts, and thus at younger cosmic times. Distant clusters of galaxies provide a natural laboratory for doing so. Just as with field galaxy surveys, however, it is advantageous to select cluster galaxies at long rest frame wavelengths in order to minimize redshift–dependent biases toward recently star–forming objects. At redshifts approaching or exceeding unity, optical photometry and colors, even in the I–band, may result in an apparent proportional enhancement of bluer, star forming galaxies, and conversely may select against galaxies exhibiting the reddest, quiescent stellar populations. Infrared (e.g. K–band) selection is a natural means of avoiding such selection effects. The long–wavelength spectra ($\lambda \gtrsim 7000$Å) of mature galaxies are dominated by light from old, evolved stars, primarily red giants, and are minimally subject to perturbation by recent star formation – nearly all normal galaxies have similar near–infrared spectra unless their stellar mass is completely dominated by a very young population. Therefore near–IR selection should choose comparable mixtures of galaxies out to quite large redshifts ($z \sim 2$), and long wavelength galaxy colors will best probe the condition of the dominant old stellar populations with minimum perturbation from small traces of recent star formation.

Infrared imaging of galaxy clusters has become possible thanks to the advent of new, large–format infrared arrays, and there is an increasing literature which studies the properties of cluster galaxies selected in this manner. It should be noted, however, that there is not as yet an unbiased sample of *clusters* selected from a purely infrared–based survey – the clusters which have been studied thus far have generally been selected optically or from flux–limited x–ray samples. The radio galaxy clusters discussed below were identified from infrared images, but may not represent an unbiased sample because of their radio selection. Truly unbiased infrared surveys to faint limits must await an order of magnitude increase in array size before they become feasible.

2.2.1 Infrared studies of distant clusters. The first studies of distant clusters using IR arrays were presented by Aragón–Salamanca *et al.* (1991,1993), who observed a sample of distant, optically–selected clusters. They found that the $V - K$ and $I - K$ colors of cluster galaxies exhibit a prominent red sequence (presumably the E/S0s), and that at high redshifts this sequence shifts toward bluer colors. At $z \approx 0.9$ they measured few, if any, galaxies as red as present–day ellipticals.

Adam Stanford, Peter Eisenhardt and I have been carrying out an extensive infrared/optical survey of galaxy clusters spanning a very wide range of redshifts. We obtain data at J, H and K, as well as in two optical bands chosen to bracket $\lambda_0 4000$Å at the cluster redshift. The images typically cover a field of view $\sim$2 Mpc on a side at the cluster redshift, and reach 2 mag below present–day L^* with $S/N > 5$. To date, we have observed more than 40 clusters, ranging from nearby Coma out to $z = 0.9$, the redshift limit of existing cluster catalogs.

In order to extend this work to still higher redshifts and earlier lookback times, Eisenhardt and I have undertaken a systematic survey for galaxy clusters around radio sources in the redshift range $0.8 < z < 1.4$. It has been known for some time that radio galaxies and radio loud quasars at $z \approx 0.5$ often inhabit rich environments (cf. Yee & Green 1984; Ellingson, Yee and Green 1991; Yates *et al.* 1989; Hill & Lilly 1991; Dickinson 1994). We may hope that this holds true at higher redshifts still, and if so then take advantage of the large number of radio galaxies presently known at $z > 1$ to provide targets for a cluster hunt. The radio galaxy also provides an *a priori* likely redshift for any cluster candidate which is found.

We have found a number of excellent cluster candidates from this survey. One of the most dramatic examples is $z = 1.206$ radio galaxy 3C 324. Many extremely red $(R - K \approx 6)$ galaxies surround 3C 324, representing a substantial excess over the number expected from the general field. I consider their colors in greater detail below; here, I simply note that they have roughly the colors expected for weakly evolved elliptical galaxies at $z \approx 1.2$. In 1994 we obtained an extremely deep HST image of the 3C 324 field which showed that these objects are, for the most part, morphologically early–type galaxies with conventional $r^{1/4}$–law surface brightness profiles. We have subsequently carried out an extensive program of spectroscopy from the Keck telescope and have confirmed that many of the surrounding galaxies do indeed share the redshift of the radio galaxy.

2.2.2 Evolution in the color–magnitude plane. Figures 1–3 present optical–infrared color–magnitude (c–m) diagrams for clusters at $z = 0.4, 0.9$ and 1.2. For all of these clusters, *HST* images allow elliptical galaxies to be selected on the basis of their morphologies; their colors can then be studied apart from the remainder of the cluster population.

In all cases, a narrow sequence of red early–type galaxies is observed in the color–magnitude plane. The scatter in the c–m diagrams remains remarkably tight at all redshifts. For E/S0 galaxies in the Coma and Virgo clusters, Bower *et al.* measured an intrinsic $\delta(V - H)$ color scatter of only 0.05 mag around the mean c–m regression line. For the $z \approx 0.4$ clusters, where our observed bandpasses very nearly match rest–frame V and H, we find an intrinsic scatter of only 0.06 mag (for galaxies brighter than L^*, where photometric errors are small). For the higher redshift clusters, the bandpasses used here sample somewhat bluer rest–frame wavelengths where star formation might have a greater effect. Nevertheless, when excluding a few unusually *red* outliers (which are very likely background objects, or perhaps heavily reddened galaxies), the scatter remains small: 0.07 mag in $I - K$ at $z = 0.9$, and 0.07 mag in $R - K$ for $z = 1.2$. This color homogeneity at all redshifts suggests a remarkable degree of coevality among cluster ellipticals. Neglecting extinction, galaxy colors reflect a combination of age and metallicity. Therefore, following the arguments of Bower *et al.* , a measured color dispersion at a given luminosity sets an upper limit to the fractional dispersion in galaxy ages $\delta t/t$. The consistently small color dispersion measured in distant clusters implies that $\delta t/t$ changes very little with increasing lookback time, despite the fact that the age of the universe (an upper limit to t) becomes smaller. Variations in metallicity or extinction would only increase the range of observed colors, requiring still tighter limits on the true dispersion in ages.

As noted by Aragón–Salamanca *et al.* , the colors of the E/S0 galaxies do become bluer with increasing redshift. At $z \approx 0.4$, this change is small – Stanford *et al.* compare the colors shown in Figure 1 to observed $(V - H)$ colors

9

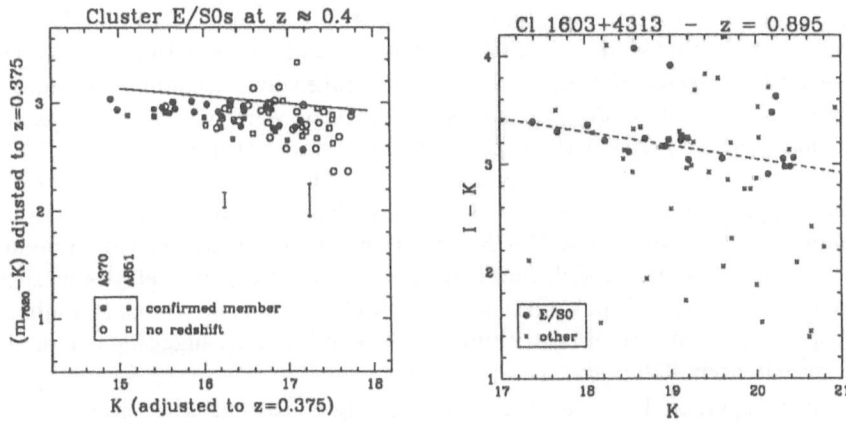

Fig. 1. *Left:* Optical–IR color–magnitude diagram for galaxies in Abell 370 ($z = 0.375$) and Abell 851 ($z = 0.407$), adapted from Stanford, Eisenhardt & Dickinson 1995. Here, only galaxies with E/S0 morphologies (classified by A. Oemler from *HST* images) have been plotted. The A851 photometry has been adjusted slightly to match the redshift of A370. Representative $\pm 1\sigma$ error bars are shown. The solid line traces the mean $(V - H)$ vs. H c–m relation for E/S0s in Coma as it would appear if redshifted to $z = 0.375$.

Fig. 2. *Right:* Color–magnitude diagram for Cl 1603+4313, a cluster at $z = 0.895$ observed with *HST* by Westphal. E/S0 galaxies are marked as filled circles. The dashed line is a fit to the E/S0 c–m relation (see text).

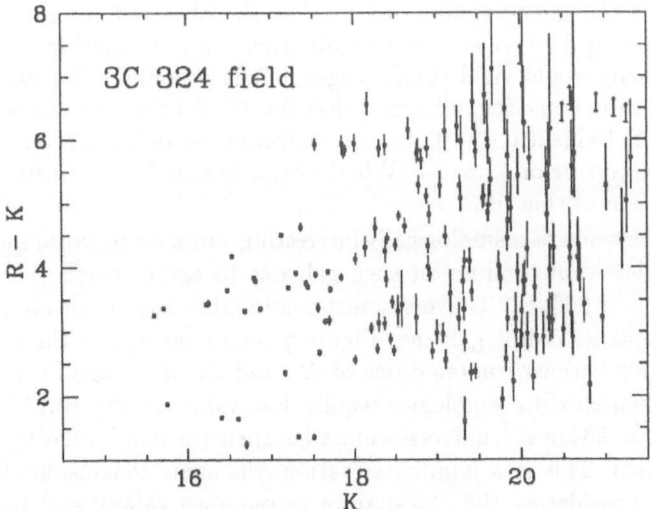

Fig. 3. C–m diagram for the 3C 324 field, showing the narrow red sequence at $R - K \approx 5.9$. Most of the galaxies in this red "finger" have simple E/S0 morphologies.

of Coma cluster galaxies and find a blueward shift of only -0.13 mag for A370 and -0.18 mag for A851 – a small shift significant at only the 2 to 3σ level when possible systematic errors are carefully taken into account. Therefore for a lookback time of $\sim 5h_{50}^{-1}$ Gyr, cluster ellipticals were nearly as red as they are today, suggesting that their formation must have taken place far earlier. At $z = 0.9$, the observed $(I - K)$ c–m relation is ~ 0.5 mag bluer than the observed colors (approximately rest–frame $B - J$) for Coma galaxies. Unfortunately we cannot directly compare the $R - K$ measurements for 3C 324 to Coma because the bandpasses do not match wavelengths where Coma photometry is presently available. However, comparison to spectral models (which do match IUE ultraviolet spectrophotometry of local ellipticals reasonably well) suggests a color shift of approximately -0.6 mag.

Taken together, these results are reasonably consistent with simple passive evolution of a simple, "single–burst" stellar population. The fact that E/S0 colors remain as red as they do, out to $z = 1.2$, strongly suggests that cluster ellipticals were *already mature* at those large look–back times. As Stephane Charlot has pointed out in his lectures at this school, age–dating a galaxy from integrated colors alone is indefinite, since color necessarily reflects a variety of parameters including age, metallicity, the star formation history, and the form of the stellar initial mass function. Therefore we cannot assign a definite age to the $z = 1.2$ galaxies we observe. However, we can reasonably assume that the fastest way to make a stellar population redden (again, barring extinction) is to let it age passively with no further star formation. Figure 4 plots predicted the $R - K$ color of a galaxy *as observed at* $z = 1.2$ as a function of age for solar metallicity models (computed using the Bruzual & Charlot 1993 code). The single–burst model reddens the most rapidly, and reaches the observed color ($\langle R - K \rangle = 5.9$) approximately 3.3 Gyr after the termination of star formation. Models with higher metallicity would yield younger ages, but it is unlikely (for galaxies with metallicities in the range found locally) that the 3C 324 galaxies can be less than a few Gyr old. Evidently, the bulk of star formation in the 3C 324 ellipticals ceased several Gyr before $z = 1.2$. Whether this is true for all clusters remains an open question at this time.

This is conclusion is cosmologically interesting, since some world models simply do not allow enough time for such galaxies to age to reach the colors we observe. At $z = 1.206$, the Universe must presumably have been old enough to accommodate as its oldest galaxies. Figure 5 shows the age of the universe at this redshift for various combinations of H_0 and q_0. If an age of 3.3 Gyr age is adopted, then closed cosmologies require low values of the Hubble parameter, while $H_0 = 80 \, \mathrm{km \, s^{-1}}$ universes must be open (or dominated by a cosmological constant). This is a familiar situation: the same conclusions have been reached from considering the oldest stars in our own galaxy and its globular clusters. Globular cluster ages are presumably more reliable (or at least better constrained) than are those of faint galaxies estimated from a single color. What is striking here is not the actual values of H_0 and q_0 which are allowable (since these require considerable faith in the choice of population synthesis models),

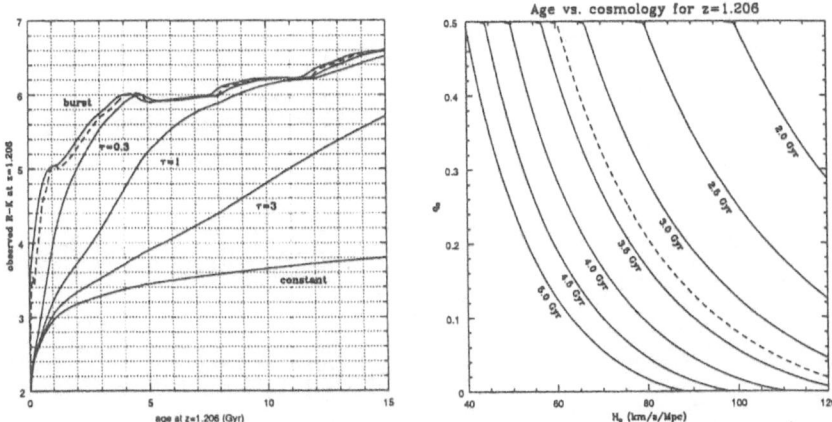

Fig. 4. (left) $R - K$ colors (as observed at $z = 1.206$) for synthetic galaxies with various star formation histories, generated with the population synthesis models of Bruzual and Charlot (1993). Models with exponentially declining declining star formation rates are labeled by their e–folding time τ in Gyr. The single burst models redden most quickly, and reach $R - K = 5.9$ after ~3.3 Gyr.

Fig. 5. (right) Loci of constant universal age at $z = 1.206$ vs. the fundamental cosmological parameters in a $\Lambda = 0$ universe. Tracks mark acceptable combinations of q_0 and H_0 which produce the labeled age. The fiducial 3.3 Gyr is marked by a dashed line.

but rather the persistence of the age problem from nearby stellar clusters out to galaxies at $z = 1.2$. It seems that we are faced either with a genuine cosmological constraint, or with the need to adjust stellar evolutionary models in order to rescale ages at all redshifts.

2.2.3 Structural parameters and luminosity evolution. *HST* images now allow us to make quantitative structural comparisons between high–z ellipticals and their present–day counterparts. Ideally, one would like to measure the fundamental plane at high redshift. This must await high signal–to–noise spectroscopy for measuring velocity dispersions: several such efforts are now underway (cf. Franx 1993). In the meanwhile, we can consider projections of the fundamental plane which depend only on morphological/photometric parameters. One example is the observed anti–correlation between galaxy size and surface brightness (Kormendy 1977). Figure 6 plots effective radii r_e and surface brightnesses $\langle \mu \rangle_e$ in the B–band for a sample of nearby elliptical galaxies compiled by Sandage & Perelmuter (1990). Data for high–z cluster ellipticals measured from *HST* images are superimposed.

The k–corrections needed to transform the photometry for the $z = 0.4$ and $z = 0.9$ clusters to B are small, since the filters through which the *HST* data were taken provide a conveniently close match to the rest–frame B–band. The

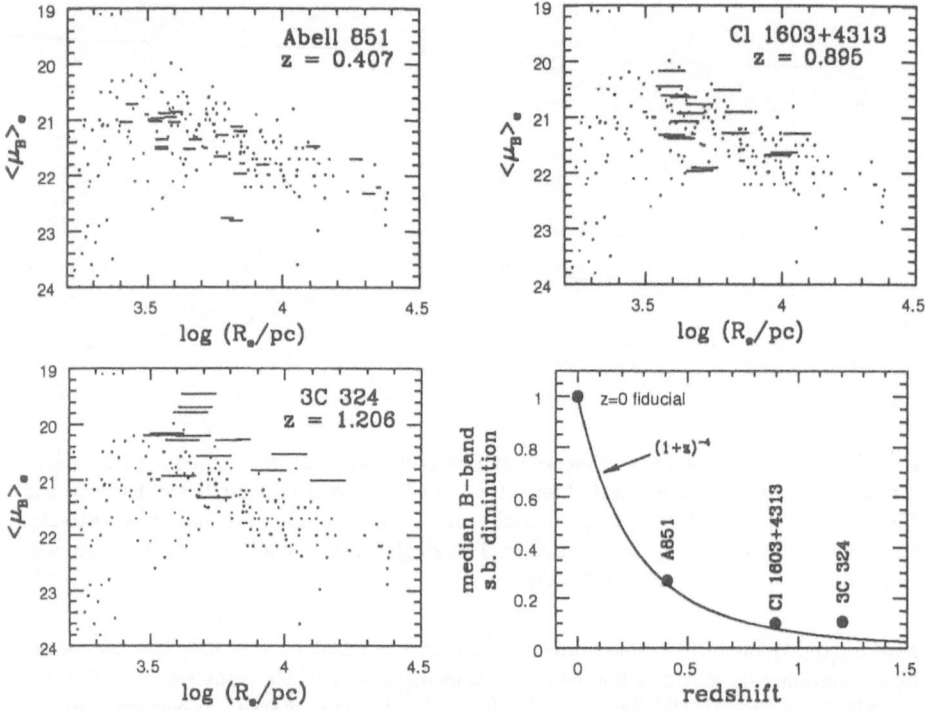

Fig. 6. Sizes and rest–frame B surface brightnesses of elliptical galaxies in high redshift clusters. The small dots are a local sample, while the dashes show the high–z galaxies, with each dash connecting R_e values for $q_0 = 0.5$ and 0. Cosmological dimming is illustrated in the lower right panel, which intercompares raw surface brightnesses from the three data sets (normalized to unity at $z = 0$).

uncertaintly is substantially larger for 3C 324 – there, the F702W filter which was used samples the galaxy spectral energy distributions at $\sim$3200Å. Correction to the B–band was made using a single–burst Bruzual & Charlot model spectrum which matches the $R-K$ colors of the galaxy. This, however, may be an incorrect assumption. A more protracted star formation history would lead to a flatter UV SED, requiring a smaller correction to the B–band. Therefore the surface brightnesses for the $z = 1.2$ ellipticals in figure 6 may have been somewhat overestimated.

At $z = 0.4$ and 0.9, the cluster galaxies fall neatly onto the locus of present-day ellipticals. By $z = 1.2$ some deviation is seen toward higher surface brightnesses (or larger sizes). In all three panels, the expected $(1 + z)^4$ cosmological dimming (Tolman 1930) has been accounted for. It is educational, however, to see this dimming displayed directly in the data. The lower right hand panel of figure 6 compares median surface brightness offsets from the normal Kormendy relation to the expected cosmological dimming. Reassuringly, the Tolman signal is strongly evident. The distant galaxies appear to lift above the expected curve, however, with the biggest departure ($\sim$ 1 mag) found at the highest redshift. This is presumably the signature of luminosity evolution. Indeed, about 1 mag of brightening is expected given the observed colors of the galaxies around 3C 324 (see §3). The greatest uncertainty comes from the k–correction for the 3C 324 galaxies, as discussed above. The deviation plotted may therefore represent an upper limit to the allowed degree of luminosity evolution rather than a direct measurement.

This indirect means of measuring luminosity evolution in cluster ellipticals agrees, broadly, with the conclusions reached by Aragón–Salamanca (1995) from a direct determination of cluster galaxy luminosity functions out to $z = 0.9$. From infrared imaging observations, he finds no significant evolution of M_K^* out to this redshift if q_0 is assumed to be 0.5. This is easily reconciled with the mild luminosity evolution at rest–frame B implied by the deviation from Tolman dimming seen in figure 6; cluster ellipticals at high redshift, as we have seen (figures 1–3), are slightly bluer than their present–day counterparts, and thus (for a given M_K) must be brighter at B. What is not yet clear is whether the apparent lack of K–band luminosity evolution, and the extreme mildness of the B–band evolution, can be squared with standard passive evolution models. Better data, and ultimately, proper determinations of the fundamental plane at high–z, are needed.

3 Galaxy Evolution from QSO Absorption–Selected Samples

Although a connection between certain classes of QSO absorption line systems and gas associated with galaxies was hypothesized long ago, the first systematic evidence supporting this was provided by Bergeron & Boissé (1991). Observing QSOs with known MgII absorption lines at $z_{abs} < z_{QSO}$, they identified galaxies near the QSO sightline and spectroscopically confirmed that their redshifts matched those of the MgII absorption doublet.

3.1 MgII–selected galaxies, $0.2 < z < 1$

Since that time, we have been carrying out surveys aimed at (1) characterizing the nature of the galaxies selected by this method, and (2) using these galaxies to study the evolution of the field galaxy population at high redshift. We refer to our sample as one "selected by gas cross–section," since a MgII rest–frame equivalent width $W_0 > 0.3$Å is essentially equivalent to a neutral hydrogen column density $N(HI) \gtrsim 10^{17}$ cm^{-2}. For the purposes of studying field galaxy evolution, this is useful primarily because the selection depends only on a robust and easily measured *rest frame* property (W_0(MgII)), and not on any *observed frame* characteristic such as apparent magnitude, color, surface brightness, etc. The method is thus free of many potential biases which affect deep magnitude–limited redshift surveys (although it may be subject to its own!). Moreover, once the gas halo cross–sections are understood (see below), the resulting sample is volume–limited, which is greatly advantageous when studying luminosity functions and the like. If the nature of absorption–selected galaxies can be firmly established, then comparison between our samples and those from the deep redshift surveys may lead to new insights about galaxy evolution.

Our first survey studied MgII absorption systems at $0.2 < z < 1.0$, and is essentially finished, with nearly complete imaging identification of candidate absorbers and spectroscopic redshift confirmation for ~80% of these candidates. We are presently pushing for 100% redshift completeness using the W.M. Keck 10m telescope. The second survey covers the range $1 < z < 2$ and is now underway using infrared and optical imaging from KPNO, Palomar and Keck, as well as selected spectroscopic follow–up from Keck. For details concerning the observations and data analysis, the reader is directed to previously published descriptions, including Steidel & Dickinson 1992, Steidel, Dickinson & Persson 1994, Steidel & Dickinson 1995, and Steidel 1995.

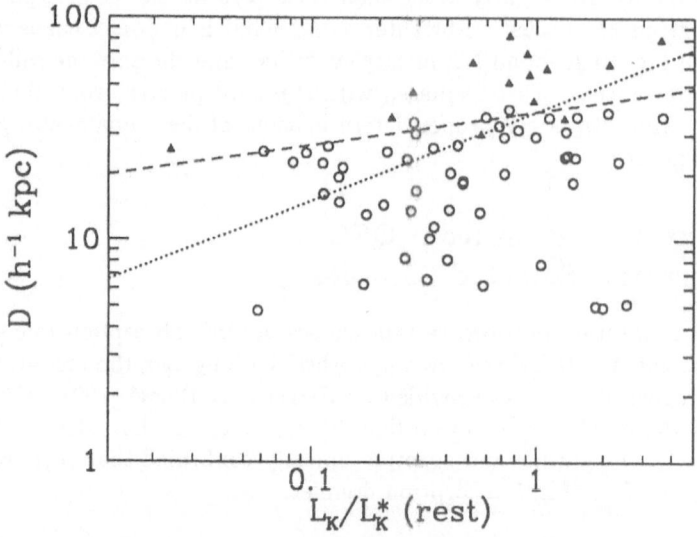

Fig. 7. Impact parameter from QSO sightline vs. infrared luminosity. The solid triangles are galaxies responsible for producing MgII absorption, while the open triangles are confirmed *non*–absorbers, i.e. galaxies which are found *not* to produce MgII absorption with $W_0 > 0.3$Å. The dashed line shows the best fit scaling relationship $D \propto L_K^{0.15}$. The dotted line traces the more conventional Holmberg scaling ($D \propto L^{0.4}$), which is strongly excluded for the MgII sample.

For the range $0.2 < z < 1.0$, we find that *all* galaxies with luminosities $L_K \gtrsim 0.05L_K^*$ are potentially MgII absorbers, provided that they fall within a particular impact parameter of the QSO sightline (see Figure 7). This impact parameter (effectively, the gaseous halo radius) scales weakly with luminosity as

$R_{halo} = 38h^{-1}(L_K/L_K^*)^{0.15}$ kpc (where $h = H_0/100\,\mathrm{km\,s^{-1}\,Mpc^{-1}}$). Conversely, we find no cases of bright intervening galaxies within this impact parameter limit which do *not* produce absorption. The only interlopers have turned out to be dwarfs, mostly very blue. Apparently, the MgII systems are dominated by "big" galaxies (within a few magnitudes of L^*). The "faint blue galaxies" which dominate deep number counts do not contribute appreciably to the gas cross-section of the Universe at $N(\mathrm{HI}) \gtrsim 10^{17}\,\mathrm{cm^{-2}}$. The K–band luminosity function of the absorbing galaxies follows a Schechter–like distribution (see Figure 8), with L^* (at $\langle z \rangle \approx 0.65$) indistinguishable from the present–day value, but with a high normalization consistent with values measured from the deep field galaxy surveys (e.g. the CFRS) at similar redshifts.

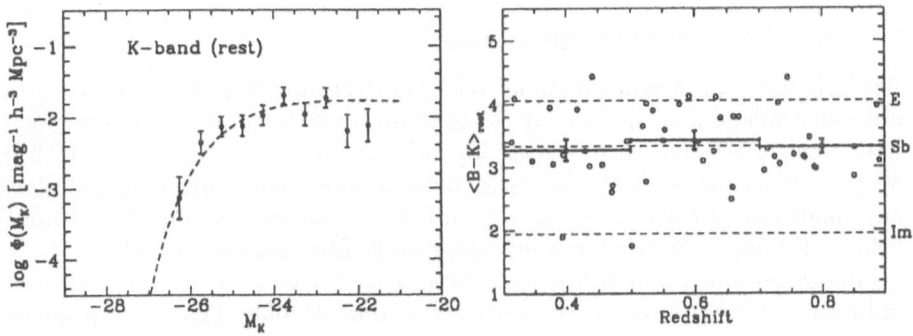

Fig. 8. *(Left:)* K–band luminosity function for MgII–selected galaxies, $0.2 < z < 1.0$. The solid line shows a Schechter–function representation of the local K–band luminosity function (from Mobasher *et al.* 1993), but normalization rescaled to match that of the MgII sample.

Fig. 9. *(Right:)* Rest–frame $(B - K)$ color vs. redshift for galaxies from the MgII–selected sample. The dashed lines mark colors of present–day E, Sb and irregular galaxies; the horizontal bars mark average colors (with errors shown) for galaxies in three redshift intervals. No significant color evolution is evident in the MgII sample.

Interestingly, the halo radius vs. luminosity scaling is better behaved when computed with K–band magnitudes (instead of optical photometry). Moreover, we see no relation between a galaxy's color and its nature as an absorber. The absorbers span the range of normal galaxy colors from flat–spectrum "Magellenic irregulars" to red ellipticals. There is no evidence for color evolution in the absorber population out to $z = 1$ (figure 9). Evidently, to first order, the presence of an extended gaseous halo does not depend on the current star formation rate in a galaxy, but rather on the luminosity of its older stellar population. Because L_K, for evolved galaxies, roughly traces the total stellar mass, we might therefore suppose that the halo diameter primarily reflects the *mass* of the galaxy.

Our extension of this survey to the redshift regime $1 < z < 2$ has only begun, but the preliminary indications are that nothing is dramatically different from the situation at $z < 1$ – the galaxy luminosities, colors, impact parameters, and space densities are roughly the same. Overall, this suggests that the "big" (massive?) galaxy population has been, for the most part, remarkably stable over a very long span of cosmic time. The apparent *absence* of luminosity evolution might seem to contradict expectations for simple passive evolution of stellar populations. However, this may be interpreted as implying a roughly constant star formation rate with redshift when averaged across the absorber population.

3.2 MgII Absorber Morphologies

Recently, we have begun obtaining *HST* WFPC2 images of MgII absorbers in order to evaluate their morphologies and to measure their orientations and inclination angles relative to the quasar line of sight. In Cycle 4 we imaged 3C 336, a $z = 0.92$ quasar with the largest number of foreground absorbers (5!) along any single line of sight in our survey. In Cycle 5, we are imaging 12 additional fields. All images are taken through the F702W filter ($\lambda_{\mathrm{eff}} \approx 7000$Å).

Figure 10 shows a montage of the absorbers imaged to date. Most are fairly ordinary galaxies spanning the range of normal Hubble Types, from bulge–dominated systems (e.g. the $z = 0.318$ and 0.660 absorbers) to late–type disks (e.g. the $z = 0.442$ and 0.723 galaxies). A few exotic objects are also found – the $z = 0.525$ absorber looks like the highly elongated, peculiar galaxies seen in many deep *HST* images and emphasized by Cowie *et al.* 1995. This galaxy, however, is extremely red, suggesting that it is a highly reddened, edge–on disk – extinction may account for its odd morphology. Several absorbers are highly inclined (e.g. the $z = 0.891$ galaxy), reinforcing other evidence that MgII absorption arises from halo material rather than from gas in the disk. The few $z > 1$ galaxies we have imaged look somewhat peculiar, but this may only reflect the fact that the R–band WFPC2 images sample their emitted–frame ultraviolet continuum.

4 Discussion

Considered together, the preceding sections seem to imply that there has been little evolution in the galaxy population out to $z \sim 1$. Cluster ellipticals have undergone little change beyond that expected from simple passive spectral evolution – perhaps even *less* evolution than passive population synthesis models would predict. Similarly, the colors and luminosity distribution of MgII absorption–selected galaxies are also virtually unchanged compared to the properties of bright galaxies ($L > 0.05L^*$) in the local universe.

Despite these measurements, one can hardly conclude that the universe of galaxies has not evolved since $z \approx 1$. The evidence from magnitude–limited field

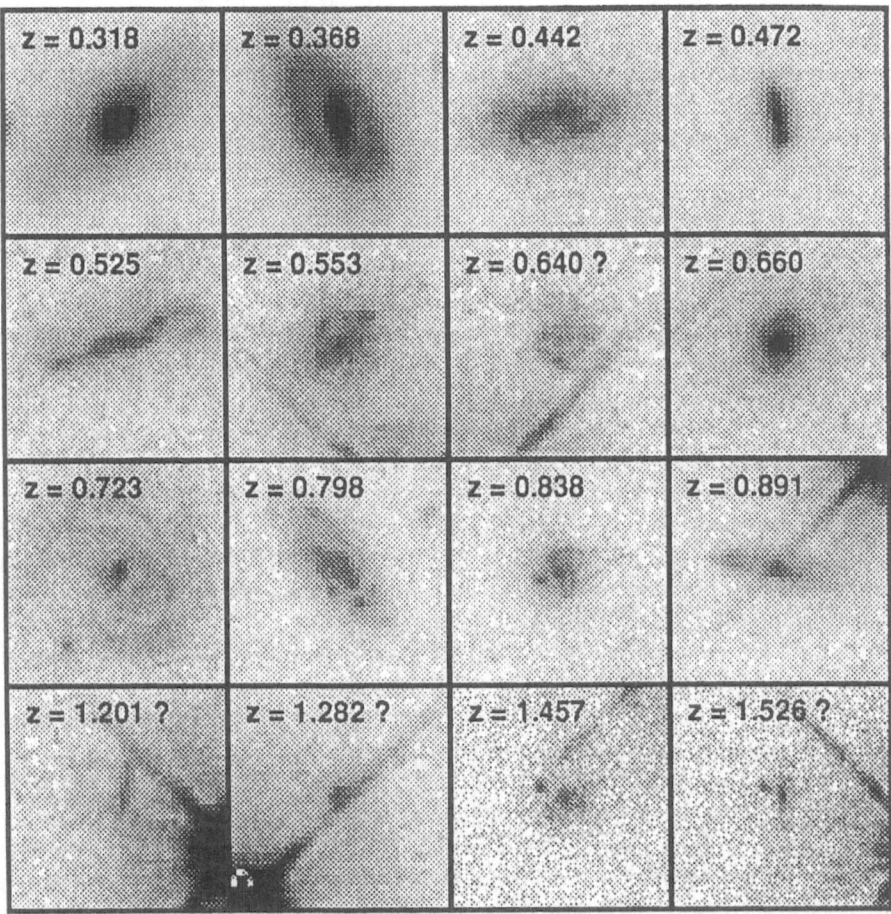

Fig. 10. Montage of *HST* WFPC2 images of MgII absorbing galaxies, arranged by redshift. Each panel is 4″ on a side. Cases where the galaxy redshifts have not been confirmed spectroscopically are indicated with a question mark.

galaxy redshift surveys strongly contradicts this. In their determination of the redshift–dependent field galaxy luminosity function from the Canada–France Redshift Survey, Lilly *et al.* (1995) find unambiguous evidence for evolution. Considering the CFRS luminosity function as a whole, they find that there is either substantial steepening of the faint–end slope, or an increase in the normalization ϕ^*, or both. Moreover, in summing the total comoving volume emissivity of galaxies in their sample, Lilly *et al.* 1996 find a dramatic increase in the luminosity density L of the universe with increasing redshift, with $\mathcal{L} \propto (1 + z)^{2.7\pm0.5}$ at rest–frame B.

Where does the discrepancy lie between the field redshift surveys and the cluster and absorption–selected samples? There are several clues. First, Lilly *et al.* find that when they divide their CFRS sample by color, selecting only galaxies redder or bluer than the color of a present–day Sbc galaxy, the evolution which they observe may be described as being entirely localized within the blue population. Like the cluster ellipticals discussed in §2.2, the red field galaxies of the CFRS show no significant change in space density of luminosity with redshift. In this regard, the field population seems similar to that in rich clusters: galaxy evolution in the latter seems to be largely confined to the blue "Butcher–Oemler" galaxies, leaving the red E/S0s relatively untouched.

This does not explain the discrepancy between the CFRS field population and the MgII absorbers discussed in §3, however. The absorber galaxies span a wide range in color, and yet show no sign of color or luminosity evolution when considered as a whole. We argue, however, that selection by gas cross–section picks out essentially *all* field galaxies with luminosities greater than $\sim 0.05 L_K^*$; therefore all galaxies in the CFRS brighter than this threshold would, potentially, be MgII absorbers were there a QSO close enough to their sightlines to serve as a source of background illumination. While some of the overall evolution in the field galaxy population might be attributed to low luminosity galaxies (i.e. fainter than $\sim 0.05 L_K^*$), there is no question that many or most of the high redshift CFRS galaxies which participate in the evolution which they observe are substantially more luminous than this "dwarf–like" threshold.

There are two possible explanations. First, Lilly *et al.* argue that there is no formal discrepancy. By dividing our MgII sample by color and fitting regression lines of absolute B magnitude vs. redshift, they find that the bluest MgII absorbers may indeed be brighter at high redshift, while the red absorbers are stable, just as is seen in the CFRS. The relatively small size of the MgII absorber sample (58 objects) makes the formal significance of this conclusion marginal, but it is nevertheless suggestive. The second explanation depends on selection effects. Despite the fact that the CFRS is selected at relatively red wavelengths (in the I–band), it is nonetheless invariably sensitive to some degree of bias toward bluer galaxies at higher redshifts, compared to the MgII–selected sample which has no photometric selection criteria. It is highly unlikely that this would account for all of the observed evolution in the CFRS luminosity function, and the overall change in the cosmic luminosity density cannot be explained by such a bias. However, together, it may be that these two explanations, taken together, together can bring the data sets into full reconciliation: the small size of the MgII sample may mask some trends with redshift, while selection biases may amplify some of the trends in the CFRS.

The *HST* Medium Deep Survey also provides evidence for differential evolution within the galaxy population. Dividing galaxies by morphological type and counting them as a function of magnitude, Driver *et al.* and Glazebrook *et al.* find that "normal" Hubble sequence galaxies follow $N(m)$ relations similar to those predicted by "no–evolution" models where the local luminosity function of galaxies is simply replicated out to high redshift. Irregular or amorphous galaxies, however, have much steeper $N(m)$. Qualitatively, this fits with the trends seen in the MgII absorber sample, where the majority of galaxies are readily recognizable members of the Hubble sequence population. Evidently, the rapidly evolving irregular population contributes little to the gas cross–section of the universe at $N(HI) \gtrsim 10^{17}$ cm^{-2}, and thus is largely excluded from samples of QSO metal line absorbers. Conversely, one would predict that the irregular galaxies from the MDS must, in general, be relatively low–mass systems, with blue colors and small K–band luminosities. Confirmation of this prediction must await infrared photometry and redshift measurements for a large enough sample to permit the construction of proper luminosity functions for the members of this population.

Overall, I would conclude at the moment that the balance of evidence weighs in favor of strong differential evolution in the galaxy population in all environments. In both clusters and the field, "red" galaxies seem to be weakly evolving, while "blue" galaxies have faded dramatically with time. Similarly, the most luminous galaxies (particularly, those with large K–band luminosities, i.e. those with the greatest mass bound up in their old stellar populations) seem to have been quite stable with time out to $z \approx 1$. Some larger future census of faint galaxy properties is still needed to confirm these evolutionary trends. At the same time, we are just beginning to probe the nature of the galaxy population at $z > 1$ through clusters, metal–line absorbers, deep field redshift surveys, and ultra–deep HST imaging. What we find there will be important for establishing the origins of the galaxy population we find around us today.

5   Acknowledgements

I am extremely grateful to my collaborators, including Peter Eisenhardt, Hy Spinrad, Adam Stanford, and Chuck Steidel, for allowing me to reproduce results from our work together, much of it prior to formal publication. Many of the results shown here stem from their efforts. I also thank the organizers of this Advanced School for a very pleasant stay in Laredo and for their financial support, and most of all the students for making the week a real pleasure both inside and outside the classroom. I gratefully acknowledge additional travel support from STScI. Portions of this research were funded by NASA/STScI GO grants nos. 5304, 5465, and 5984.

6   References

Aragón–Salamanca, A., 1995, in *Fresh Views on Elliptical Galaxies*, ASP Conference Series, eds. A. Buzzoni, A. Renzini & A. Serrano, p. 265.

Aragón–Salamanca, A., Ellis, R., and Sharples, R., 1991, MNRAS, 248, 128.

Aragón–Salamanca, A., Ellis, R., Couch, W.J., and Carter, D. 1993, MNRAS, 262, 764.

Bergeron, J., and Boissé, P. 1991, A&A, 243, 344.

Bower, R.G., Lucey, J.R., and Ellis, R.S. 1992, MNRAS, 254, 601.

Bruzual, G., and Charlot, S. 1993, ApJ, 405, 538.

Butcher, H.R., and Oemler, A. 1978, ApJ, 219, 18.

Butcher, H.R., and Oemler, A. 1984, ApJ, 285, 426.

Couch, W.J., Ellis, R.S., Sharples, R.M., and Smail, I. 1994, ApJ, 430, 121.

Cowie, L.L., Hu, E.M., and Songaila, A. 1995, Nature, 377, 603.

Dickinson, M. 1994, Ph.D. thesis, U.C. Berkeley.

Dickinson, M. 1995, in *Fresh Views on Elliptical Galaxies*, ASP Conference Series, eds. A. Buzzoni, A. Renzini & A. Serrano, p. 283.

Dressler, A., Oemler, A., Butcher, H.R., and Gunn, J.E. 1994, ApJ, 430, 107.

Dressler, A., Oemler, A., Sparks, W.B., and Lucas, R.A. 1994, ApJ, 435, L23.

Driver, S.P., Windhorst, R.A., Windhorst, A., Ostrander, E.J., and Griffiths, R. 1995, ApJ, 449, L23.

Ellingson, E., Yee, H.K.C., and Green, R.F. 1991, ApJ, 371, 49.

Ellis, R., Colless, M., Broadhurst, T., Heyl, J., and Glazebrook, K. 1996, MNRAS (in press).

Franx, M. 1993, PASP, 691, 1058.

Glazebrook, K., Ellis, R.S., Santiago, B., and Griffith, R. 1995, MNRAS, 275, L19.

Gunn, J.E., Hoessel, J.G., and Oke, J.B. 1986, ApJ, 306, 30.

Hill, G.J., and Lilly, S.J. 1991, ApJ, 367, 1.

Kormendy, J. 1977, ApJ, 218, 333.

Hill, G.J., and Lilly, S.J. 1991, ApJ, 367, 1.

Lavery, R.J., and Henry, J.P. 1988, ApJ, 304, L5.

Lilly, S.J., Tresse, L., Hammer, F., Crampton, D., and Le Fèvre, O. 1995, ApJ, 455, 108.

Lilly, S.J., Le Fèvre, O., Hammer, F., and Crampton, D. 1996, ApJ (in press).

Mobasher, B., Sharples, R.M., and Ellis, R.S. 1993, MNRAS, 263, 560.

Moore, B., Katz, N., Lake, G., Dressler, A., amd Oemler, A. 1996, Nature (submitted).

Oemler, A., Dressler, A., and Butcher, H.R. 1996, ApJ(submitted).

Sandage, A., and Perelmuter, J-M, 1990, ApJ, 361, 1.

Stanford, S.A., Eisenhardt, P.R.M., and Dickinson, M. 1995, ApJ, 450, 512.

Steidel, C.C., and Dickinson, M. 1992, ApJ, 394, 81.

Steidel, C.C., Dickinson, M., and Persson, S.E. 1994, ApJ, 437, L75.

Steidel, C.C., and Dickinson, M. 1995, in *Wide Field Spectroscopy & the Distant Universe*, eds. S. Maddox and A. Aragón–Salamanca, World Scientific, p. 349.

Steidel, C.C., in *QSO Absorption Lines*, ed. G. Meylan, Springer, p. 139.

Tolman, R.C. 1930, *Proc. Nat. Acad. Sci.*, 16, 511.

Worthey, G., Trager, S.C., and Faber, S.M. 1995, in *Fresh Views on Elliptical Galaxies*, ASP Conference Series, eds. A. Buzzoni, A. Renzini & A. Serrano, p. 203.

Yates, M., Miller, L., and Peacock, J. 1989, MNRAS, 240, 129.

Yee, H.K.C., and Green, R.F. 1984, ApJ, 280, 79.

The Lyα Clouds as Tracers of the Evolution of the Universe

Jose Luis Sanz

Instituto de Fisica de Cantabria (CSIC-UC),
Universidad de Cantabria, 39005-Santander

Abstract: The Lyα clouds are a useful tool for Cosmology because they sample the history of the universe to high redshift ($z \simeq 5$). Their physical properties at any time (column densities, velocity dispersions, chemical abundances) and their statistical ones (number density, spatial correlation) give relevant information about the intergalactic medium. On the other hand, the relative lack of clouds near QSOs can be used to constrain the UV-background at high-z. It is now claimed that the origin of the clouds is tightly related to the process of structure formation in the universe and maybe a high proportion are in fact related to galaxies.

1 Introduction

One of the most interesting, relevant and unsolved problem in Cosmology is related to the origin, formation and evolution of galaxies. The history of the universe at intermediate redshift ($0 < z < 1$) can be sampled through normal galaxies and new relevant information is coming up from HST and the Keck telescopes (see the review by Mark Dickinson in these Proceedings). At high redshift ($1 < z < 5$), apart from powerful radiogalaxies , QSOs and a few Lyα emitters, the only information comes from the absorbers that are in the line-of-sight of QSOs. At very high redshift ($z \simeq 10^3$), the cosmic microwave background decoupled from the matter and show us now a picture of the universe (through its spectrum and anisotropies as detected by the COBE satellite) at that time. At the big-bang ($z \simeq 10^{10}$), indirect information about that epoch is obtained through the nucleosynthesis of light elements.

To be published in *The Universe at High-z, Large-Scale Structure and the Cosmic Microwave Background*, ed. E. Martinez-Gonzalez and J. L. Sanz (Springer Verlag, Heidelberg)

From the observational point of view, we are now seeing in imaging (HST) and spectroscopy (Keck) how the galaxies were up to $z \simeq 1$(Casertano et al. 1995). Moreover, at higher-z ($z > 1$), one finds serendipitiously a few Lyα - emitters near some QSOs (Lowenthal et al. 1991; Francis et al. 1996; Hu et al. 1996) that could be protogalaxies in their first burst of star formation. Defining the protogalaxies at high-z is complicated because either the star formation process is so slow that we are not able to detect such weak emission with the present technology or dust shrouds the light which comes out from these emerging protogalaxies.

The absorbers in the l.o.s. of QSOs play a fundamental role defining the progenitors through absorption processes. They show many absorption lines, most of which are identified with H-Lyα absorption and the rest of the lines are identified as metal transitions (MgII, CIV,...). It is customary to differenciate between Lyα clouds (with column density $12 < logN_{HI} < 17$ and no traces of metals), metal line absorbers (MgII-systems and CIV-systems with $logN_{HI} > 17$) and damped Lyα systems ($logN_{HI} > 20$). The Lyα clouds were historically associated to intergalactic clouds of primordial origin (Lynds 1971; Sargent et al. 1980), although CIV is associated with strong lines (see section 4) whereas the MgII, CIV-absorbers and damped Lyα systems are associated to a different population: normal galaxies or proto-galaxies.

Regarding the Lyα clouds at low-z, we emphasize that some of them have been identified with galaxies very recently (Lanzetta et al. 1995) but other clouds are not (Le Brun et al. 1995). In the first case maybe the correct interpretation is that then the l.o.s. to the QSO is passing through the outer halo of a galaxy. In other cases, the possibility that the clouds are the by-products (substructure) of the structure formation process has been delineated taking into account faillure to detect them observationally and numerical simulations dealing with standard scenarios of galaxy formation.

On the other hand, a systematic study of the MgII-systems (Bergeron and Boissé 1991; Steidel et al. 1994) lead to the identification of the absorbers through direct imaging and spectroscopy. The conclusion is that this type of metal line absorbers are directly linked to the halo of normal galaxies. Nobody has identified by direct imaging and spectroscopy any CIV-system at high-z with no MgII absorption (although there is a recent claim by Benitez et al. 1996).

It is now widely accepted that the damped Lyα - systems ($N_{HI} \simeq 10^{20} cm^{-2}$) are disk (or protodisk) galaxies and recent observations with the Keck seems to test such hypothesis from the kinematics of a system at $z = 2.309$ in the l.o.s. of the QSO PHL 957 (Wolfe et al. 1994). No clear identification of a damped Lyα system by direct imaging and spectroscopy has been done to date (temptative detections in imaging have recently claimed by Steidel et al. 1996a, b for three damped systems at low-z, $z_D = 0.3950, 0.6922, 0.8596$).

Taking into account all of that, one can ask the following interesting question: where are the progenitors of the present galaxies? The answer is not a trivial matter because the first problem is that from the observational point of view the question is not well defined: we really do not know how a galaxy at high-z (e.g.

$z > 3$) looks-like, although some light have been recently obtained from HST observations up to $z \simeq 1$ and from the study of the QSO absorbers. A recent high-light, is the discovery by Steidel and collaborators (1996) of a population of normal star-forming galaxies at redshifts $3.0 < z < 3.5$. The spectra are similar to those of nearby star-forming galaxies and maybe the high-z counterparts of the spheroid component of present-day liuminous galaxies has been identified.

In this review, we will comment about the Lyα clouds population, making emphasis on the observational aspects. The nature and composition of the clouds is one of the most relevant open problems in cosmology. There are also some interesting aspects related to them: one can get information about the intergalactic medium and the process of structure formation up to $z \simeq 5$ (and this is the most oustanding problem for cosmology). On the other hand, their low chemical abundance can give insight about the primordial universe because maybe such a population is the less evolved (as compared with metal and damped-systems). Finally, it is interesting to remark the possible influence of the clouds on the diferent background radiations (e.g. microwave photons).

2 Physical properties of Lyα clouds

Large telescopes (4m) and more recently the Keck one and fine instrumentation (e.g. echelle spectrographs and TEX CCDs detectors) have allowed spectral information for more than 15 QSOs, at intermediate and high-z, with high-resolution (FWHM$\simeq 10 \, km/s$) and high SNR (> 10) during the last years. On the other hand, the HST has allowed to take spectral information of some QSOs at low-z but at a lower resolution. In many cases, the absorption lines appear isolated (there is a blending problem that is more relevant for high-z) and can be fitted to Voigt profiles. Usually, it is assumed that the shape of the profile is generated by natural broadening plus thermal motions inside the cloud although deviations of this shape can be generated in the collapse phase of structure during the evolution of the universe (McGill, C. 1990; Rauch, M. 1995). Through the fitting process, two observational parameters can be obtained: the HI column density (N_{HI}) and the velocity dispersion ($b = 2^{1/2}\sigma$). In the case of pure thermal motions: $b^2 = 2kT/m$ or $b(km/s) \simeq 13T_4^{1/2}$, where T_4 is the temperature in units of $10^{4 \, \circ}K$.

Recently, some global Lyα clouds samples (at high resolution) have been considered in the literature. Sample 1 (de la Fuente *et al.* 1995) contains 654 lines -from 7 QSOs- covering a wide range in redshift $1.85 < z < 4.32$. The QSOs have been selected taking into account a similar SNR (> 10) and resolution ($R > 12500$) and the lines are restricted to satisfy strict criteria to obtain an homogeneous sample: $b \geq 15 \, km/s$, $log N_{HI} \geq 13.3$, $\Delta b \geq 2 \, km/s$ and $\Delta log N_{HI} \geq 0.03$ (originally more than 1000 lines are in the set and finally only 654 survive). Sample 2 (Cristiani 1995) contains 1100 lines -from 10 QSOs- covering approximately from $2 < z < 4$. These QSOs have been selected from the ESO key-programme (6 QSOs with, SNR > 20 and R> 20000) and complemented with other 4 from the literature with similar SNR. Sample 3 (Hu *et al.*

1995) is based on 4 QSOs observed with the HIRES spectrograph on the Keck telescope. The SNR> 50 and R> 36000 spectra define a set of 1056 lines. A primary analysis has been done in the column density range $12.3 < logN_{HI} < 14.5$ with 990 lines.

2.1 The column density distribution

The HI-column density distribution $p(logN_{HI})$ for the sample 1 mentioned above is characterized by a plateau in the range $13.4 < logN_{HI} < 14.0$ and a strong fall-off for $logN_{HI} > 14.0$. A least squares fit to a double power-law gives a break at 13.99 and slopes -0.006 and -1.1, respectively. Similar results have been obtained with the sample 2 in the range $13.0 < logN_{HI} < 16.0$. Possible biases that alter this distribution have been quantified via numerical simulations (Cristiani 1995) and the result is that the slope below the break increases to $\simeq -0.4$ but the distribution is still inconsistent with a single power-law because there is a deficit of lines in the range $14 < logN_{HI} < 17$. On the other hand, confirming some of the previous results but at higher resolution, the sample 3 have been analized in the range $12.3 < logN_{HI} < 14.5$ and corrected for incompletness via numerical simulations, the result can be fit by a power law with slope -0.46 (or a 95% confidence interval of (-0.37, -0.51)). Summing up, below a break at $logN_{HI} \simeq 14.0$ there is a plateau that maybe can be corrected for incompletness to a slope of $\simeq -0.4$ whereas above the break a slope of $\simeq -1$ is more apprpriate. It is a matter of debate whether Lyα clouds with $logN_{HI} < 14$ and $logN_{HI} > 14$ comprise a single population (see also the differences in chemical composition in the next section).

2.2 The velocity dispersion distribution

The b-distribution has a bell-shape form with a strong tail. The bulk of the lines have Doppler parameters in the range $(20, 40)\,km/s$ with only a few cases with $b < 10\,km/s$. For sample 1 one finds $< b >= 31.6\,km/s$, $b_{med} = 33\,km/s$ and $\sigma_b = 13\,km/s$. For sample 2 similar values are found, only 15% (2%) of the lines have $b < 20\,km/s$ ($b < 10\,km/s$) whereas for sample 3 less than 1% of the lines have $b < 20\,km/s$.

2.3 The correlation $b - N_{HI}$

This has been a controversial topic during the last years (Pettini *et al.* 1990; Carswell *et al.* 1991; Sanz *et al.* 1993; Rauch *et al.* 1993; Cristiani *et al.* 1994; Rodriguez-Pascual *et al.* 1995). The $b - N_{HI}$ diagram shows a deficit of lines in the upper-left part due to a selection effect (low-column density lines are hidden in the noise) but there is also a deficit of strong lines ($logN_{HI} > 14$) with small b. This is clear from samples 1 and 2 but also in individual QSOs. From sample 1, a weak but noticeable correlation between b and N_{HI} is observed in the data (a two-dimensional KS-test gives a probability of 0.047 for the observed

$p_{ob}(b, logN_{HI})$ distribution being a realization of the uncorrelated model given by $p(b, logN_{HI}) = p_{ob}(b) p_{ob}(logN_{HI}))$ and supporting this view is the result that the b-distributions below and above the cut-off point at $logN_{HI} \simeq 14$ have a similar asymmetric bell-shape form but they show significative differences in mean and median values (in fact, a KS-test applied to both b-distributions gives a probability of $p \simeq 10^{-6}$ for coming from the same parent population). We comment that the "migration diagrams" in the $b - N_{HI}$ plane found by Rauch et al. (1993) -due to finite SNR- have no practical importance in the region $14 < logN_{HI} < 14.5$ and they do not strongly affect the arguments mentioned above. Obviously, the data from the sample 3 with $logN_{HI} > 14$ will help to shed light on this relevant issue. Summing up, we can say that below $logN_{HI} \simeq 14$ the mean value for the Doppler parameter is $b \simeq 25\,km/s$ and there is no appreciable correlation with N_{HI} whereas above $logN_{HI} \simeq 14$ the mean value is bigger $b \simeq 35\,km/s$ and there exits a noticeable trend with the column density. These results are consistent with the absorbing clouds being very flattened, highly ionized clouds, gravitationally confined by dark matter, where the observed $b - logN_{HI}$ trend is due to orientation effects (Rodriguez-Pascual et al. 1995). This simplified model is also able to reproduce the observed broken power-law distribution of HI-column densities and the b-distribution. Finally, we remark that a weak correlation between b and N_{HI} is predicted through the numerical simulations performed in the CDM+Λ model including an ionizing background (Miralda-Escudé et al. 1995).

2.4 Size

Spectroscopic observations along the two l.o.s. to projected pairs or gravitational lensed QSOs allow to probe the size, geometry and spatial clustering of Lyα clouds. We remark that "cloud size" can be interpreted as defining a coherent structure but maybe there exist correlated but distinct structures with the same properties. Pioneer studies of the Lyα forests of QSO pairs at separations of $\simeq 1'$ (Shaver & Robertson 1983; Crotts 1989) found weak correlations and showed that the size of the clouds is less than $\simeq 300h^{-1}kpc$. Work on the gravitationally lensed QSO UM 673 (Smette et al. 1992) showed that the size of the clouds is larger than $6h^{-1}kpc$. Recent work (Bechtold et al. 1994; Dinshaw et al. 1994) on the QSO pair Q1343+2640A/B ($9''.5$ separation) implies a firm model-independent lower-limit on the size of the clouds of $40h^{-1}kpc$ at $z \simeq 1.8$ if $\Omega = 1$. Moreover, under the assumption of uniform radius spherical absorbers one gets the upper-limit on the size of $\simeq 280h^{-1}kpc$ with 98% confidence. The median value for the size is $\simeq 90h^{-1}kpc$, all the numbers scaling by a factor 1.44 and 1.85 for $\Omega = 0.1$ and $\Omega = 0.1, \Lambda = 0.9$, respectively. They also detect differences in the column density and radial velocity of the absorbing gas. A characteristic radius of $> 70h^{-1}kpc$ is incompatible with the two popular models for the clouds: gas confined by minihalos of CDM (Rees 1986) and confination by a hot IGM (Sargent et al. 1980). The characteristic velocity difference of $\simeq 60km/s$ is also a problem for models that assume either clouds in the disks or in the virialized halos of normal galaxies that would give velocity dispersions above $200 km/s$. An extension of this analysis to more general geometries has been considered (Fang *et al.* 1995) proposing a picture of Lyα clouds at $z \sim 2$ as non-linear, gravitationally collapsing objects and suggesting a connection with faint blue galaxies at $z \sim 1$. Simulations of double l.o.s. (Charlton *et al.* 1995) assuming various mass distributions, geometries and kinematics indicate that only a smooth disk/slab-like model with systematic velocities remains consistent with the inferred properties of a single population of Lyα absorbers. Similar results have been recently obtained for the two components (3''.0 separation) of the gravitational lens candidate HE 1104-1805 (Smette *et al.* 1995): lines are strongly correlated and there are not differences in velocities within $\sim 10 km/s$. They derive a 2σ lower-limit at $z \simeq 2$ of $50 h^{-1}kpc$ for spherical clouds if $\Omega = 1$. At lower-z ($z < 1$), the study of the pair Q0107-025A/B lead a strong lower-limit on the size of the clouds: $170 h^{-1}kpc$ at $z \sim 0.7$ (Dinshaw *et al.* 1995), but the statistics of the coincident lines (4) is very small. If low and high-z clouds represent a single population, this increasing in size with time is expected in a hierarchical structure formation scenario.

2.5 Internal structure

Recent data of the QSO Q0302-003 observed at high-resolution with HIRES at Keck (Cowie *et al.* 1995) have shown for the first time that the structure of the strong clouds ($log N_{HI} > 14.5$) is very complex. Half of the lines have CIV (see subsection 4.3) and their kinematic structure show that the clouds are not isolated but blended complexes of cloudlets (1 to 3 components) individually having temperatures of $T \simeq 35000 - 70000 K$ and velocity dispersions of $\simeq 18 km/s$. These complexes require a strong gravitational potential, so either the clouds are embedded in the halos of galaxies or in pancakes in the IGM. These results showing velocity sub-structure in the Lyα clouds have been confirmed with another quasar, Q0636+6801 ($z_{em} = 3.178$), by Tytler *et al.* (1995).

2.6 The 2-point correlation function

The 2-point correlation function is usually computed in the velocity space, estimated by: $\xi(v) = -1 + N_{obs}/N_{exp}$, where N_{obs} and N_{exp} are the observed number of line pairs at separation v and the number of pairs expected from a random distribution in redshift. For sample 2, 10^3 numerical simulations have been performed to get N_{exp} and the result is that a weak but noticeable signal is present with amplitude $\simeq 0.2$ in the $100 km/s$ bin (the existence of correlation on small scales was suggested by Webb 1987). Moreover, the correlation dissapears for weak lines with $log N_{HI} < 13.6$ and increases for higher N_{HI} values to $\simeq 0.6$ in the same bin with 7σ confidence (an effect first noted by Chernomordik 1995, based on a reanalysis of 3 QSOs spectra at high-resolution with lines in the range $2.7 < z < 3.7$ and $log N_{HI} > 14$). All of this is in agreement with the results derived from sample 3: no correlation is found at $v > 150 km/s$ for $12.8 < log N_{HI} < 14.5$ but there is an excess of 0.17 ± 0.045 at $50 < v < 150 km/s$

and also the correlation increases with column density. All of this seems to suggest that the clouds are correlated with clouds of similar column density.

On the other hand, the Lyα clouds that exhibit CIV absorption (see subsection 4.3) are clustered on a scale $\leq 10^3 km/s$ (Tytler et al. 1995), suggesting the existence of two populations: one primordial that shows no correlation and other with CIV that cluster with an amplitude that is weaker than the one corresponding to metal line systems and galaxies.

At low-z, low resolution spectra for 35 QSOs with $0.5 < z < 1.4$ taken with the HST (The HST QSO absorption line key project) have been analized and more than 80 Lyα clouds identified but they do not show significant evidence of clumping (Bahcall et al. 1993). Of course, a bigger sample is needed to reach a definitive conclusion.

3 Evolutionary properties of Lyα clouds

The study of the evolution of the different physical parameters (b, N_{HI}, size) analized in the previous section and others (chemical abundance, dark matter content, UV-radiation field) is relevant to understand not only the process of formation and evolution of the clouds themselves but also the history of the universe and the process of structure formation in general. It is not possible, at least from samples 1 and 3, to distinguish any trace of evolution of the column density distribution. We will comment on b in the following subsection.

3.1 Evolution of the velocity dispersion b

In subsection 2.3, we have remarked that the strong ($log N_{HI} > 14$) and weak ($log N_{HI} < 14$) absorption clouds show different properties in the sample 1. A representation of the mean value of b for each QSO of the sample against z show that b is approximately constant for the weak systems, whereas for the strong ones seems to be evolution (b increases with time) at least for the five QSOs with $2.5 < z < 4$. In the simplified model of flattened clouds outlined by Rodriguez-Pascual et al. (1995), strong broad absorption lines are drawn from tilted systems in which the large HI-column density is due to a larger absorption path and the larger width is due to a gravitational component of the velocity dispersion in the plane of the slab. Within this framework, the average increase found in b for strong lines at low-z cooud be associated to an increase in the size of the clouds (Dinshaw et al. 1995). A more detailed explanation of this fact can be given due to the difficulty to explain large b values as thermal broadening (Press and Rybicki 1993): on the one hand, Webb (1987) finds that the lines are more clustered at low-z (in small scales) and Cristiani et al. (1995) that the clustering tends to be higher for larger column densities. So, if we assume (Hu et al. 1995) that each Lyα cloud is a blend of components of comparable N_{HI} and that the observed b is produced by such a blend (this assumption follows from the structure of the CIV lines found by Cowie et al. 1995), then the three arguments lead to an increase of b with time.

3.2 The evolution of size

Regarding size, we have commented at the end of subsection 2.3 that there is an apparent increase in size with time: $R \simeq 90\,h^{-1}kpc$ at $z \sim 1.8$ (Bechtold *et al.* 1994; Dinshaw*et al.* 1994) and $R \simeq 170h^{-1}kpc$ at $z \sim 0.7$ (Dinshaw*et al.* 1995), which is expected in hierarchical scenarios.

3.3 Number density $N(z)$

Regarding the number of systems per unit interval of redshift, dN/dz, and relaxing the criteria (de la Fuente *et al.* 1995) applied in the previous section for sample 1 (considering only the first two criteria, then 708 lines survive), a fit to a power law $(dN/dz \propto (1+z)^\gamma)$ gives the value $\gamma = 2.35 \pm 0.22$ in the redshift range $1.9 < z < 4.3$ but a higher slope $\gamma = 3.22 \pm 0.41$ is obtained in the range $2.1 < z < 3.7$ (in this case one avoids two QSOs of the sample, one at low-z and the other at high-z with $z_{em} = 2.14$ and $z_{em} = 4.51$, respectively). For sample 2 the value $\gamma = 2.7$ has been quoted (Cristiani 1995). These values can be compared with the ones obtained from previous large samples at intermediate resolution: Lu *et al.* (1991) obtained $\gamma = 2.75 \pm 0.29$ for $1.7 < z < 3.8$ and Röser (1995) quoted $\gamma = 2.44 \pm 0.44$ for $1.7 < z < 3.3$. All of these results indicate a clear evolution in the clouds for $2 < z < 4$ ($\gamma = 0.5, 1, 2$ for no evolution and $(\Lambda, \Omega) = (0,1), (0,0), (1,0)$, respectively). On the other hand, observations with HST (Morris *et al.* 1991; Bahcall *et al.* 1993) suggest a flattening in the line distribution at low-z: $\gamma = 0.79 \pm 0.37$. More recently, Bahcall *et al.* (1996) have quoted the value $\gamma = 0.58 \pm 0.50$ on the basis of 135 lines identified as extragalactic Lyα clouds for $z_{ab} \leq 1.3$. So, to what extent evolution in the Lyα clouds ceases at low-z or the clouds at low $(z < 2)$ and high-z $(z > 2)$ constitute a different population is a current matter of debate.

3.4 Evolution of the 2-point correlation function

The evolution of the 2-point correlation function with redshift has been addressed by Cristiani (1995). Based on the sample 2 with $logN_{HI} > 13.8$, it is obtained that the amplitude of the correlation at $100\,km/s$ decreases with increasing z from 0.85 ± 0.14 at $1.7 < z < 3.1$, to 0.74 ± 0.14 at $3.1 < z < 3.7$ and 0.21 ± 0.14 at $3.7 < z < 4.0$. Of course, this is what it is expected in hierarchical scenarios. Data from the HST are still at too low-resolution to get a more definitive answer on evolution but the previous analysis suggests a picture where rapid evolution of clustering took place at high-z ($z \simeq 3.5$).

4 Chemical abundances in Lyα clouds

Until very recently, the Lyα clouds were thought to be constituted by primordial matter H and He. Now, thanks to the capability of HST sensitive to UV radiation, He has been found. On the other hand, the Keck telescope has allowed the first tentative detection of Deuterium outside our Galaxy, at high redshift ($z = 3.3$) and also the detection of CIV in some Lyα clouds with strong HI column density. Our understanding of the clouds can dramatically change with these findings but also all of this is a test for the big-bang nucleosynthesis and a challenge for theories of structure formation.

4.1 Helium

The absorption of neutral helium, HeI, is at 584Å and as singly ionized helium, HeII, at a shorter wavelength 304Å, so until recently He had not been detected in the Lyα clouds because the corresponding redshifted absorption lines fall into the UV region. Jakobsen et $al.$ (1994) obtained a spectrum of the QSO Q0302-003 ($z_{em} = 3.286$) with the HST FOC objetive prism and they found a gap in the flux below the redshifted 304Å line. An estimation of the optical depth leads to $\tau > 1.7$ (90% confidence level) at $z = 3.2$. On the other hand, Tytler et $al.$ (1995), using HST with FOS, reported a similar absorption for the QSO Q1935-6914 ($z_{em} = 3.185$) but the estimated $\tau = 1.0 \pm 0.2$. The discrepancy between the two values for τ at the same z could be due to inhomogeneities in the IGM. More recently, Davidsen et $al.$ (1995) have used the HUT on the ASTRO-2 to study the QSO HS 1700+6416 ($z_{em} = 2.72$) and estimated $\tau = 1.0 \pm 0.07$ averaged in the range $2.2 < z < 2.6$. The spectra obtained to date do not have enough resolution to distinguish whether the HeII absorption is due to discrete Lyα clouds or is produced by a more diffuse IGM between the clouds. However, the recent detection (Songaila et $al.$ 1995) of a population of weak Lyα clouds with HI column density down to $log N_{HI} = 12.3$ for the same QSO (using the HIRES spectrograph at Keck, $R = 36000$), shows that there is no need to invoke the diffuse IGM but that such a population of clouds can account for the HeII absorption.

4.2 Deuterium

The first tentative detection of deuterium, D, outside our Galaxy was made by Songaila et $al.$ (1994), using Keck, and Carswell et $al.$ (1994), using MMT. A spectrum of the QSO Q0014+813 ($z_{em} = 3.42$) shows a HI cloud with an absorption feature at the precise wavelength predicted for cosmic D at $z = 3.32$. The implied abundance $D/H = 2.5 \times 10^{-4}$, although compatible with the BBN predictions, is much higher than the abundance derived from models of Galactic evolution (in our Galaxy today is less than 2×10^{-5}, Linsky et $al.$ 1993).

On the other hand, Tytler and Fan (1995) have found in the spectrum of QSO 1937-1009 ($z_{em} = 3.78$) a cloud at $z = 3.57$, with $log N_{HI} > 17.8$ and weak metal lines, showing D aborption but the abundance is an order of magnitude lower $D/H = 2 \times 10^{-5}$. More recent results supporting a higher abundance include: Carswell et al. (1995) quote $D/H \geq 2 \times 10^{-5}$ with a best guess of $D/H = 2 \times 10^{-4}$, in a system in Q0420-388, Rugers and Hogan (1995) find $D/H = 1.1 \times 10^{-4}$ analyzing the spectrum of GC0636+68.

A possible explanation of the D feature in some cases would be HI absorption by a small cloud placed at the appropriate z in the forest, but Rugers and Hogan (1995) have made a new reanalysis of the Keck spectrum of Q0014+813 assuming two components to represent the absorbing cloud. Spectral fits to several Ly lines give $log N_{HI} = 16.76, 16.90$, respectively. The main results are that the D features are too narrow to be interlopers (i.e. HI clouds), good agreement with temperature and redshift with their HI counterparts and both components give the same abundance: $D/H = (1.9 \pm 0.5), (1.9 \pm 0.4) \times 10^{-4}$, respectively, with an independent lower-limit on the sum of 1.3×10^{-4}. If one accepts the D explanation with an abundance of $(1.9 \pm 0.4) \times 10^{-4}$ as primordial, then standard BBN gives a baryon to photon ratio $\eta = (1.7 \pm 0.2) \times 10^{-10}$ and this implies $\Omega_b h^2 = (6.2 \pm 0.8) \times 10^{-3}$, i.e about the value estimated from observations of luminous matter in the universe. The previous value for Ω_b is in agreement with estimates of primordial abundances of ^{4}He and ^{7}Li. In particular, this result imply that the massive galaxy haloes must be made of non-baryonic matter.

4.3 Metals

Until very recently it has been an open question whether the Lyα clouds are chemically primordial or have some weak metallicity. A tentative detection of CIVλ1548 at $\geq 99.99\%$ confidence has been reported (Lu 1991), using stacking of many metal absorption line spectra corresponding to many Lyα lines. He estimated a relative abundance of $[C/H] \simeq -3.2$ (the abundance is defined relative to solar, i.e. $[M/H] = log(M/H) - (log(M/H))_\odot$) for cloud populations with $log N_{HI} \simeq 15.7$, suggesting that not all Lyα clouds are primordial. However, a similar search by Tytler and Fan (1995), using a spectra of QSO HS1946+7658 at $10\,km/s$-resolution, leads to an upper-limit of $[C/H] < -2$ but now the more common systems are weak clouds with $log N_{HI} \simeq 13 - 14$.

More recently, Cowie et al. (1995) have got the spectrum of Q0302-003 with very high SNR and resolution ($R = 36000$) using HIRES at Keck. They found that approximately half of the Lyα clouds with $log N_{HI} > 14.5$ have CIV lines with $log N_{CIV} > 12$, then giving a metallicity of $[C/H] \simeq -2$. Consistent with this estimate is the result by Tytler et al. (1995) with the spectrum of Q0636+6801 ($z_{em} = 3.178$) taken at Keck, they find that 60% of the clouds show CIV and abundance $[C/H] > -2.5$ for clouds with $log N_{HI} \geq 14.5$. Womble et al. (1995), based on a spectrum at Keck of the gravitational lens Q1422+2309 ($z_{em} = 3.63$), find that $40 - 45\%$ of the clouds with $log N_{HI} > 14.3$ are enriched with heavy elements and estimate a typical abundance of $[C/H] \simeq -2.3$ at $< z > \simeq 3.1$. All of this has left open the question of whether the Lyα population is bimodal: one big primordial population with no trace of metals contaminated by weak metal line clouds (which could be associated with the outer halos of galaxies or protogalaxies).

5 Identification of Lyα clouds

The possible association (or lack) of Lyα absorbers with galaxies or proto-galaxies is one of the most interesting topics. It was unexpected that low-z clouds were found in sufficient numbers to enable studies related not only to evolution but also to their relationship to galaxies. There are two approaches: one consists in studying the clustering properties of the clouds and galaxies and the other is the possible identification of a cloud with something one can call a "galaxy". In the last case, the projected separation and velocity difference of the galaxy with respect to the cloud are relevant quantities, togheter with the equivalent width of the absorption line, that define such association or not.

No clear identification of Lyα clouds at high-z with galaxies has been done to date as far as we know. Francis *et al.* (1996) have recently reported the discovery of of a group of galaxies at $z = 2.38$, detecting to Lyα-emitting galaxies, which supports the interpretation of a supercluster of QSO absorption lines (at least $10\,Mpc$ in comoving size) at that redshift. In the low-z case, the following observations have been made: the l.o.s. to the QSO 3C273 shows 7 Lyα absorption lines, from HST observations, below $10^4\,km/s$. Salpeter and Hoffman (1995) have correlated them with 22 disk galaxies that are near the l.o.s. to the QSO and found that in 3 cases there is an individual galaxy close enough in position (between $\simeq 200 - 250\,kpc$) and in velocity ($< 120\,km/s$) which may produce the absorption. Moreover, for a four line there is a small group of galaxies nearby and for the rest no galaxy is seen.

Lanzetta *et al.* (1995) have made imaging and spectroscopy of faint galaxies (46 in the range $0.07 < z < 0.5526$) in fields of HST centered on QSOs. They found 9 galaxies coincident in z with Lyα clouds and they conclude that at $z < 1$ most luminous galaxies are surrounded by extended gaseous haloes of $\simeq 160\,h^{-1}kpc$ and the fraction of absorbing clouds arising in luminous galaxies could be very high (maybe more than 50%).

More recently, Le Brun *et al.* (1995) have considered a sample that includes 3 new fields of QSOs from the HST and combined those of previous studies. They found 32 absorber-galaxy associations versus 11 cases with no absorption, no clear anti-correlation between rest-frame equivalent width and impact parameter, no correlation between galaxy luminosity and impact parameter and that the relative velocity distribution of the associations are consistent with either galaxy rotation velocities or the local velocity dispersion in large-scale structure.

All of this surveys favors the existence of two populations among the Lyα clouds: one which is physically associated with normal galaxies and another one invisible and more primordial. Maybe low-surface brightness galaxies, dwarfs or primordial gas clouds could constitute the last population. The recent observations (Barcons *et al.* 1995) of two low-luminosity spiral galaxies, with z coincident with those of absorption lines ($z_a = 0.09, 0.075$) in the spectra of two QSOs (Q1704+6048 and Q2135-1446) at impact parameters of $62, 48\, h^{-1} kpc$, respectively, and the fact that the gas responsable for the absorption lines appears to take part of the rotation of the galaxies, suggest that the haloes of galaxies are dominated by dark matter and also indicates that low-surface brightness galaxies can play a certain role defining the second population.

Finally, the Lyα clouds can reside in the large-scale structure of the universe (filaments, walls or void-like regions). This has been recently reported by Stocke *et al.* 1995 and Shull *et al.* 1995. They have undertaken a spectral search, with HST, for low-z absorbing clouds in the l.o.s. to bright QSOs behind well mapped voids and superclusters. In the first paper, 8 absorption clouds towards 3 QSOs are found, 7 of which are located in supercluster structures and 1 absorber is in a void, whereas in the second paper, they identify 10 clouds (plus 1 probable) toward 4 targets and conclude that at least 3 lie within voids.

6 The UV-background

The number of clouds per unit redshift increases with redshift (see section 3.3) but this rise is less rapid as one approaches to the QSO. This effect, first discoverd by Carswell *et al.* (1982), is usually called the "inverse effect" or the "proximity effect", and it has been atributed to photoionization of clouds by the bright QSOs in whose spectra they are observed and which are less likely to exceed the theshold in neutral hydrogen column density needed for detection (Bajtlik *et al.* 1988). Assuming such an interpretation, the proximity effect provides an indirect measure of the intensity of UV ionizing radiation from all sources because the relative importance of photoionization by a QSO is inversely proportional to such an intensity.

6.1 High-z

Bajtlik *et al.* (1988) interpreted the inverse effect as a proximity effect. They assumed a model in which the clouds are highly ionized by the general background of UV radiation with a frequency dependence above the Lyman limit which is the same as that of the QSO radiation. Then, data at intermediate-resolution imply that the UVB radiation is constant for $1.7 < z < 3.8$ and given by $log J_{LL} = -21.0 \pm 0.5$, $J_{LL}(erg\, cm^{-2} s^{-1} Hz^{-1} sr^{-1})$ is the Liman limit intensity of the UV radiation. This represents the first indirect measurement of the UV background at high-z, but -as the authors pointed out- it is not completely clear where this UV radiation is coming from. Bechtold (1994), on the basis of 34 QSO spectra taken at intermediate-resolution, concludes that the evolution of the UVB is not well constrained in the interval $1.6 < z < 4.1$ and assuming it constant obtains $log J_{LL} = -20.53$, although uncertainties do not elliminate QSOs as the main source to the UVB. Giallongo *et al.* (1993) have observed the QSO PKS 2126-158 ($z_{em} = 3.27$) at high-resolution ($\simeq 14\, km/s$), obtaining from the proximity effect $logJ_{LL} = -21.15$, so not far from the QSO contribution. More recently, Cristiani (1995) using the high-resolution sample 2 (see section 2) has estimated $logJ_{LL} = -21.32 \pm 0.08$ for the intensity of the UVB, through a "maximum likelihood analysis" that includes not only J but also β (the slope of the column density distribution) and γ (the slope of the cosmological evolution rate) and this value is close to the one estimated for the integrated contribution of all the QSOs (Haardt and Madau 1995). He also concludes that the UV background is consistent with no evolution in $2 < z < 4$. On the other hand, the influence of a foreground QSO on the Lyα forest of another high-z QSO has been recently inveetigated for 3 objects at $2 < z < 2.7$ (Fernández-Soto et al. 1995). The results are consistent with the existence of a proximity effect due to the foreground QSO and find a best value of $logJ_{LL} = -20.5$ and a lower-limit of $logJ_{LL} > -21.8$ (2σ level) for the intensity of the UVB at those redshifts.

The reionization of the IGM by quasars at high-z has been discused in several papers (Bechtold et al. 1987; Miralda-Escudé and Ostriker 1990; Madau 1992; Meiksin and Madau 1993). In the last paper, the authors obtained that the integrated QSO contribution to the UVB over the range $3 < z < 5$ to the flux maybe as large as $J_{LL} \simeq 3[(1+z)/4.5]^{1/2} \times 10^{-22}$ and this level is compatible with 1σ upper-limits to the G-P effect at $z = 3$ if the diffuse component of the IGM satisfies $\Omega_{IGM}h^2 \leq 0.005$ (see next section).

6.2 Low-z

Kulkarni and Fall (1993), based on observations of 13 QSOs by Bahcall et al. (1993) with the FOC on HST, have reported for the first time the detection of the proximity effect at low-z. They have estimated $logJ_{LL} = -23.22$ at $z \simeq 0.5$. At lower redshift, Vogel et al. (1995) have observed the SW component of the Haynes-Giovanelli cloud HI 1225+01 which is optimal to measure the local metagalactic flux and found the upper-limit $logJ_{LL} < -23.10$.

Madau (1992) has computed the spectrum of the diffuse UVB, taking into account the opacity of Lyα clouds and Ly Limit systems, from observed QSOs as a function of redshift and obtained that J_{LL} increases by a factor 25-35 between $z = 0$ and $z \simeq 2$, which is to be compared with previous estimations at low and high-z.

7 The Gunn-Peterson test

7.1 HI G-P test

Gunn and Peterson (1965) suggested that a uniform distribution of HI in the intergalactic medium would generate a depression in the continuum of any QSO in the blue (i.e. shortward of Lyα emission) due to Lyα absorption, the so-called "Gunn-Peterson effect". The optical depth at redshift z is given by $\tau_{GP} = 4.14 \times 10^{10}h^{-1}n_{HI}(1+z)^{-1}(1+\Omega z)^{-1/2}$, where $n_{HI}(z)$ is the number density of neutral hydrogen. There have been several observational attempts to detect a uniform medium (Steidel & Sargent 1987; Jenkins & Ostriker 1991;

Webb *et al.* 1992; Giallongo *et al.* 1992), leading to strong upper-limits on n_{HI}. More recently, Giallongo *et al.* (1994) have obtained the value $\tau_{GP} = 0.02 \pm 0.03$ in the interval $4.1 < z < 4.3$ from the high-resolution spectrum ($\simeq 40\,km/s$) of the QSO 1202-0725. For a highly ionized IGM, τ_{GP} can be expressed as a function of the IGM density if one assumes ionization equilibrium, where the heating is dominated by photoionization of the UV-background. Then, the previous upper-limit constrains the IGM baryon density to be $\Omega_{IGM} \leq 0.01$ (practically independent of the temperature of the medium) if the QSOs are the major contributors to the ionizing background (the UV flux is assumed to be constant for $z > 3$ with intensity $J_{-21} = 0.3$ and $J \propto \nu^{-0.7}$). So, much of the baryon density derived from the big-bang nucleosynthesis ($\Omega_b \simeq 0.05$ for $H_o = 50$) remains to be explained, and maybe most of the baryons are already in bound systems (Lyα clouds, metal line systems and damped Lyα systems) at high-z. A similar result has been found by Williger *et al.* (1994), in their study of the $z = 4.5$ QSO BR1033-0327 the continum optical depth is estimated to be $\tau_{GP} \leq 0.1$ at $z \simeq 4.3$.

A potential problem could arise if one assumes that the density of the IGM is close to the baryonic density required by primordial nucleosynthesis, because then the intensity of the ionizing background (required to avoid the G-P through) could be in conflict with the constraint derived from the proximity effect. However, Reisenegger and Miralda-Escudé (1995) have considered a more realistic scenario taking into account the evolution of structure by gravity, in such a case the IGM is inhomogeneous on scales larger than the Jeans length and τ_{GP} would fluctuate. For models of structure formation with *rms* mass fluctuation on the Jeans scale $\sigma_J(z = 4) = 1$ (typical of low-Ω CDM models), this implies a median value of τ_{GP} that is consistent with the estimated intensity from the proximity effect $log J_{LL} \simeq -21$.

7.2 He G-P test

On the other hand, if there exists either neutral helium HeI or singly ionized HeII in the IGM, then an analogous G-P through should be seen shortward of the redshifted wavelength $\lambda 584$Å , or $\lambda 304$Å , respectively. Previous attempts to detect such G-P HeI effect with IUE (Tripp *et al.* 1990) and HST (Beaver *et al.* 1991) failed, giving 3σ upper-limits of $n_{HeI}(z = 1.722) \leq 7 \times 10^{-11} h\,cm^{-3}$ and $n_{HeI}(z = 2.148) \leq 1.1 \times 10^{-10} h\,cm^{-3}$, respectively, for $\Omega = 1$. However, there is a recent claim of detection of absorption below HeII in the quasar Q0302-003 with $\tau > 1.7$ at the 90% c.l. (Jakobsen *et al.* 1994). With the low-resolution of the FOC observations ($R = 100$), it is imposible to distinguish if the absorption is due to small Lyα clouds or to an IGM. Recently, Songaila *et al.* (1996) have studied the same QSO with high-resolution ($R = 36000$) at the Keck, showing that there are many low-column density Lyα clouds that could explain the observed absorption seen with HST without requiring the existence of the IGM.

7 Models of Lyα clouds

Two competing models describing the structure of isolated clouds have been developped: a) pressure confined model (Sargent *et al.* 1980) and b) minihalo model (Rees 1986). More recently, numerical simulations (incorporating gravity and gas dynamics) have been performed in the context of popular models of galaxy formation (CDM), so dealing with the problem of origin, formation and evolution of the clouds.

7.1 Pressure confined models

Sargent *et al.* (1980) interpreted the Lyα systems as intergalactic clouds confined by the pressure of a general intergalactic medium (IGM) whose physical properties were determined on the basis of theoretical and observational constraints ($n_{IGM} \simeq 10^{-5} cm^{-3}, T_{IGM} \simeq 3 \times 10^5 K$). Numerical simulations of spherical, optically thin, pressure confined clouds in an adiabatically evolving intercloud medium lead to discrepancies with observational data (Williger and Babul 1992). They found a deficiency of clouds with $log N_{HI} > 15.5$, the column density distribution for a resticted range of redshift around $z \simeq 2.7$ has a strong cut-off not seen in the data and also the number density of lines evolves with z inconsistently with the observations. A possible way out is to assume non-spherical systems to describe the clouds. Anyway, the properties that are necessary to assume on the IGM are a "fine tuning", the confinement by pressure should emerge through the evolution of the universe and not be fixed a priori.

7.2 Mini-halo model

The minihalo model (Rees 1986; Ikeuchi 1986) assumes that any Lyα cloud consists of stable photoionized gas surrounded by a sperical mini-halo of cold dark matter which confines gravitationally the baryonic component. For uniform density, baryionic density $\Omega_b = 0.1$ and temperature $T = 3 \times 10^4 K$, one obtains a virial velocity for the gas of $16 - 34\,km/s$, that is in agreement with the observational data.

However, the physical properties of the clouds (column densities and velocity dispersions, see section 2.3) indicate that they are very flattened structures if gravitation plays the dominant role defining Lyα clouds as isolated structures (Rodriguez-Pascual *et al.* 1995; Haehnelt 1995). This can be considered as a variant of the previous sperical model, so a cloud can be represented by a sheet-like struture of photoionized gas (baryonic component) surrounded by a dark matter halo which contains the 99% of the mass. Of course, it is interesting to see to what extent this type of ad-hoc structure is generated in the process of evolution of the matter fluctuations in the cosmological context and this will be considered in the next section.

7.3 Numerical simulations

Recently, several numerical simulations have been performed in the context of models of galaxy formation, incorporating gravity and gas dynamics, thus avoiding the idealizations assumed in the analytical calculations.

Petitjean *et al.* (1995), Mücket *et al.* (1995) have studied the evolution of Lyα clouds in a CDM model, normalized to $\sigma_{16} = 1$, performing numerical simulations with a PM code using 128^3 particles. They found that the clouds trace the dark matter filaments and that the number density of lines fits well the data at any redshift for $logN_{HI} > 14$. At low-z, the 2-point correlation function for the clouds has a slope comparable to the one of galaxies, being a 25% of the clouds associated with galaxies.

Hernquist *et al.* (1995) considered an $\Omega = 1$ cold dark matter model (CDM) with baryon density $\Omega_b = 0.05$, normalization $\sigma_{16} = 0.7$ and an ionizing background approximately constant between $2 < z < 3$ in a volume represented by a periodic cube of comoving size $22.222\,Mpc$, using 64^3 SPH particles and 64^3 dark matter particles with masses $1.45 \times 10^8 M_\odot$ and $2.8 \times 10^9 M_\odot$, respectively. The main result is that the Lyα clouds can develop as a natural by-product in such a hierarchical theory of structure formation with an ionizing bakground at high-z. Absorbers with column densities $logN_{HI} \simeq 13 - 15$ are generated in different ways (filaments of warm gas, caustics, high-density halos of hot gas, layers of cool gas and local fluctuations of the IGM) with temperatures in the range $10^4 - 10^5\,K$, being the typical absorber a flattened structure of low-density (with density fluctuation $\delta \simeq 0 - 10$) and b-parameters often dominated by peculiar motions rather than thermal broadening, so no sharp distinction between clouds and smoth IGM can be done. Gravitational and pressure confinement both play a certain role but most of the clouds are far from thermal or dynamical equilibrium. With this type of semi-realistic model, some observational properties of the Lyα clouds (HI and b distributions) can be qualitatively reproduced, but the fact that changing parameters (Zhang *et al.* 1995 assume a COBE normalization and a higher intensity for the ionizing bakground) or assuming a low-density flat model (Cen *et al.* 1994; Miralda-Escudé *et al.* 1995), can also reproduces the same observational properties suggest that such tests are not strong enough to constraint models.

Potential problems of the CDM model include: a deficit of low-column density Lyα clouds and Ly Limit systems, non-Voigt profiles for the lines and differences in the lines along the l.o.s. of double QSOs.

8 Summary

Low-column density systems ($N_{HI} < 10^{17}$) absorbing in the l.o.s. of QSOs define a population called Lyα clouds. Maybe a high proportion of them (50%) are related to the outer-halo of normal galaxies and the rest are tight to the substructure that emerges in the process of galaxy formation.

In the case of $logN_{HI} > 14.2$, they contain metals (CIV) reflecting a certain level of enrichment of the nearby medium or the cloud itself by the on-going star formation process.

Their physical properties (column densities and velocity dispersions) indicate that they are very flattened structures if gravitation plays the dominant role defining Lyα clouds as isolated structures.

The typical size of a Lyα cloud is $\simeq 90\,h^{-1}kpc$ at $z = 1.8$ and $\simeq 170\,h^{-1}kpc$ at $z = 0.7$.

On the other hand, recent high-resolution studies with Keck seem to imply substructure for $logN_{HI} > 14.3$, each Lyα cloud being constituted by 1-3 components.

The correlation function is appreciable on scales up to $\simeq 150\,km/s$ and is compatible with 0 for higher velocity separations, that indicates that the global Lyα population has correlation properties very different from metal line absorbers and galaxies. The population of clouds that exhibit CIV absorption show clustering on scales of $\leq 10^3 km/s$.

The number density of clouds increase with redshift above the expectation in standard Friedmann models, what can be interpreted as evolution driven by merging. This is typical in any hierarchical theory where the merging produces structure from small substructure at any time.

An evolution is observed in the velocity dispersion with redshift that can be interpreted as an increase in the cloud size.

The study of the proximity effect in the neighborhood of the QSOs allows to obtain constrains on the UV-background. The amplitude is approximately constant (or compatible with it) in the range $1.5 < z < 3.5$. This coincides with the era of maximal QSO activity with a declinig below and above such a region in the number of optically detected QSOs.

Recent numerical hydro-dynamical simulations in the context of standard scenarios of galaxy formation (CDM, CDM+Λ) lead to the conclusion that small structure lying in the filaments and pancakes at high-z could be the origin of at least a part of the Lyα cloud population.

Summing up, some light has been shed during the last years taking into account high resolution data taken with 4m telescopes, HST and more recently with the Keck, but relevant questions like:

i) What is the origin, formation and evolution of the Lyα clouds?, what about the relationship to the structure formation problem?

ii) Is there a single population?, or do we need to distinguish between low and high-z populations, weak and strong clouds?

iii) What is the relationship with other absorbers (metal line systems, Lyman limit systems, damped Lyα systems)?

iv) What is the relationship with galaxies and the large-scale structure of the universe?

v) What fraction of baryonic and dark matter is in the clouds?, what is their contribution to the global density of the universe?

vi) What is the chemical composition of the clouds?, is it primordial?

vii) What is the internal structure of the clouds?

viii) Is there an intergalactic medium?

ix) What is the origin and the properties of the ionizing background at high-z?

need to be explored in more detail in the near future, taking into account the capabilities of 8m telescopes, HST and new UV telescopes like FUSE, in order to get a more definitive answeer about the nature and composition of the Lyα clouds.

9 Acknowledgements

I thank my collaborators and colleagues, including E. Martinez-González, P. M. Rodriguez, A. de la Fuente and N. Benitez, for many discussions of the topics presented here. This research has also been supported by a grant (PB92-0741) from the Spanish DGICYT.

References

Bahcall, J. N., Bergeron, J., Boksenberg, A. et al. 1993, ApJS, 87, 1.

Bahcall, J. N. et al. 1996, ApJ, 457, 19.

Bajtlik, S., Duncan, R. C. and Ostriker, J. P. 1988, ApJ, 327, 570.

Barcons, X., Lanzetta, K. M. and Webb, J. K. 1995, Nature, 376, 321.

Beaver et al. 1991, ApJ, 377, L1.

Bechtold, J. 1994, ApJS, 91, 1.

Bechtold, J., Weymann, R., Lin, Z. and Malkan, M. A. 1987, ApJ, 315, 180.

Bechtold, J., Crotts, A. P. S., Duncan, R. C. and Fang, Y. 1994, ApJ, 327,570.

Benitez, N., Martinez-González, E., Sanz, J. L., Aguirre, A. and Alises, M., 1996, submitted to ApJ.

Bergeron, J. and Boissé, P. 1991, A&A, 243, 344.

Carswell, R. F. et al. 1995, MNRAS,278, 506.

Carswell, R. F., Lanzetta, K. M., Parnell, H. C. and Webb, J. K. 1991, ApJ, 371, 36.

Carswell, R. F., Rauch, M., Weymann, R. J., Cooke, A. J. and Webb, J. K. 1994, MNRAS, 268, L1.

Carswell, R. F., Whelan,, J. A. J., Smith, M. G., Boksemberg, A. and Tytler, D. 1982, MNRAS, 198, 91.

Casertano, S., Ratnatunga, K. U., Griffiths, R. E., Im, M., Neuschaefer, L. W., Ostrander, E. J. and Windhorst, R. A. 1995, ApJ, in press.

Cen, R., Miralda-Escudé, J., Ostriker, J. P. and Rauch, M. 1994, ApJ, 437, L9.

Charlton, J. C., Churchill, C. W. and Linder, S. M. 1995, ApJ, 452, L81.

Chernomordik, V. V. 1995, ApJ, 440,431.

Cowie, L. L., Songaila, A., Kim, T.-S. and Hu, E. M. 1995, AJ, 109, 1522.

Cristiani, S. 1995, astro-ph/9512086.

Cristiani, S., D'Odorico, S., Fontana, A., Giallongo, E. and Savaglio, S. 1995, MNRAS, 273, 1016.

Crotts, A. P. S. 1989, ApJ, 336, 550.

Davidsen, A. et al. 1995, AAS meeting, Pittsburgh, June 1995.

Dickinson, M. 1996, in this volume.

Dinshaw, N., Impey, C. D., Foltz, C. B., Weymann, R. J. and Chaffee, F. H. 1994, ApJ, 437, L87.

Dinshaw, N., Foltz, C. B., Impey, C. D., Weymann, R. J. and Morris, S. L. 1995, Nature, 373, 223.

de la Fuente, A., Rodriguez-Pascual, P. M., Sanz. J. L. and Recondo, M. C. 1995, preprint.

Fang, Y., Duncan, R. C., Crotts, A. P. S. and Bechtold, J., 1995, astro-ph/9510112.

Fernández-Soto, A., Barcons, X., Carballo, R. and Webb, J. K. 1995, MNRAS, 277, 235.

Francis et al. 1996, ApJ, 457, 490.

Giallongo, E., Cristiani, S. and Trevese, D. 1992, ApJ, 325, L12.

Giallongo, E., Cristiani, S., Fontana, A. and Trevese, D. 1993, ApJ, 416, 137.

Giallongo et al. 1994, ApJ, 425, L1.

Haardt, F. and Madau, P. 1995, astro-ph/9509093.

Haehnelt, M. G. 1995, astro-ph/9512034.

Hernquist, L., Katz, N., Weinberg, D. H. and Miralda-Escudé, J. 1995, astro-ph/9509105.

Hu, E. M., Kim, T., Cowie, L. L., Songaila, A. and Rauch, M. 1996, AJ, 110, 1526.

Hu, E. M., McMahon, R. G. and Egami, E. 1996, preprint.

Ikeuchi, S. 1986, Astr. Sp. Sci. 118, 509.

Jakobsen, P., Boksenberg, A., Deharveng, J. M. et al. 1994, Nature, 370, 35.

Jenkins, E. B. and Ostriker, J. P. 1991, ApJ, 376, 33.

Kulkarni, V. P. and Fall, M. 1993, ApJ, 413, L63.

Lanzetta, K. M., Bowen, D. V., Tytler, D. and Webb, J. K. 1995, ApJ, 442, 538.

Le Brun, V., Bergeron, J. and Boissé, P. 1995, preprint.

Linsky, J. L., Diplas, A., Wood, B. E., Brown, A., Ayres, T. R. and Savage, B. D. 1995, ApJ, in press.

Lu, L. 1991, ApJ, 379, 99.

Lu, L., Wolfe, A. M., and Turnshek, D. A. 1991, ApJ, 367,19.

Lynds, C. R. 1971, ApJ, 164, L73.

Lowenthal, J. D., Hogan, C. J., Green, R. F., Caulet, A., Woodgate, B. E., Brown, L. and Foltz, C. B. 1991, ApJ, 377, L73.

Madau, P. 1992, ApJ, 389, L1.

Meiksin, A. and Madau, P. 1993, ApJ, 412, 34.

McGill, C. 1990, MNRAS, 256, 43P.

Morris, S. L., Weymann, R. J., savage, B. L., and Guilliland, R. 1991, ApJ, 377, L21.

Miralda-Escudé, J. and Ostriker, J. P. 1990, ApJ, 350, 1.

Mücket, J., Petitjean, P., Kates, R. E. and Riediger, R. 1995, astro-ph/9508129.

Petitjean, P., Mucket, J. and Kates, R. E. 1995, A&A, 295, L9.

Pettini, M., Hunstead, R. W., Smith, L. J. and Mar, D. P. 1990, MNRAS, 246, 545.

Press, W. H. and Ribicki, G. B. 1993, ApJ, 418, 585.

Rauch, M. 1995, astro-ph/9512002.

Rauch, M. Carswell, R. F., Webb, J. K. and Weymann, R. J. 1993, MNRAS, 260, 589.

Rees, M. J. 1986, MNRAS, 218, 25.

Reisenegger, A. and Miralda-Escudé, J. 1995, ApJ, 449,476.

Rodriguez-Pascual, P. M., de la Fuente, A., Sanz, J. L., Recondo, M. C., Clavel, J., Santos-Lleo, M. and Wamsteker, W. 1995, ApJ, 448, 575.

Röser, H. 1995, A&A,299,641.

Rugers, M. and Hogan, C. J. 1995, preprint.

Salpeter, E. E. and Hoffman, G. L. 1995, ApJ, 441, 51.

Sanz, J. L., Clavel, J., Naylor, T. and Wamsteker, W. 1993, MNRAS, 260, 589.

Sargent, W. L. W., Young, P., Boksemberg, M. and Tytler, D. 1980, ApJS, 42, 41.

Shaver, P. A. and Robertson, J. G. 1983, ApJ, 268, L57.

Shull, J. M., Stocke, J. T. and Penton, S. 1996, AJ, 111, 72.

Smette, A., Surdej, J., Shaver, P. A. et al. 1992, ApJ, 389, 39.

Smette , A., Robertson, J. G., Shaver, P. A., Reimers, D., Wisotzki, L. and Kohler, Th. 1995, A&AS, 113, 199.

Songaila, A., Cowie, L. L., Hogan, C. J. and Rugers, M. 1994, Nature, 368, 599.

Songaila, A., Hu, E. M. and Cowie, L. L. 1995, Nature, 375, 324.

Steidel, C. C. and Sargent, W. L. W. 1987, ApJ, 318, L11.

Steidel, C. C., Dickinson, M. and Persson, S. E. 1994, ApJ, 437, L75.

Steidel, C. C., Pettini, M., Dickinson, M. and Persson, S. E. 1996a, MIT preprint No. CSR-94-24.

Steidel, C. C., Bowen, D. V., Blades, J. C. and Dickinson, M. 1996b, preprint.

Steidel, C. C., Giavalisco, M., Pettini, M., Dickinson, M. and Adelberger, K. L. 1996, preprint.

Stocke, J. T., Shull, J. M., Penton, S., Donahue, M. and Carilli, C. 1995, ApJ, 451, 24.

Tripp, T. M., Green, R. F. and Bechtold, J. 1990, 364, L29.

Tytler, D. and Fan, X.-M. 1994, ApJ, 321,49.

Tytler, D., Fan, X.-M., Burles, S. et al. 1995, in *QSO Absorption Lines*, Proc. ESO Workshop, Munich 1995; ed. G. Meylan, Springer-Verlag, Heidelberg, Germany.

Vogel, S. N., Weymann, R., Rauch, M. and Hamilton, T. 1995, ApJ, 441, 162.

Webb, J. K. 1987, in IAU Symp. 124, Observational Cosmology, ed. A. Hewitt, G. Burbidge and L. Z. Fang (Dordrecht: Reidel), 803.

Webb, J. K., Barcons, X., Carswell, R. F., and Parnell, H. C. 1992, MNRAS,255, 319.

Williger, G. M. and Babul, A. 1992, ApJ, 399, 385.

Williger, G. M. et al. 1994, ApJ, 428, 574.

Wolfe, A. M., Fan, X.-M., Tytler, D., Vogt, S. S., Keane, M. J. and Lanzetta, K. M. 1994, ApJ, 435, L101.

Womble, D. S., Sargent, W. L. W. and Lyons, H. S. 1995, preprint.

Zhang, Y., Anninos, P. and Norman, M. L. 1995, preprint.

High–Redshift Radio Galaxies

George Miley

Leiden Observatory, Postbus 9513, 2300 RA Leiden, The Netherlands

Abstract: This is a summary of a personally-biassed review that I gave of our present knowledge about high–redshift radio galaxies.

1 Introduction

During these two lectures I shall give you my personal view of the present state of play in the high-redshift galaxy game. After a general introduction to the subject, summarizing the main differences between distant radio galaxies and their nearby counterparts, I shall concentrate on the intriguing question of what makes the UV/optical/IR continuum radiation in these objects. I shall then describe some recent work which sheds light on this question. We shall see that our knowledge of the responsible radiation mechanism is (to paraphrase Sean O' Casey) "in a state of chassis". I shall conclude with some words about the environment in which high-redshift radio galaxies are located. For those of you who are interested in pursuing some of the topics in more detail I commend you to the recent excellent review by Pat McCarthy and the references therein (McCarthy 1993).

Radio galaxies are unique cosmological probes. As with radio-loud quasars, the presence of luminous radio continuum and optical line emission enable radio galaxies to be observed and recognized at large distances, up to $z = 3.8$. However, unlike the situation for most quasars, their optical emission can be spatially resolved from the ground and studied in detail.

Progress in detecting distant radio galaxies has been rapid in recent years. If someone had told me only a decade ago that I would be standing here discussing galaxies having $z > 2$, I would not have believed them. However, the use of CCDs and the exploitation of new selection criteria has reaped dramatic rewards. Now, more than 60 radio galaxies are known with $z > 2$. More than half of these have been found by our group by concentrating on radio sources with the steepest spectra, most of these in a "Key Programme" of the European Southern Observatory. Although several people contributed to this Key Programme, most of the work was done by Huub Röttgering, who presented his Ph.D thesis in January and Rob van Ojik, who succeeded him.

Redshifts of 1.5 to 4 correspond to a time when the Universe was 10% - 20% of its present age. This was a crucial period in history when galaxy formation must have been rampant. It corresponds to the AGN era, a two-billion year "delta function" in the population evolution of luminous quasars and radio galaxies, when their space-density rose to a value several hundred times larger than the present density before the species mysteriously and suddenly became almost extinct.

Radio galaxies are extremely valuable for studying this period of the Universe because they emit **three components**, (IR-optical-UV continuum, emission lines and radio continuum) that are all **highly luminous and spatially extended**. Not only can different and complementary sets of diagnostics be derived from each of these components, but studies of the relationships between these various diagnostics and the interaction between the different components provides yet additional knowledge about the physical conditions. Such data are relevant for understanding the formation and evolution of galaxies, active nuclei and radio sources.

2 General Properties

The various spectral regimes of high-redshift radio galaxies have several properties that differ from the corresponding components in low-redshift radio galaxies. Although these changes give us information about the high-redshift Universe, we must be cautious of interpreting them as pure evolution. High-redshift radio galaxies are among the most luminous objects in the Universe at almost all wavebands and (as yet) Malmquist bias makes it difficult to disentangle the effects of redshift from those of luminosity.

2.1 The Radio Continuum

High-redshift radio sources have smaller linear sizes than low-redshift ones, indicating the effect of a denser confining medium in the early Universe. There is conflicting evidence that distant radio sources are more bent than nearby ones. Although this effect appears to be present for the radio counterparts of quasars with normal radio spectra (Barthel and Miley 1988), it is absent for radio galaxies with the steepest radio spectra.

2.2 UV/Optical/IR Emission Lines

The associated galaxies have a rich emission-line spectrum, with enormous luminosities (Ly α typically $10^{43.5}$ - $10^{45.5}$ erg/s). The emitting gas is spatially extended in a giant halo, up to 100 kpc, with a characteristic velocity range of 10^3 km/s (FWHM). The usual rituals of interpreting spectroscopy sugest that the source of ionization is anisotropic photoionization from a hidden quasar nucleus and yield characteristic densities of 10^{-2} cm^{-3} and ionized gas masses of /sim 10^{10} M$_\odot$. As was pointed out by McCarthy in his thesis, the ionized gas is morphologically related to the radio sources for $z > 0.1$. In some cases, this can be explained as alignment due to the effects of anisotropic photoionization along the radio axes. In other cases, there is clear evidence for physical interaction between the radio jet and the ionized gas as is seen at lower redshifts most prominently in 3C 277.3 (Coma A) (van Breugel et al. 1985).

2.3 UV/Optical/IR Continuum

The identified objects have R-magnitudes which are typically between 21.5 and 24. Until a few years ago, this radiation was believed to be produced by an old (/sim 10^9 y) population of stars in a "normal" giant elliptical galaxy. The main evidence for this was the remarkably low scatter and continuity in the infrared Hubble diagram (K vs. log z). A cottage industry was devoted to interpreting the spectral energy distributions of such objects in terms of evolving stellar population mixes and using the Hubble diagram to determine H_0 and q_0. The field was as elegant as the beautiful attempts to unify active galaxies on the basis of orientation that we heard so much about earlier today. Unfortunately, it was a theology built on a house of cards! Even though we now know that high-redshift galaxies are not simply elliptical galaxies whose optical properties are unaffected by their radio sources, the continuity and low scatter in the IR Hubble diagram pointed out by Lilly and Longair (1984) are important results that must be explained by any viable model for high-redshift radio galaxies.

A spanner was thrown in the works by the discovery that not only are the giant Ly α halos related to the radio sources, but the UV/optical/IR continua are also preferentially aligned along the radio sources for objects having $z > 0.7$ (Chambers et al. 1987; McCarthy et al. 1987; Chambers et al. 1988; Eisenhardt and Choski 1990; Rigler et al. 1992). We are clearly dealing with more complex systems than giant elliptical galaxies, objects that are significantly different from low-redshift galaxies.

3 Models of the Optical Continuum

Several classes of models have been proposed or considered to account for the alignment effect. Starburst models, interpret the alignment as due to star formation stimulated by the radio jet as it propagates outward from the nucleus (Rees 1989; De Young 1989; Begelman and Cioffi 1989; Bithell and Rees 1990). Two distinct scenarios are invoked. The first postulates a 10^7 y starburst superimposed on an old 10^9 y conventional elliptical galaxy. The second explains the spectral energy distribution in terms of a single relatively young (few x 10^7 y) stellar population whose formation is triggered by the jet.

Another class of models was stimulated by the discovery that, at least in a few cases, the extended emission is polarized in the UV. This prompted the suggestion that the aligned component is due to scattering of light from a hidden quasar by electrons (Fabian 1989) or dust (Tadhunter et al. 1989).

I should mention two additional possibilities that have been proposed to explain the alignment effect. Daly (1992) suggested that inverse Compton scattering of the cosmic background photons by the radio-emitting relativistic electrons might well be the culprit. However, such a model had previously been considered by Chambers et al. (1988) and discounted because of the non-coincidence of the optical and radio morphologies and the huge departures from equipartition of the radio source energetics that would be needed (up to nine orders of magnitude).

The final model I draw to your attention ia a recent proposal by Eales (1992) that the alignment is primarily a statistical effect caused by enhancement of radio emission during interaction of the jet with the medium in the surrounding galaxy. This postulates (i) the radio luminosity increases with ambient density and (ii) distant galaxies have an anisotropic density distribution, elongated along their major axes. I shall come back to this later.

4 Relevant New Results

4.1 Contamination of Continuum by Emission Lines

Most of the evidence for alignment of the UV/optical/IR continuum with the radio emission is based on broad-band images. IR spectroscopy by Eales and Rawlings (1993) indicate that a substantial part of the broad-band K flux is affected by the presence of emission by $H\alpha$ or [OIII] in the band. This casts doubt on the reality of the reported radio/K-band alignments. Table 1 illustrates the problem. This shows the suspect redshift ranges for the four strong lines that dominate the visible spectra. All other lines blueward of Ly α have equivalent widths much smaller than 100Åand therefore contribute less than a few percent to the broad-band fluxes. It is clear that the strong lines are indeed a serious contaminant in the IR, but for $z > 1.2$, they have a negligible effect in the R-band images. Even discounting the lowest redshift objects, there is no doubt in my mind that a strong radio/optical alignment effect exists in the large remaining sample. A similar conclusion is reached by McCarthy (1993). The radio/IR alignment effect is more suspect. However, a beautiful recent image of 4C 41.17 at z = 3.8 taken by the Keck telescope in K' band excludes emission lines, but shows that the light emitted near 5000Å is clearly aligned along the radio axis (Graham et al. 1994).

Table 1. Redshifts of Contamination by Bright Emission Lines

Line	λ (Å)	Optical (R-Band)	Infrared (K-Band)
$Ly\alpha$	1216	$3.8 < z < 5.5$	
$[OII]$	3727	$0.6 < z < 1.2$	
$[OIII]$	5007		$2.8 < z < 3.8$
$H\alpha$	6563		$1.7 < z < 2.7$

4.2 HST Observations

Because the HST can provide information about sub-kiloparsec structure in such objects, Ken Chambers, Wil van Breugel, Duccio Macchetto and I therefore instigated a pilot project to observe two of the most distant galaxies known, 4C 41.17 at z = 3.8 and 4C 28.58 at z = 2.9. (Miley et al. 1992; Miley 1993). Despite the HST aberration, the pilot project was successful. The rest frame ultraviolet continua of both galaxies were well detected and exhibit clumpiness on a scale of a few hundred parsec, optimally suited to mapping with the HST and the observed morphologies were used to constrain models of distant radio galaxies. I will here mention some of the implications:

(i). Clumpiness. Both galaxies observed emit a large fraction of their fluxes (30% to 50%) in clumps. Independent of detailed models, the clumpiness implies that considerable inhomogeneities exist in the inner regions of distant radio galaxies. It is possible that the same inhomogeneities which are responsible for the HST structure are also connected with the clouds responsible for the heavy element absorption systems that sometimes occur close to the emission redshifts of some quasars. If the clumps are stellar, stellar population models would imply a mass in stars of $> 10^{10}$ M_0 contained in each 500 pc clump.

(ii). Connection with Radio. For 4C 41.17 the HST data show clearly that the alignment effect can occur strongly on the sub-kiloparsec scale. The detailed results have implications for models of the radio/optical alignment. The similar radio optical clumpiness on the circumnuclear scale is difficult to reconcile with the model of Eales (1992), for which the radio/optical alignment is a statistical effect whereby the radio emission is enhanced when the jet is emitted along the major axis of the galaxy. The similarity in the radio and optical curvature is suggestive. If the radio/optical bending is observed to be a common property in radio galaxies, it would be an argument against simple scattering models.

We have now reobserved 4C 41.17 with the repaired HST and the new observations confirm our earlier results. In addition, we observe 20 to 30 fainter clumps that are likely to be delineate the formation of this remarkable galaxy.

4.3 Evidence for Young Stars

There are two recent results which provide evidence for the presence of a young stellar population in high-redshift radio galaxies. The first is work by Hammer et al. (1993), who coadded deredshifted spectra of ten 3C radio galaxies with $0.75 < z < 1.1$. Although the resulting spectrum *shows clear absorption features indicating stars*, there is no sign of the well-known $4000\mathring{A}$ break that would be expected from an old stellar population. Hammer et al. conclude that any old population of stars cannot be responsible for more than 15% of the flux from these objects. This really dents the old conservative viewpoint that we are observing elliptical galaxies.

The other relevant observation is the spectrum of TX0211-122, a z=2.3 galaxy discovered during our ESO Key Programme. The spectrum is highly anomalous with a flux ratio of Lyα to NV $\lambda1240$ which is a factor of 30 smaller than for a typical high-redshift radio galaxy. The is unlike that of any other extragalactic object except that of the $z = 2.3$ IRAS galaxy F10214+4724, which is also a substantial radio source. In both objects **depression of Lyα appears to be accompanied by simultaneous enhancement of NV**. The relatively large NV/CIV ratio implies that the line-emitting gas is overabundant in nitrogen. The most feasible explanation for the spectra is that both objects have undergone a period of intense star formation. The most massive stars would evolve quickly, producing (i) the dust responsible for attenuating the Lyα and (ii) a relative overabundance of nitrogen.

4.4 Evidence for Hot Electrons

Although there are energetic problems in explaining the alignment effect purely in terms of electron scattering (Eales and Rawlings, 1990; McCarthy 1993), it is significant that ROSAT has detected X-rays from 3C 356 at $z = 1.079$ (Crawford and Fabian 1993). the X-Ray source has a luminosity of approximately 2.5 x 10^{44} erg s^{-1}, implying the existence of a hot approx. (10^7K) dense ($/\mathrm{sim}10^{-2}$ cm^{-3}) gas cooling at a rate of about 500 - 1500 M$_0$ per year.

4.5 Evidence for Dust

The optical polarization work discussed by Cimatti et al. (1993) shows that high-redshift aligned radio galaxies have polarizations with a strong tendency to be oriented perpendicular to the radio axes. The wavelength dependance of the polarization fraction is unclear, although there is a slight tendency for the polarization fractions to be larger in the UV. Cimatti et al. interpret the data in terms of dust scattering. A dust mass of approximately 10^8 M$_0$ is required. Additional evidence for the presence of dust in at least one object is provided by the small Lyα/NV$\lambda1140$ flux ratio in TX0211-122 described in 4.3.

5 Nature of the Optical Continuum

In the light of the various recent observations, it is useful to reconsider the present state of the game regarding mechanisms for producing the optical continuum. After discarding the Compton scattering and anisotropically enhanced radio emission mechanisms for the reasons discussed above, let us examine problems which we encounter if we take each of the remaining models as sole explanations for the observed IR/optical/UV continua.

5.1 Dust Scattering

Objections to this model are:

(i) Absence of broad emission lines in the scattered light. Although one might speculate that the hidden "god" responsible for all the light is a BLLac object rather than a quasar, given the luminosities required, this would be very contrived.

(ii) Radio-optical Alignment persists into the red and IR. Because of the wavelength dependance of dust scattering, the alignment should only be present in the UV rest frame. Although the line contamination has rendered the IR imaging results unclear, the Keck continuum picture of 4C 41.17 suggests that the alignment effect persists out to at least $5000\mathring{A}$.

(iii) The IR Hubble Diagram. It is difficult to explain the continuity and small scatter in the K vs log z diagram.

(iv) The luminosities of the "hidden gods" would have to be typically at least 16th magnitude at $z = 2$, which is brighter than most known quasars.

5.2 Electron Scattering

Objections to this model are:

(i) The increase of polarization to the UV, if confirmed, would create problems. More data are needed here.

(ii) It is difficult to create enough scattered light without swamping it by stars formed in the cooling flows.

(iii) The IR Hubble Diagram (see 5.1).

(iv) High quasar luminosities are required (see 5.1).

5.3 Old Stars with Young Burst

Objections to this model are:

(i) Where is the $4000\mathring{A}$ break?

(ii) Cannot explain the polarization.

5.4 Young Stars

An objections to this model is that it cannot explain the polarization.

As I warned you at the beginning of my talk, the situation is in one big mess. In my view it is unlikely that any one of the proposed explanations, by itself, can account for the UV/optical/IR continuum or the alignment effect. If I were a betting man, I would back the young star model **together with** one or both of the scattering hypotheses as being **jointly** responsible.

6 The Environment

There are several recent results which provide intriguing information about the region environment of $z > 2$ radio galaxies. I wish to devote the remainder of my talk to discussing these.

6.1 Companions and Clusters

A large number of high-redshift radio galaxies appear to be double, with the axis of the optical double preferentially oriented along the radio axes. Examples are 3C 65, 68.2, 194, 356, 368, TX 0828 +193 and 1436+ 157. A statistical analysis of this effect was made by Röttgering (1993), who considered 31 high-redshift radio galaxies with radio sizes of between 4" and 20". He found that the tendency of "companions" to lie close to the radio axes is significant at the 5% level. Moreover, of the two high-redshift galaxies observed in the HST pilot study, one of them 4C 28.58 is double on the sub-kiloparsec scale.

How can the doubleness of the optical images be explained? One possibility is that we are witnessing galaxy merging. An initial scenario proposed to explain the alignment effect postulated that the axis of the radio source was oriented along the direction of galaxy-galaxy interactions (Djorgovski 1987), but no good mechanism was proposed to explain why this should be so. However, a connection between the radio axis and the merger axis is unnecessary. Suppose that mergers were common during the $z=2$ epoch and that either the starburst or scattering mechanism for enhancing luminosity occurs. When the merging galaxies are located along the beam, their luminosity would be enhanced compared with other merging galaxies and their fluxes become comparable with the parent radio galaxies.

According to such a composite starburst/merger picture a gas-rich companion is in the late stages of merging with the host galaxy of 4C 28.58. The jet responsible for the outer radio emission encounters the nuclear region of this companion, located about 3 kpc from the nucleus of the main galaxy. Shocks from the jet compress the gas, triggering an enormous episode of star formation. Mechanisms similar to those proposed to be occurring in the host galaxy by e.g. Rees (1989) and Begelman and Cioffi (1989) would operate in the companion. Scattering by dust and/or gas in the companion galaxy could also contribute to the observed flux.

Some evidence that the highest redshift galaxies are indeed surrounded by a cluster or group of less luminous galaxies is provided by the Keck picture of 4C 41.17 (Graham et al. 1994). A speculative person might conjecture on this basis that high-redshift radio galaxies are the progenitors of cD galaxies at the centres of still-forming rich clusters. Our recent 10-orbit image with the repaired HST shows more that 4C 41.17 is comprised of more than 20 continuum clumps within a region of several tens of kiloparsec. The image is as might be expected from current models of galaxy formation.

6.2 Ly α Absorption

For the last 25 years one of the most important methods for investigating the early Universe has been to study quasar absorption lines. There are two limitations of such studies due to the fact that the light being absorbed originates in quasars. First, the quasars are unresolved, so in general no information can be obtained about the spatial scale of the absorbers. Secondly, the absorption lines having z_{em} /sim to z_{abs}, i.e. those close to the quasar, are attenuated (presumably ionized) due to their proximity to the QSO. Absorption studies of radio galaxies would be hampered by neither of these effects. We (Hunstead, Röttgering, Miley, van Ojik and Wieringa) therefore recently instigated a project to search for absorption lines in high-redshift radio galaxies. Our initial targets were selected from galaxies having the brightest Ly α. It is in the region of this strong line that we would have the most chance of seeing absorption. Evidence for possible absorption in 4C 41.17 has been given by Hippelein and Meisenheimer (1993).

We have observed two objects with the AAT at a resolution of 1.4Å(FWHM) and are excited about the results. There appears to be several absorption features in both targets. In 0943-242 we see a sharp deep absorption trough that reaches zero intensity and has a half-width of more than 400 km/s. The probability of obtaining such a system by a chance on the basis of the statistics of quasar absorption lines is less than 1 in 1000. The absorbers must therefore be intimately connected with the galaxy itself, or the environment of the galaxy, e.g. a protocluster. The depth of the absorption trough implies that the absorbing cloud at least covers the complete Lyα emission region (> 30 kpc). Although the exact column density involved is dependant on the number of absorbing clouds involved, the value will lie between 10^{18} and 10^{20} cm^{-2}, implying a total mass in HI of approximately 10^7 M_0. There appears to be a velocity shift along the slit of roughly 10 km/s which, if interpreted as rotation would imply a dynamic mass of 10^9 M_0. A radio jet impacting on an object of such a mass could easily be deflected. It is tempting to speculate that the culprit responsible for the Lyα absorption is similar to one of the companion objects visible in the Keck picture of 4C 41.17. This first high-resolution absorption study is certainly enough to whet the appetite for more!

7 Conclusions and the Future

Taking all the various observations mentioned above, I believe that a little imagination coupled with a whiff of speculation and might lead a reasonable astronomer to the following conclusions:

(i) High-redshift radio galaxies are the progenitors of giant elliptical galaxies, located near the centres of protoclusters, maybe at the centres of giant cooling flows.

(ii) They are undergoing vigorous star formation which is producing dust.

(iii) The radio/optical alignments are produced by a combination of jet-induced star formation and scattering of light from a hidden quasar, both in the parent galaxy and in nearby companions.

(iv) The galaxies are located in highly inhomogeneous regions of the $z = 2$ Universe.

High-redshift radio galaxies will provide one of the most important tools for studying the early Universe, using the present and next generations of large telescopes. Their scales are particularly well-suited to study with the HST. Selecting objects with extreme radio spectra from the large-sky surveys at present underway with the Westerbork, VLA and Molonglo telescope will provide a rich harvest for many decades in the future.

Acknowledgements: I thank my collaborators, particularly Huub Röttgering, Richard Hunstead and Rob van Ojik for many useful chats.

References

Barthel, P.D., and Miley, G.K., 1988, Nature, 333, 319.

Begelman, M.C., and Cioffi, D.F., 1989, Ap.J. (Letters), 345, L21.

Bithell, M. and Rees, M.J., 1990, Mon. Not. Roy. Astro. Soc. 242, 570.

Chambers, Miley and van Breugel, 1987, Nature, 329, 604.

Chambers, K.C., Miley, G. K., and Joyce, R.R., 1988, Ap. J. 329, L75.

Cimatti, A., di Serego Alighieri, S., Fosbury, R.A.E., Salvati, M. and Taylor, D., 1993, Mon. Not. Roy Astro. Soc., 264, 421.

Crawford, C.S., and Fabian, A.C., 1993, Mon. Not. Roy. Astro. Soc. , 260, L15.

Daly, R., 1992, Ap. J. 399, 426.

De Young, D. S., 1989, Ap. J. (Letters), 342, L59.

Djorgovski, S., 1987, in Proc. Santa Cruz Astrophysics Workshop on "Nearly Normal Galaxies" ed. S. Faber, Springer Verlag.

Eales, S.A., and Rawlings, S., 1990, Mon.Not.Roy.Astro.Soc., 243, 1p.

Eales, S.A., 1992, Ap. J., 397, 49.

Eales, S. A. and Rawlings, S., 1993, Ap. J., 411, 67.

Eisenhardt, and Chokshi, A., 1990, Ap.J. (Letters), 351, L9.

Fabian, 1989, Mon. Not. Roy. Astro. Soc., 243, 1p.

Graham, J.,et al., 1994, Ap.J., 420,5.

Hammer, F., Le Fèvre, O., and Angonin, M.C., 1993, Nature, 362, 324.

Hippelein, H. and Meisenheimer, K. 1993, Nature 362, 224.

Lilly, S. J., Longair, M. S. 1984, Mon. Not. Roy. Astron. Soc. 211, 833.

McCarthy, P.J, van Breugel, W.J.M., Spinrad,H., & Djorgovski, S. 1987, Ap. J., (Letters), 321, L29.

McCarthy, P.J., 1993, Ann. Rev. Astron. Astrophys., 31, 639.

Miley, G. K., Chambers, K. C., van Breugel, W. and Macchetto, D., 1992, Ap. J. (Lett), 40, L69.

Miley, G. K. 1993, Distant Galaxies in Proc. ST-ECF / STScI Workshop "Science with the Space Telescope" eds. P. Benevenuti and E. Schreier, ESO Conf. and Workshop Proc. No. 44, p1.

Rees, M.J., 1989, M. N. R. A. S., 239, 1p.

Rigler, M.A., Lilly, S.J., Stockton, A., Hammer, F., and Le Fèvre, O.,1992, Ap. J., 385, 61.

Röttgering, H., 1993, Ph.D Thesis, University of Leiden.

Tadhunter, C.N., Fosbury, R.A.E., and di Sergio Aligheri, S., 1989, "Proc. of the Como Conference on BL Lac Objects: 10 Years After", ed. L. Marasci

van Breugel, W.J.M., Miley, G. K., Heckman, T., Butcher, H. and Bridle, A., 1985, Ap. J. 290, 496.

Spectral Evolution of Galaxies

Stéphane Charlot [1,2]

[1]Institut d'Astrophysique du CNRS, 98 bis Boulevard Arago, F-75014
Paris, France
[2]Kitt Peak National Observatory, 950 N. Cherry Avenue, Tucson, AZ
85726, U.S.A.

Abstract: We review models of spectral evolution of galaxies and current constraints
set by observations at high redshifts on the formation and evolution of galaxies. These
lectures assume previous basic knowledge of the general properties of stars and galaxies.

1 Stellar Population Synthesis

1.1 Introduction

The integrated light from star clusters and galaxies should reflect the distribution
of stellar masses, ages, and metallicities within them, providing us with important clues on the past history of star formation. Stellar population synthesis, the
modeling of the spectral energy distribution emitted by specific populations of
stars, is a natural approach to identifying such clues. Studies in this field have
led to the development of population synthesis models, usually termed "evolutionary" (Crampin & Hoyle 1961; Tinsley 1978; Bruzual 1983; Arimoto & Yoshii
1987; Guiderdoni & Rocca-Volmerange 1987; Buzzoni 1989; Bruzual & Charlot
1993, 1996; Bressan, Chiosi, & Fagotto 1994; Fritze von Alvensleben & Gehrard
1994; Worthey 1994; Mayya 1995; Weiss, Peletier, & Matteucci 1995). In such
models the detailed physical processes affecting gas properties and star formation efficiency in a galaxy are generally all reduced to few crude assumptions
about the stellar birthrate (the relative efficiency of the various processes involved are extremely difficult to determine, even in nearby star-forming regions;
e.g., Silk 1996). The main adjustable parameters are then usually the stellar
initial mass function (IMF), the star formation rate, and in some cases the rate
of chemical enrichment. For a given set of these parameters one computes the
time-dependent distribution of stars in the theoretical Hertzsprung-Russell diagram (hereafter H-R diagram), from which the integrated spectral evolution of
the stellar population can be obtained. These models are widely used to study

stellar systems which are too far away for individual stars to be discerned. In particular, they constitute an essential tool of observational cosmology.

We first briefly mention the basic ingredients and simplifying assumptions underlying population synthesis models (see, e.g., Tinsley 1980 for more details). The IMF, noted $\phi(m)$, is defined in such a way that $\phi(m)dm$ is the number of stars born with masses between m and $m + dm$. We use here the normalization

$$\int_{m_L}^{m_U} dm \, m\phi(m) = 1 \, M_\odot. \tag{1}$$

The lower and upper mass cutoffs are still subject to debate (e.g., Scalo 1986, and references therein). Standard values are $m_L = 0.1 \, M_\odot$ and $m_U = 100 \, M_\odot$. Salpeter (1955) has shown from star counts that the IMF in the solar neighborhood is well represented by a single power law

$$\phi(m) = m^{-(1+x)}, \tag{2}$$

with a slope $x = 1.35$. More recent studies suggest that the IMF probably has a flatter slope at lower masses, and that it might even be bimodal, with low-mass and high-mass stars forming in different environments and at different rates (e.g., Larson 1986; Scalo 1986; see Silk 1996 for a review of current theories of star formation).

In general, in population synthesis models the IMF is assumed to be constant in time, space, and metallicity and to be similar to that in the solar neighborhood. The other main free parameters, the star formation and chemical enrichment rates, are then to be constrained by observations. The star formation rate is defined as the mass of gas transformed into stars per unit time,

$$\psi(t) = -dM_g/dt. \tag{3}$$

Stars form heavy elements and return enriched matter into the interstellar medium, especially at the end of their lives. Hence, chemical enrichment is a natural product of stellar evolution in galaxies. The chemical enrichment rate is defined as the increase per unit time in the mass fraction Z of all elements heavier than helium in the gas phase,

$$\chi(t) = dZ/dt. \tag{4}$$

The parameter Z is usually called "metallicity" of the gas, and its value in the solar neighborhood is $Z \approx 0.02$ (the mass fractions of hydrogen and helium are $X \approx 0.70$ and $Y \approx 0.28$, respectively; hence $X + Y + Z = 1$). A full understanding of galaxy evolution therefore requires that chemical and spectral evolution be treated consistently. However, the connection between ψ and χ is not trivial, as it is likely to depend on location, age, and environment. Hence, it appears more tractable at first, if not necessary, to study separately models of chemical and spectral evolution of galaxies. By combining the two types of models, one can hope to eventually arrive at a rough understanding of galaxy evolution.

Spectral evolution of galaxies can be investigated without prior knowledge of chemical evolution because galaxies with any star formation histories can be expanded in series of instantaneous bursts, each having fixed metallicity. Thus, the spectral energy distribution $F_\lambda(t)$ at age t of a stellar population with arbitrary star formation rate and chemical enrichment rate can be modeled by means of a simple convolution integral of the spectra $F_Z(t')$ of instantaneous-burst populations with age t' and metallicity $Z(t - t')$ as,

$$F_\lambda(t) = \int_0^t d\tau \psi(t - \tau) F_{Z(t-\tau)}(\tau). \tag{5}$$

The first goal of population synthesis models is therefore to build accurate spectral evolution models $F_Z(t)$ for instantaneous-burst stellar population with fixed Z, the influence of chemical enrichment in the determination of $Z(t)$ appearing only as the next order complication. This generalized approach to the spectral evolution of galaxies is known as "isochrone synthesis" (Charlot and Bruzual 1991; see §1.2 below). It also implies that the properties predicted for any stellar population can be traced back to those of isochrones predicted for instantaneous burst populations with fixed metallicities. In §1.2 below, we describe the general principles of stellar population synthesis models and review standard results about the spectral evolution of galaxies with different star formation histories. Then, in §1.3, we investigate the main uncertainties underlying the predictions of current population synthesis models and prospects for future improvement.

1.2 Concepts of Spectral Evolution

In this section we describe the various steps involved in the modeling of spectral evolution of galaxies and present the most significant results usually inferred from population synthesis models. To perform isochrone synthesis we must first compute accurate isochrones in the theoretical H-R diagram for instantaneous-burst stellar population with fixed metallicities and in a wide range of ages. An isochrone at age t is defined by the loci in the H-R diagram of stars born coevally at $t = 0$ with masses distributed according to the IMF. This can be computed from the predictions of stellar evolution theory for the evolutionary tracks followed in the H-R diagram by individual stars with given initial mass and metallicity.

We recall that stars can generally be arranged into three main mass ranges, within which the various stages of evolution are similar. *Massive* stars ($m \gtrsim 8\ M_\odot$) go through all nuclear burning phases until the formation of an iron core. Their subsequent fate is either a type II or Ib supernova, followed by the formation of a neutron star or a black hole. *Intermediate-mass* stars have initial masses in the range $M_{\mathrm{HeF}} \lesssim m \lesssim M_{\mathrm{up}}$. Here, $M_{\mathrm{HeF}} \approx 1.9\ M_\odot$ is the minimum mass limit for quiet He-ignition while $M_{\mathrm{up}} \approx 5 - 7 M_\odot$ is roughly the maximum mass limit for degenerate C-ignition (the values of these characteristic masses are somewhat model dependent). After core-He exhaustion, intermediate-mass stars ignite helium in a shell which expands outward, and causes the surrounding hydrogen shell to extinguish. This constitutes the early AGB phase, since stars then undertake their second ascent of the giant branch in the H-R diagram.

Although the detail of the evolution along the AGB depends on initial mass, all stars re-ignite their hydrogen shell at some point and enter a double-shell burning phase characterized by the periodic thermal pulses of the He-shell. This constitutes the thermally pulsing AGB phase. Intermediate-mass stars then lose their envelope, leading to the formation of a planetary nebula nucleus (PNN) that evolves into a white dwarf (WD). The evolution of stars with masses in the range $5 - 7 \lesssim m \lesssim 8\ M_\odot$ is uncertain. They most likely undergo non-degenerate (off-center) C-ignition at the end of their early AGB evolution, and they could either become WDs if they lose enough mass or collapse as neutron stars. Finally, *low-mass* stars $(0.08 M_\odot \lesssim M_{\mathrm{HeF}})$ ignite helium degenerately (the "He-Flash") at the tip of the red giant branch (RGB), afterwhich their evolution is similar to that of intermediate-mass stars.

Figure 1 illustrates the evolutionary tracks followed in the theoretical H-R diagram by stars in the main mass ranges described above. The tracks are from the solar metallicity computations of Bressan et al. (1993). The evolutionary lifetime along the tracks, 90% of which is spent burning hydrogen on the main sequence, is a strong decreasing function of initial mass, roughly $t_{\mathrm{evo}} \approx 10^{10}(m/M_\odot)^{-2}$ yr.

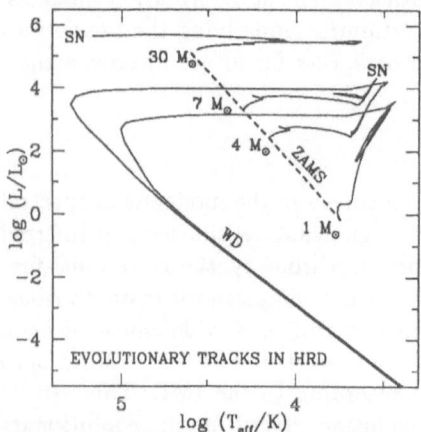 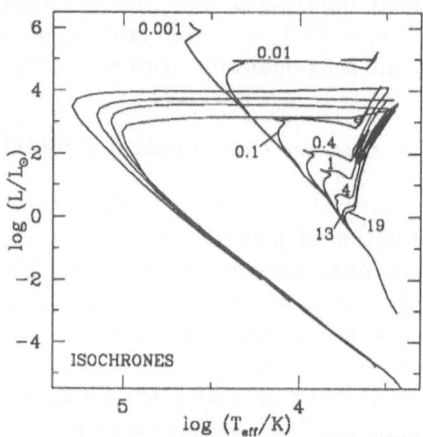

Fig. 1. Evolutionary tracks of stars with solar metallicity in the theoretical H-R diagram (from Bressan et al. 1993). The tracks start on the zero-age main sequence (ZAMS). Low- and intermediate-mass stars terminate their evolution as white dwarfs (WD) and massive stars as supernovae (SN).

Fig. 2. Theoretical isochrones interpolated at various ages, as described in the text, from a complete set of evolutionary tracks of stars with solar metallicity and initial masses in the range $0.1 \le m \le 100\ M_\odot$. Ages are indicated in Gyr next to the isochrones.

We note that the "turnoff" effective temperature and luminosity of a star roughly scale with initial mass as $T_{\mathrm{eff}} \propto m^{0.5}$ and $L \propto m^3$ (a characteristic turnoff marks the end of core-H burning in the H-R diagram when the star leaves the main sequence and starts to cool; both scaling relations flatten at large m). To construct accurate isochrones at various ages for an instantaneous burst stellar population, one must first identify stages of equivalent physical significance (e.g., the end of core-H burning) among a sample of evolutionary tracks for a complete set of initial masses, some stages being defined only in a restricted mass range (e.g., the He-flash for low-mass stars). An isochrone that corresponds to a continuous distribution of initial masses is then defined by interpolation from the evolutionary tracks at a fixed age t of the loci in the H-R diagram of stars in all evolutionary stages. Figure 2 shows examples of isochrones computed using a full set of 28 evolutionary tracks for solar-metallicity stars from Bressan et al. A total of 310 evolutionary phases were identified to define the isochrones. The number of stars in each evolutionary phase along the isochrones is given by the IMF weight, $\phi(m)$, and the (interpolated) time spent in that phase.

The integrated spectral energy distribution of the stellar population at a given age is simply the sum of the spectra of stars along the isochrone at that age. Before we proceed with a detailed modeling, it is worth noting that the behavior of the integrated optical spectrum of a stellar population can be guessed from a simple reasoning neglecting the contribution from advanced stages of stellar evolution (evolved stars contribute mainly at infrared wavelengths; see below). We first remark that the spectrum of a star can be roughly approximated by that of a blackbody of same temperature and luminosity related by the Stefan-Boltzmann law, $L = 4\pi R^2 \sigma T_{\mathrm{eff}}^4$.[1] Therefore, the wavelength of maximum intensity of a stellar spectrum is, according to Wien's law, around $\lambda_{\mathrm{max}} = 2.9\,(T_{\mathrm{eff}}/1000\mathrm{K})^{-1}\,\mu\mathrm{m}$ (if intensity is defined per unit wavelength). Since massive stars have larger luminosities and higher temperatures than low-mass stars, their spectra are brighter and peak at shorter wavelengths. Photometrically, this implies that massive stars are bluer than low-mass stars (see eq. [6] below). Thus, if stars of all masses are assumed to be born at once and with an IMF similar to that in the solar-neighborhood (eq. [1]), the fractional contribution to the integrated spectrum by stars of mass m will scale initially as $L(m)\phi(m) \propto m^{0.65}$. We then expect blue, short-lived massive stars to strongly dominate the optical spectrum at early ages. After these stars disappear, the optical spectrum must fade rapidly and redden as less massive, cooler stars dominate the light.

We now investigate in more detail the spectral evolution of an instantaneous burst stellar population. To begin with, we show in Figure 3 how the actual spectrum of a star (a solar-metallicity, main-sequence star with $\log T_{\mathrm{eff}} = 3.6$) compares with that of a pure blackbody of same temperature. Although the two spectra have similar shapes, line blanketing by many atomic transitions in the stellar atmosphere reduces substantially the emerging radiation with respect to a pure blackbody model. Therefore, stellar spectra cannot be reliably approximated by unblanketed blackbody spectra, and one of the main difficulties in

---

[1] The present argument is useful mostly to provide a schematic idea of the optical spectral properties of a stellar population. In reality, the radius entering the Stefan-Bolztmann relation is not well defined for evolved (post-main sequence) stars with extended thick envelopes, in which radiation emerges from different radii at different wavelengths. In such cases, T_{eff} can be considerably larger than the actual radiation temperature at most wavelengths.

building a population synthesis is to assemble a complete library of individual spectra for stars in wide ranges of temperature, luminosity, and metallicity.

Figure 4 shows the spectral evolution of an instantaneous burst stellar population with solar metallicity computed using such a complete library of stellar spectra and summing the spectra of individual stars along the isochrones shown in Figure 2 (from Bruzual & Charlot 1996). As expected, at 10^6 yr the spectrum is entirely dominated by short-lived, young massive stars on the main sequence and peaks in the ultraviolet. After 1×10^7 yr, the most massive stars have evolved off the main sequence and become red supergiants, making the ultraviolet light decline and the near-infrared light rise. From a few times 10^8 yr to $\gtrsim 1 \times 10^9$ yr, the AGB stars maintain a high near-infrared luminosity, and the ultraviolet light continues to drop as the turnoff mass decreases on the main sequence. After a few times 10^9 yr, the RGB takes over the production of the near-infrared light. The rise in the far ultraviolet after $4-10$ Gyr is produced by low-mass stars in their post-AGB evolution. The most remarkable feature in Figure 4 is the nearly unevolving shape of the optical to near-infrared spectrum at ages from 4 to 19 Gyr. The reason for this is that low-mass stars evolve from the main sequence to the end of the AGB in a small range of temperatures (see Fig. 2).

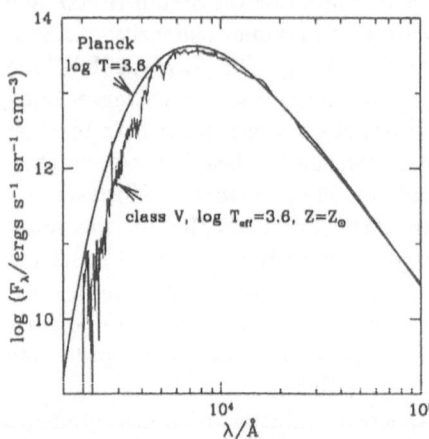

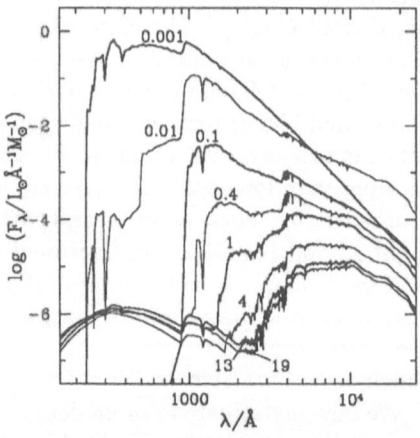

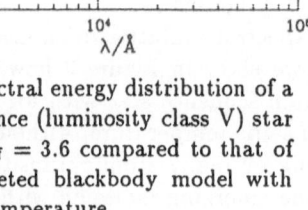

Fig. 3. Spectral energy distribution of a main sequence (luminosity class V) star with $\log T_{\text{eff}} = 3.6$ compared to that of an unblanketed blackbody model with the same temperature.

Fig. 4. Spectral evolution of an instantaneous burst stellar population with solar metallicity and a Salpeter IMF. Ages (in Gyr) are indicated next to the spectra.

For galaxies about which detailed spectral information is not available, colors indicating the flux emitted in broad spectral bands represent a useful alternative. Several systems can be chosen to define colors from a spectral energy distribution. In the standard color system, an $M_i - M_j$ "color index" (or simply "color") between two spectral bands M_i and M_j centered around two wavelengths $\lambda_i < \lambda_j$ is defined by

$$M_i - M_j = \text{zeropoint} - 2.5 \log(F_i/F_j), \tag{6}$$

where $F_k = \int d\lambda R_{k,\lambda} F_\lambda$ (in units of erg s^{-1} cm^{-2}) is the convolution of the spectrum F_λ by the response function $R_{k\lambda}$ of the kth filter. Hence, by definition, a color is bluer when it is more negative. The zeropoint is defined in such a way that for the standard A0V star Vega, $M_i - M_j$ is always zero and the V magnitude is 0.03 (see Thuan & Gunn 1976; and Oke & Gunn 1983 for definitions of other photometric and spectrophotometric systems). For reference, the effective wavelengths (and bandwidths) of the standard U, B, V, R, I, J, K, and L filters are, in μm, 0.359 (0.065), 0.441 (0.104), 0.547 (0.118), 0.652 (0.087), 0.827 (0.211), 1.33 (0.285), 1.71 (0.280), 2.15 (0.394), and 3.50 (0.540) respectively. As an application, Figure 5 shows the evolution of the $U - B$ and $V - K$ colors and mass-to-visual light ratio, M/L_V, for the instantaneous burst stellar population of Figure 4. As expected, the colors are bluest at early ages and redden progressively as the most massive stars terminate their evolution. However, after only a few billion years, the evolution is very slow. The mass-to-light ratio increases steadily with time because while the total mass of the stellar population remains constant, the V-band luminosity decreases together with the turnoff mass.

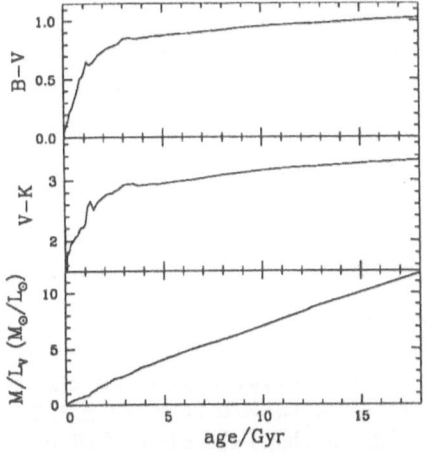
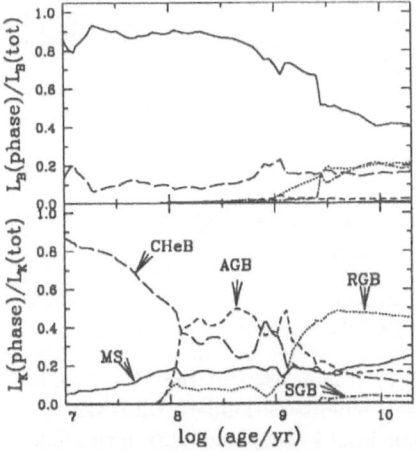

Fig. 5. Evolution of the $B - V$ and $V - K$ colors and mass-to-visual light ratio M/L_V for an instantaneous burst stellar population with solar metallicity and a Salpeter IMF.

Fig. 6. Fractional contributions to the B and K light of the stellar population in Fig. 5 by stars on the main sequence and various post-main sequence evolutionary phases (see text for acronyms).

As an instructive complement to Figure 5, Figure 6 illustrates separately the evolution (here as a function of log age) of the contributions to the integrated B and K light by stars on the main sequence (MS) and in various post-main sequence evolutionary stages: subgiant branch (SGB; for low-mass stars), RGB, core-He burning phase (CHeB; which includes the supergiant phase for massive stars), and AGB. Main-sequence stars (essentially near the turnoff) dominate the integrated B light all ages. This justifies a posteriori the arguments used previously to describe spectral evolution in the optical from rough scaling relations. In the K band, however, main sequence stars never account for more than 20% of the integrated light. The emission is dominated first by supergiants, and then by AGB descendants of intermediate-mass stars from about 1×10^8 to 1×10^9 yr. At later ages, the RGB progeny of low-mass stars accounts for nearly half the integrated infrared light.

The photometric evolution of galaxies with continuous star formation rates can be readily understood from equation (5) and the photometric evolution of an instantaneous burst stellar population in Figure 4. We fix for the moment the metallicity to the solar value and compute the spectral evolution of stellar populations with exponentially declining star formation rates $\psi(t) = \exp(-t/\tau)/\tau$, with timescales of star formation $\tau = 1, 2, 4, 7$ Gyr, and ∞ (corresponding to $\psi(t) =$constant). The results are shown in Figure 7, along with the mean observed colors of nearby galaxies of various morphological types (E to Irr).

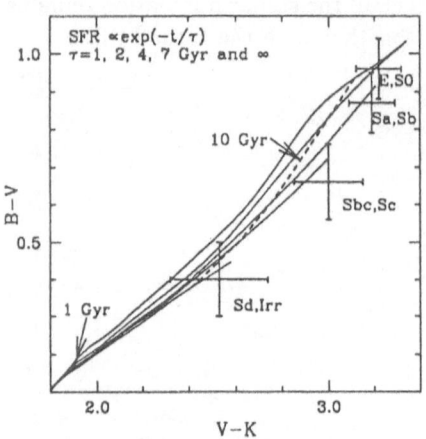 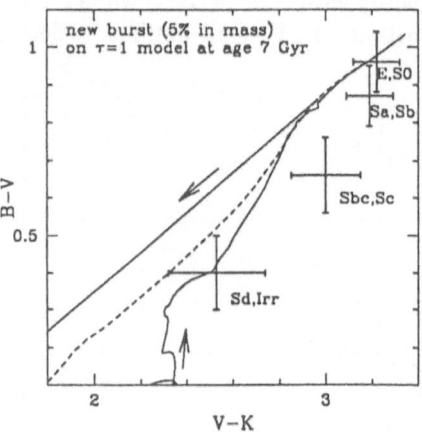

Fig. 7. Comparison of the $V - K$ and $B - V$ colors of models with various exponentially declining star formation rates with the observed colors of nearby galaxies of different morphological types. Timescales of star formation (in Gyr) are indicated next to the curves. Dashed lines show the location of the models at 1 and 10 Gyr.

Fig. 8. The dashed line corresponds to the model with $\tau = 1$ Gyr in Fig. 7. The solid line shows the effect of adding a new burst of duration 10^8 yr on this model at an age of 7 Gyr, in which 5% of the final galaxy mass in stars is formed. Arrows indicate the sense of the rapid blueing and subsequent reddening that follow burst.

The models start in the blue region of the BVK plane (lower left) and evolve redward. The two dashed lines join the locations of the various models at fixed ages of 1 and 10 Gyr. All models follow a similar path in the BVK plane until about $\lesssim 1$ Gyr. Then, models with longer timescales of star formation remain relatively blue, evolving slowly toward the colors of later-type Spirals, whereas rapidly star-forming models redden faster and lead to the colors of early-type E/S0 galaxies. In fact, ongoing star formation in models with large τ's can maintain blue colors for as long as the stellar population is replenished in massive stars. The modest reddening undergone by these models, which is more pronounced in $V - K$ than in $B - V$, is caused mainly by the accumulation old, low-mass RGB stars. As soon as star formation drops significantly, at ages $t > \tau$, the most massive stars disappear and colors redden rapidly.

The strong dominance of young massive stars in any stellar population has important consequences for determining the history of star formation in galaxies from their observed spectral energy distribution. As an introduction to this issue, we show in Figure 8 the evolutionary path in the BVK plane of a model galaxy with a short star formation timescale, $\tau = 1$ Gyr, on which a new burst of duration 10^8 yr and involving 5% of the final galaxy mass is superimposed at an age of 7 Gyr. The new addition of massive stars blues instantaneously the colors. In fact, less than 1 Gyr after the burst, the model galaxy has colors typical of a nearby irregular galaxy. Then, when massive stars terminate their evolution, the colors of the model galaxy in Figure 8 redden rapidly again, and less than 3 Gyr after the burst the galaxy appears virtually identical to what it would in the absence of recent star formation.

Figures 7 and 8 together suggest that, even at fixed metallicity, colors are degenerate indicators of the the past history of star formation in galaxies. Within less than 0.1 mag in $B - V$ and $V - K$ colors, an irregular galaxy can be modeled by either a 10 Gyr old stellar population with constant star formation rate, or a younger stellar population with a declining star formation rate, or even by an old, passively evolving stellar population with a small amount of recent star formation. Furthermore, E/S0 galaxies appear to have colors essentially characteristic of any stellar population in which star formation has ceased for at least a few billion years. This degeneracy of the appearance of galaxies relative to the past history of star formation remains even when the whole spectrum from the optical to infrared is considered instead of simply broad-band colors (e.g., Bruzual & Charlot 1993). This leads us to one of the most important conclusions about spectral evolution: the spectral continuum shape of a galaxy depends essentially only on the ratio of the present to past-averaged star formation rates. This may be roughly estimated as

$$\frac{\psi(\text{present})}{\langle \psi \rangle} \approx \frac{\psi(\text{present}) \times \text{age}}{M_*}, \tag{7}$$

where M_* is the total mass in stars. Evidently, $\psi(\text{present})/\langle \psi \rangle$ does not provide us with any information on absolute ages nor on the details of the past history of star formation. It is remarkable to note, however, that this ratio correlates well with galaxy morphological type. The observed values range from 0 for E/S0 galaxies to 0.1 for Sa/Sab galaxies, 1 for Sbc/Sc galaxies, and up to a few for Sm/Im galaxies (Gallagher, Hunter, & Tutukov 1984; Kennicutt, Tamblyn, & Congdon 1994).

Another useful prediction of population synthesis models is mass-to-light ratio, as this can ultimately help us to estimate the mass of a distant galaxy from its observed luminosity. Since very low-mass stars on the lower main sequence are very faint, they can contribute significantly to the mass but not to the integrated light of a whole stellar population. Therefore, the lower cutoff of the IMF is a key uncertain parameter in determinations of mass-to-light ratios. The upper cutoff is a less critical parameter if the slope of the IMF is similar to that in the solar neighborhood. The reason for this is that, near the cutoff ($m \gtrsim 50\,M_\odot$), the luminosity depends less strongly on main sequence mass ($L \propto m^2$ instead of m^3 at turnoff). Hence the integrated mass and luminosity both depend very weakly on m_U, roughly as $\partial M/\partial m_U \sim m_U \phi(m_U) \propto m_U^{-1.35}$ and $\partial L/\partial m_U \sim m_U^2 \phi(m_U) \propto m_U^{-0.35}$, respectively.

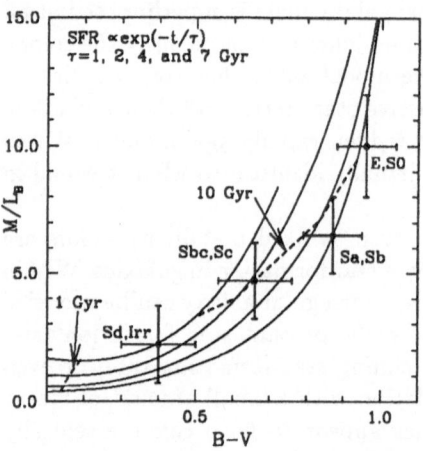

Fig. 9. Mass-to-blue light ratio versus $B - V$ color for the same models as in Fig. 7, compared to the values observed in nearby galaxies of various morphological types.

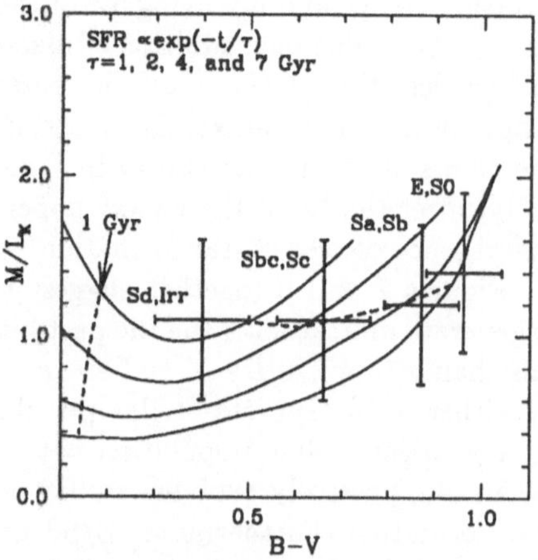

Fig. 10. Mass-to-infrared light ratio versus $B-V$ color for the same models as in Fig. 7, compared to the values observed in nearby galaxies of various morphological types.

Figures 9 and 10 show M/L_B and M/L_K as a function of $B - V$ color for models with the same range of star formation histories as in Figure 7. Traditionally, such ratios are expressed in units of $M_\odot/L_\odot$, where $L_\odot$ stands for the solar luminosity in the band under consideration (i.e., $L_{\odot B}$ and $L_{\odot K}$ in Figs. 9 and 10, respectively). As expected from Figure 7, the mass-to-blue light ratio increases with increasing color because, as a stellar population ages, the optical luminosity declines and the colors redden. However, Figure 10 shows that the mass-to-infrared light ratio is almost insensitive to the star formation history (note that the vertical scale is 5 times smaller in Fig. 10 than in Fig. 9). This is all the more remarkable in that different types of stars dominate the infrared light at different ages in an evolving stellar population (supergiants, AGB, and RGB stars; see Fig. 8). This interesting property renders mass estimates from the infrared light of galaxies more reliable than estimates from the optical light, although the infrared light does not traces only old (RGB) stars, as is often assumed. We also note that, again, models with timescales of star formation ranging from 1 to 7 Gyr are found to reproduce the mass-to-light ratios observed in nearby galaxies of early to late morphological types, respectively. One of the reasons for this success is that observations refer to the mass within the Holmberg radii of the galaxies, i.e., where the photographic (essentially V) surface brightness reaches 26.5 mag arcsec^{-2}. Within this radius, the contribution to the mass by dark matter halos is not expected to be larger than 25% (Ostriker & Caldwell).

1.3 Uncertainties in the Modeling of Stellar Populations

The use of population synthesis models to recover the mixture of stars in a galaxy from observations of the integrated light suffers from various uncertainties. As we have seen above, part of the problem is that spectral evolution depends essentially only on the ratio of the present to past-averaged star formation rates. Another problem is that both age and metallicity have a similar effect on integrated colors and line strengths. As emphasized by Worthey (1994), if two stellar populations differ in age and metallicity by $d\log\mathrm{age}/d\log Z \approx 3/2$, they will appear virtually identical in all colors and most optical spectral indices (such as the characteristic absorption lines of prominent atoms and molecules). To illustrate this, we show in Figure 11 the evolution of the $B - V$ and $V - K$ colors and M/L_V mass-to-light ratio for instantaneous-burst stellar populations with fixed IMF and different metallicities. The main effect of increasing metallicity is to redden the colors and increase M/L_V at fixed age, which is similar to the effect of increasing age at fixed metallicity. The reason for this is that, at higher metallicity, stellar effective temperatures and luminosities are lower (see, e.g., Schaller et al. 1992; Fagotto et al. 1994).

Unfortunately, this "age-metallicity degeneracy" is further aggravated by significant differences among existing models in the predicted spectral properties of a stellar population with fixed age and metallicity. These discrepancies are caused by uncertainties in the main underlying assumptions of population synthesis models: the stellar evolution theory used to predict the distribution of stars in the theoretical H-R diagram and the library of spectra assigned to stars as a function of temperature, luminosity, and metallicity from which colors are calculated. Figure 12 shows the $B - V$ and $V - K$ colors and M/L_V ratio predicted for an instantaneous burst stellar population with solar metallicity and fixed IMF according to several recent population synthesis models (Bruzual & Charlot 1996 using the Geneva [B&C] and Padua [B&C alt.] tracks; Bertelli et al. 1994 [BBCFN]; Worthey 1994 [GW]). The deviations of 0.05 mag in $B - V$ color, 0.25 mag in $V - K$ color, and 25% in M/L_V are large compared to typical observational uncertainties, and they imply alarming ambiguities in the interpretation of galaxy colors. Charlot, Worthey, & Bressan (1996) have investigated in detail the sources of disprepancies in the predictions of current population synthesis models. They conclude that the main source of disagreement is the stellar evolution theory that relies on several critical factors which are either not sufficiently understood or cannot yet be determined uniquely from comparisons with observations (opacities, heavy element mixture, helium content, convection, diffusion, mass loss, rotational mixing). Differences in spectral calibrations appear to have smaller consequences, although some major limitations are the difficult spectral modeling of cool stars and the unavailability of calibration stars for metal-rich populations and populations with altered chemical mixes.

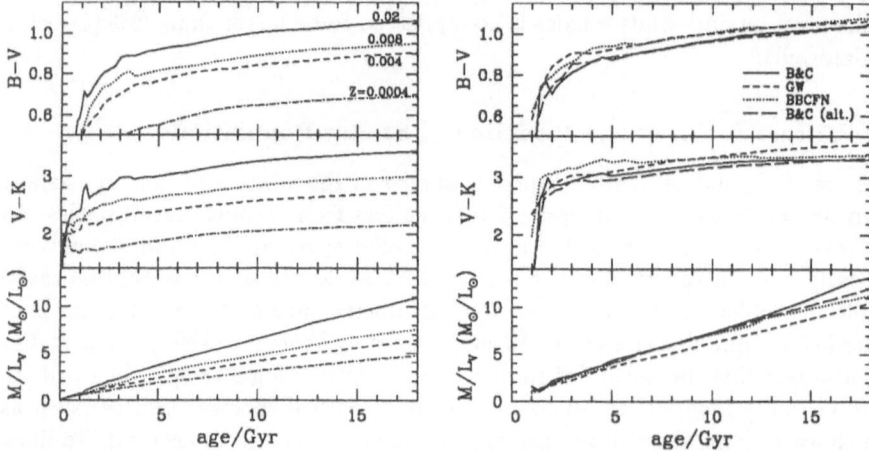

Fig. 11. Evolution of the $B-V$ and $V-K$ colors and M/L_V ratio for instantaneous burst stellar populations with different metallicities and a Salpeter IMF. Increasing metallicity reddens the colors and lowers the luminosity at fixed age.

Fig. 12. Evolution of $B-V$, $V-K$, and M/L_V for an instantaneous burst stellar population with solar metallicity and a Salpeter IMF, according to various recent population synthesis models (see text for sources).

It is worth noting that the significant differences in the predictions of the various models in Figure 12 cannot be straightforwardly resolved by means of comparisons with observations. In fact, the models all compare to the observed colors of star clusters and galaxies approximately in a similar way. This is shown in Figure 13, in which the $B-V$, $U-V$, and $V-K$ colors of the solar-metallicity models of Figure 12 are plotted for ages 1 to 18 Gyr against integrated colors for Galactic and M 31 globular clusters (Burstein et al. 1984), LMC clusters older than about 1 Gyr (van den Bergh 1981; Persson et al. 1983), and elliptical and S0 galaxies (Frogel et al. 1978; Peletier 1989). The general agreement between model and observed color sequences is all the more remarkable in that the data, unlike the models, include a wide range of metallicities. Globular clusters have metallicities in the range $0.0001 \lesssim Z \lesssim 0.006$ and are presumably uniformly old (e.g. Hesser 1993). Elliptical and S0 galaxies probably contain several generations of stars (see also §2.4 below), although they usually are not expected to have experienced large recent bursts of star formation. According to González (1993), the dominant stellar populations in these galaxies would have metallicities slightly less than twice solar. The younger LMC clusters in Figure 2 (with ages $\gtrsim 1$ Gyr) have $Z \approx 0.01$, but the older clusters are probably as metal-poor as Galactic globular clusters (e.g., Cohen 1982).

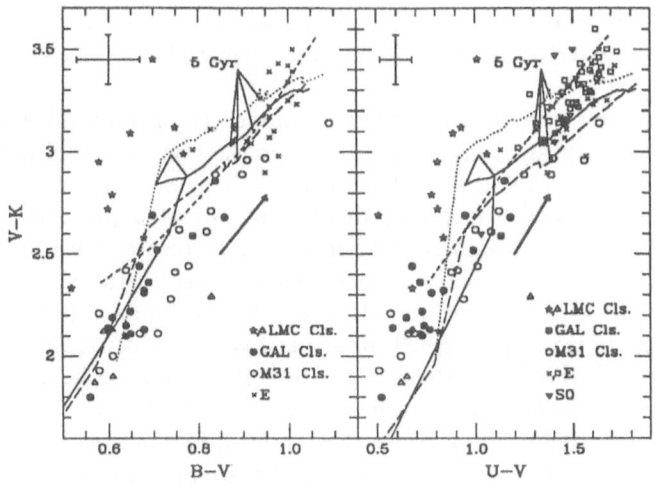

Fig. 13. Comparison of the $B - V$, $U - V$, and $V - K$ colors predicted by the solar-metallicity models of Fig. 12 at ages 1 to 18 Gyr with observed colors for star clusters and galaxies (all corrected for reddening). Model loci at 5 Gyr are also marked. The observations include Galactic and M 31 globular clusters (Burstein et al. 1984), LMC clusters (van den Bergh 1981; Persson et al. 1983), and nearby elliptical and S0 galaxies (*crosses*: Peletier 1989; *squares* and *upside down triangles*: Frogel et al. 1978). The LMC clusters are divided into two categories: upright triangles refer to the oldest clusters (type VII in the Searle, Wilkinson, & Bagnuolo 1980 classification), and stars refer to younger clusters (types IV, V, and VI). Conservative maximum observational error bars are indicated on the top left of each panel. Arrows representing the effect of reddening for $E(B - V) = 0.1$ mag indicate how extinction can further aggravate the age-metallicity degeneracy.

Figure 13 illustrates the lack of clear diagnostic diagrams in integrated broad-band colors for population synthesis models. For example, while at fixed metallicity colors tend to redden with age (Fig. 5), the higher metallicity of the younger LMC clusters in Figure 13 causes their colors to be redder than those of the old, more metal-poor clusters. We note that the scatter in $V - K$ color for the younger LMC clusters is caused by the high sensitivity of the integrated light to the presence of even a few extremely cool TP-AGB carbon stars in these objects. Furthermore, the relatively blue colors in Figure 13 of one S0 galaxy of nearly solar metallicity which recently underwent a burst of star formation (the local group galaxy NGC 205 with $V - K = 2.1$ and $U - V = 0.8$; e.g., Davidge 1992) coincide with the colors of the most metal-poor, oldest clusters in the Galaxy. This provides dramatic empirical evidence of the severity of the age-metallicity degeneracy. Also, the overall match by solar-metallicity models to the color sequences of star clusters and early-type galaxies is another example of this degeneracy. Extinction can further aggravate the age-metallicity degeneracy, as reddening vectors are parallel to the age/metallicity color sequences in Figure 13.

The main interest in population synthesis models is generally the determination of age, metallicity, and mass of galaxies from their observed spectral energy distributions. The uncertainties associated to such determinations can be estimated from the dispersion in the predictions of current population synthesis models relying on different input stellar models and spectra. Such an analysis indicates that, for idealized galaxies containing a single generation of stars and no dust, properties derived from broad-band colors and most optical spectral indices are accurate by roughly $\pm 35\%$ in age at fixed metallicity, 25% in metallicity at fixed age if the heavy element mixture is assumed to be scaled-solar, and 35% in mass at fixed metallicity and fixed IMF (Charlot et al. 1996). Even for idealized galaxies defined in this way, the uncertainties in age and metallicity determinations will increase, as expressed by the age-metallicity degeneracy, if both quantities are allowed to vary. For real galaxies containing several generations of stars with different metallicities (and perhaps even dust), determinations of age, metallicity, and mass of the various components will be far more uncertain. Finally, we note that in the gas-rich environments of star forming galaxies, nebular emission by recombination of hydrogen, helium, and heavy elements ionized by young stars is another major source of uncertainties as it can contribute significantly to the integrated light (Charlot 1996; García-Vargas, Bressan, & Leitherer 1996).

2 The Appearance of Galaxies at High Redshifts

2.1 Introduction

The expected widespread population of high-redshift analogs or progenitors of present-day disk galaxies has still not been observed in emission. This leaves large uncertainties on the appearance of young galaxies (see White 1989; Pritchet 1994). Early models predicted that "primeval galaxies" undergoing strong starbursts at redshifts $z \gtrsim 2$ should be detectable at magnitudes $R \sim 21 - 23$, but deep spectroscopic surveys to $B \lesssim 22.5$, $I \lesssim 22.1$, and $K \lesssim 20$ have not revealed such a population of forming galaxies (Partridge & Peebles 1967; Meier 1976; Colless et al. 1993; Cowie et al. 1994; Lilly et al. 1995). Partridge and Peebles (1967) pointed out that Lyα emission could be the most prominent and easily detectable signature of primeval galaxies. The reason for this is that the ionizing radiation from young stars in galaxies should lead to a strong and narrow Lyα line by recombination of the hydrogen in the ambient interstellar medium, which would be more readily visible than continuum radiation against the sky noise. A few galaxy-like objects have been discovered at redshifts $z \lesssim 4$, which occasionally show strong Lyα emission (Pritchet 1994; Spinrad 1989; Macchetto et al. 1993; Rowan-Robinson et al. 1993). However, most of these objects are peculiar, and their connection to present-day galaxies is not at all clear. In fact, all blank sky searches for Lyα emission from ordinary galactic disks at high redshifts have given null results, suggesting that young galaxies form stars slowly or in large volumes, or that Lyα photons are absorbed by dust (see Baron & White 1987, and references therein).

Independently, much has been learned on the distribution of HI in the universe at redshifts $z \lesssim 3.5$ from absorption-line studies of distant quasars (see for example Petitjean et al. 1993). The strongest absorption lines are attributed to the damped Lyα systems, that are generally interpreted as the best candidates for ordinary galactic disks at high redshift (Lanzetta et al. 1991). These have observed HI column densities $N_{HI} \gtrsim 2 \times 10^{20}\,\mathrm{cm}^{-2}$, and their abundances in heavy elements and dust at $z \approx 2.5$ amount to about 10% of the values in the Milky Way (Pei et al. 1991; Pettini et al. 1994). The damped Lyα systems do not show Lyα emission at the level expected from young, dust-poor disk galaxies (see Pettini et al. 1994). Alternatively, the evolution with redshift of the gas density integrated over all HI absorbers, $\Omega_{HI}(z)$, provides some constraint on the global depletion of cold gas though star formation in the universe since $z \approx 3.5$ (Lanzetta et al. 1995; Pei & Fall 1995). Absorption-line systems of distant quasars have also been used successfully as a way to select normal galaxies in emission out to $z \approx 1.6$ against the population of relatively nearby blue galaxies that dominate galaxy number counts at faint magnitudes (Bergeron 1988; Steidel & Dickinson 1995; Steidel et al. 1995; Aragón-Salamanca et al. 1994). These studies have shown that field galaxies with luminosities around L^* exhibit only little evolution in their space density, luminosity, and optical/infrared colors at redshifts $0.2 \lesssim z \lesssim 1.6$.

Recent observations therefore seem to indicate that the formation and evolution of normal disk galaxies has been less spectacular than originally thought. In what follows, we explain how the apparent lack of Lyα emission from young galaxies at high redshift is probably mainly a consequence of the relatively brief periods in which primeval galaxies are dust-free, and hence Lyα-bright. In fact, most present observational constraints on young galaxies appear to be in agreement with the predictions of theories based on hierarchical clustering, in which galaxies form slowly and relatively recently. The very blue, primeval galaxy phase expected at the onset of star formation would then be faint and short-lived. Since at redshifts $z \gtrsim 2$ galaxies are expected to be difficult to detect, one may think of using population synthesis models to trace back the early history of star formation from observations at lower redshifts. We also discuss below the limitations of this approach.

2.2 Lyman-Alpha Emission from Young Galaxies

The observed Lyα emission from a young galaxy depends on the star formation rate and IMF, but also on several other factors: the contributions by supernova remnants and active galactic nuclei, the orientation of the galaxy, and absorption by dust. The contribution to the Lyα emission by stars can be estimated using stellar population synthesis models. We assume for the moment that circumstellar HII regions are the only sources of Lyα photons and that the column density of the ambient HI is large enough that case B recombination applies

($N_{HI} \gtrsim 10^{17}\,\text{cm}^{-2}$) but otherwise ignore the effects on the interstellar medium on the transfer of Lyα photons. Under these "minimal" assumptions and using recent population synthesis models, Charlot & Fall (1993) have shown that the Lyα emission from a galaxy depends sensitively on the age and IMF slope, even when the star formation rate is constant. The dependence on the IMF upper cutoff and metallicity, on the other hand, are much weaker. Thus, only a rough estimate of the Lyα equivalent width of a young, dust-free galaxy is permitted, about $50 - 120$ Å.

We now briefly review the other factors that can affect the observed Lyα emission from a young galaxy (see Charlot & Fall 1993 for more details). Shull & Silk (1979) have computed the time-averaged, Lyα luminosity of a population of Type II supernova remnants using a radiative-shock code with low metallicity. Their results indicate that the contribution to the Lyα emission by supernova remnants is always less than the contribution by stars (typically 10% for a solar-neighborhood IMF) and can therefore be neglected. Active Galactic Nuclei (AGNs) are another potential source of ionizing radiation in a galaxy. We assume for simplicity that the spectrum of an AGN can be approximated by a power law $f_\nu \propto \nu^{-\alpha}$ with an index blueward of Lyα in the range $1 \lesssim \alpha \lesssim 2$. If we also assume that the AGN is completely surrounded by HI, that case B recombination applies, and that absorption by dust is negligible, then the Lyα equivalent width is $827\alpha^{-1}(3/4)^\alpha$ Å, or 600 Å for $\alpha = 1$ and 200 Å for $\alpha = 2$. The fact that most bright quasars have observed Lyα equivalent widths in the range $50 - 150$ Å could reflect a partial covering of the AGNs by HI clouds in the broad-line regions (in fact, some ionizing radiation escapes from quasars), attenuation of the Lyα emission by dust, or orientation effects (see below). Thus, in principle, AGNs can produce higher Lyα equivalent widths than stellar populations. However, the presence of an AGN in a galaxy is usually revealed by other readily identifiable signatures: strong emission lines of highly ionized species (CIV, HeII, etc.) and broad emission lines with velocity widths several times larger than those expected from the virial motions within galaxies.

The Lyα photons produced in galaxies will suffer a large number of resonant scatterings in the ambient neutral atomic hydrogen. In the absence of dust, this would lead to no net enhancement of the angle-averaged Lyα emission from a galaxy or of the total Lyα emission from a sample of randomly-oriented galaxies. However, since the Lyα line is emitted more isotropically than the continuum, the Lyα equivalent width of an individual galaxy will decrease as it is viewed more nearly edge-on. For example, in the idealized case of a plane-parallel slab, the ratio of the observed to angle-averaged Lyα equivalent width will decrease from 2.3 to 0 for viewing angles to the normal ranging from $0°$ to $90°$. The resonant scattering of Lyα photons by HI also increases enormously their chances of absorption by dust grains. The attenuation is expected to be important when the dimensionless dust-to-gas ratio, defined in terms of the extinction optical depth in the B band by $k \equiv 10^{21}(\tau_B/N_{HI})\,\text{cm}^{-2}$, exceeds the critical value $k_{\text{crit}} \approx 0.01(N_{HI\perp}/10^{21}\,\text{cm}^{-2})^{-4/3}(\sigma_V/10\,\text{km s}^{-1})^{2/3}$ (see Charlot & Fall 1993). In this expression, $N_{HI\perp}$ and σ_V are the face-on column density and line-of-sight velocity dispersion of HI. For reference, the dust-to-gas ratio in the Milky Way and Large and Small Magellanic Clouds are, respectively, $k \approx 0.8$, $k \approx 0.2$, and $k \approx 0.02$, and the face-on HI column densities within the optically visible regions of most spiral galaxies lie in the range $10^{20} \lesssim N_{HI\perp} \lesssim 10^{21}\,cm^{-2}$. Thus, we expect $k \gtrsim k_{crit}$ unless the dust-to-gas ratio is much smaller than the value in the Milky Way. In particular, some attenuation of the Lyα emission by dust is expected in the damped Lyα systems, since $N_{HI} \gtrsim 2 \times 10^{20}\,cm^{-2}$ and $k \approx 0.1$ (although there may be a large dispersion around this value; see Pei et al. 1991; Pettini et al. 1994). Moreover, the attenuation of Lyα emission by dust depends sensitively on the structure of the interstellar medium in a galaxy. In the case of a multiphase medium, the transfer of Lyα photons will depend largely on the topology of the interfaces between HI and HII regions (see Spitzer 1978; Neufeld 1991). As a result of these complications, it is nearly impossible to deduce star formation rates from the observed Lyα emission of a galaxy.

The above arguments may be used to interpret the observations of and searches for Lyα emission from nearby star-forming galaxies, damped Lyα systems, blank sky, and the companions of quasars and damped Lyα systems (see Charlot & Fall 1993). It appears that, when Lyα emission is weak or absent, as is the case in most star-forming galaxies at low redshifts and in damped Lyα systems at high redshifts, the observed abundance of dust is sufficient to absorb most of the Lyα photons. On the other hand, when Lyα emission is strong, the presence of highly ionized species, large velocity widths, or nearby quasars indicate that much of the ionizing radiation may be supplied by AGNs. The hope has always been that the searches for Lyα emission at high redshifts would reveal a population of primeval galaxies, in which the abundances of heavy elements and hence dust were low enough that most of the Lyα photons could escape. Such a population may exist at some redshifts. However, since the Lyα emission is attenuated when the dust-to-gas ratio exceeds $1 - 10\%$ of the value in the Milky Way, a typical galaxy probably spends only the first few percent of its lifetime in a Lyα-bright phase. We therefore expect primeval galaxies, as defined above, to be relatively rare at most redshifts, consistent with the null results of all searches to date.

2.3 What Should Young Galaxies Look Like?

If galaxies undergoing their first episodes of star formation in the young universe resemble nearby examples of starburst galaxies, one would expect them to exhibit very blue continua and strong Balmer emission lines. However, the most distant counterparts of nearby disk galaxies found by association with quasar absorption-line systems do not show such extreme signatures (Steidel & Dickinson 1995; Aragón-Salamanca et al. 1994). Instead, out to $z \approx 1.6$, normal field galaxies appear to display strikingly little evolution in their space density, luminosity, and optical/infrared colors (most of the evolution appears to be confined in the faint end of the galaxy luminosity function). Other constraints on the history of star formation in galaxies, and hence on their appearance, may be obtained from the evolution of the integrated neutral gas density of the universe. Recent surveys indicate that the total amount of gas in damped Lyα systems at a redshift $z \approx 3.5$ could be nearly as high as the total density of luminous material in present-day galactic disks (Lanzetta et al. 1995; Wolfe 1995). This would imply that most stars have formed relatively recently, consistent with the expectation from theories of galaxy formation based on hierarchical clustering (although a precise interpretation of the observed Ω_{HI} evolution requires including the effect of quasar obscuration by dust in damped Lyα systems; see Pei & Fall 1995).

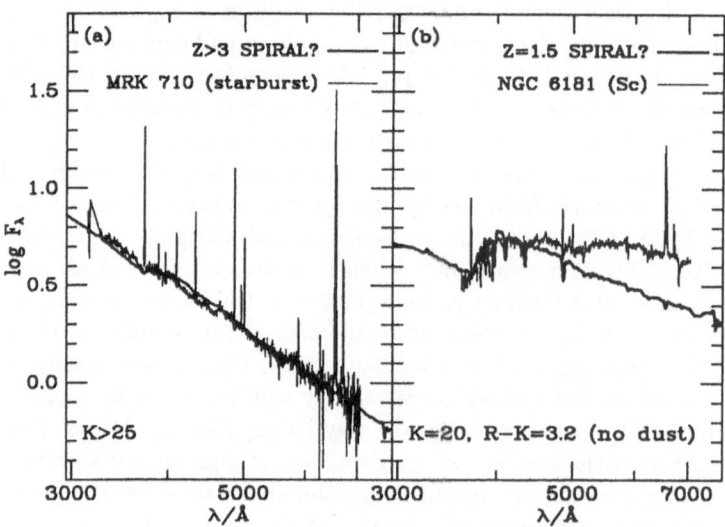

Fig. 14. (a) Possible appearance of a progenitor spiral galaxy at $z > 3$ satisfying the current observational constraints on the evolution of $\Omega_{HI}(z)$ and the observed properties of the Milky Way in a standard cold dark matter universe (thick line) compared to the observed spectrum of the nearby starburst galaxy Mrk 710 (thin line). The model spectrum does not include emission lines (see text for more details). (b) Same model galaxy as in (a) viewed at $z = 1.5$ and compared to the observed spectrum of the nearby spiral galaxy NGC 6181. The spectra in (a) and (b) are in the rest frames of the galaxies, and the predicted apparent K magnitudes (and $R - K$ color at $z = 1.5$) are indicated at the bottom.

A natural question to ask, then, is what do models accommodating present observational constraints predict for the appearance of very young galaxies? To answer this question, we have computed the early spectral evolution of a spiral galaxy using a combination of the Kauffmann et al. (1993) semi-analytic model of galaxy formation and the Bruzual & Charlot (1993) population synthesis code. The star formation rate can be adjusted to reproduce both the evolution of $\Omega_{HI}(z)$ at $z \lesssim 3.5$ in a standard cold dark matter universe and the spectral energy distribution of a typical nearby spiral galaxy at $z = 0$ (see Kauffmann & Charlot

1994). The early spectral evolution of this model is presented in Figure 14. Figure 14a shows that the galaxy undergoes an extremely blue phase at $z >$ 3, during which the spectrum resembles the observed spectrum of the nearby starburst galaxy Mrk 710 with strong H-Balmer and oxygen emission lines (from Vacca & Conti 1992). At this time, the young galaxy could qualify as a "primeval galaxy" as defined earlier. However, the phase is short-lived ($\lesssim 10^7$ yr) and very faint ($K > 25$). Then, the onset of evolved supergiant, asymptotic giant branch, and red giant branch stars reddens the spectrum substantially. At $z = 1.5$, twenty percent of the stars present at $z = 0$ have formed, and the model galaxy in Figure 14b has $K \approx 20$ and $R - K \approx 3.2$, i.e., a spectrum only moderately bluer than that of the nearby spiral galaxy NGC 6181 (from Kennicutt 1992). These predicted colors, which ignore reddening by dust, are interestingly close to the observed $K \approx 19.5$ and $R - K \approx 4$ of galaxies discovered at $z \approx 1.5$ in association with quasar absorption-line systems (Steidel & Dickinson 1995). Hence, the present results would reinforce the suggestion that the spectra of normal disk galaxies have evolved only moderately for much of their lifetime. The extremely blue phase at the onset of star formation, which might coincide with a Lyα-bright phase, is expected to be, on average, much fainter and short-lived (this initial phase may be brighter for elliptical galaxies if these formed most of their stars in an initial strong burst of star formation).

2.4 Tracing Back the History of Star Formation in Galaxies

Since young galaxies may be hard to detect at redshifts beyond $z \sim 2$, an alternative is to try and trace back the earlier history of star formation from observations at lower redshifts. The conventional approach to this problem is to use stellar population synthesis models and search for the evolution of the star formation rate that will reproduce the observed spectral characteristics of galaxies. As we have seen in §1, the degeneracy of galaxy spectra in age, metallicity, and history of star formation hampers the dating of passively evolving stellar populations such as elliptical galaxies from the single knowledge of the continuum spectrum, or equivalently, of broad-band colors. Furthermore, from the continuum spectra of spiral galaxies, one can at best obtain ratios of present to past-averaged star formation rates.

Some additional information on the past history of star formation in galaxies may be learned from the stellar absorption lines of hydrogen and of other prominent atoms and molecules (such as Mg, Mg$_2$, Fe, Ca, Na, Sr, and CN), even though most optical spectral indices appear to be degenerate in age and metallicity. For example, main-sequence A and B stars (that have lifetimes $\lesssim 2$ Gyr) are expected to strengthen the H-Balmer series and weaken the prominent metallic lines in the integrated galaxy spectrum. This can be most simply illustrated by considering the case of intermediate-age stellar populations in early-type galaxies. There is growing photometric and dynamical evidence that many E/S0 galaxies have formed stars only a few billion years ago (Pickles 1989; Rose 1985; Schweizer & Seitzer 1992; and references therein). The prototypical example is the dwarf elliptical galaxy M 32, which is believed to have undergone substantial star formation until only about 5 Gyr ago (e.g., O'Connell 1980).

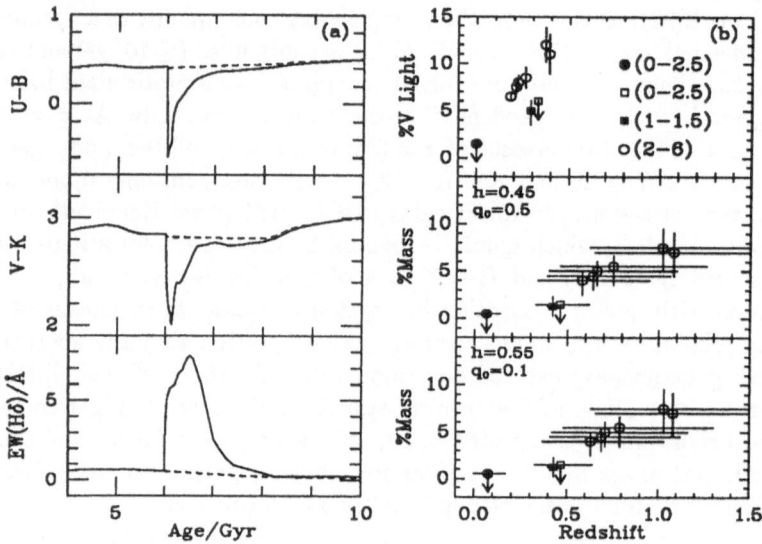

Fig. 15. (a) Evolution of the $U - B$ and $V - K$ colors and Hδ absorption equivalent width of a model elliptical galaxy formed in a single burst at $t = 0$ (dashed line) and of a similar model galaxy on which a new burst involving 10% of the final mass is added at an age of 6 Gyr (solid line). (b) Upper panel: observed contributions by intermediate-age stars to the V luminosity of E/S0 galaxies at $z \approx 0$ and in low-redshift clusters inferred from stellar absorption-line studies (see text for sources). Different symbols correspond to different studies or different ranges of intermediate ages (indicated in Gyr). Lower panels: mass fraction of stars formed in the progenitors of E/S0 galaxies as a function of redshift derived for two cosmologies from the observations shown in the upper panel. The horizontal error bars follow from the uncertainties on the ages of stars detected in low-redshifts E/S0 galaxies. The vertical error bars follow from the uncertainties on the determination of the contribution by these stars to the mass for the allowed range of ages.

Figure 15a illustrates how spectral absorption features such as the Balmer Hδ equivalent width can be used to detect late bursts of star formation in early-type galaxies with colors otherwise typical of old, passively evolving stellar populations. The solid lines correspond to a model elliptical galaxy formed in a major burst at age $t = 0$, on which a new burst involving 10% of the final mass is added at an age of 6 Gyr. At ages $t > 6.5$ Gyr, the $U - B$ and $V - K$ colors of this model differ by less than 0.1 mag from the values in the absence a second burst (shown by the dashed lines). However, the Hδ equivalent width continues to evolve significantly for nearly 1 Gyr because of the presence of A and B stars.

Spectral Evolution of Galaxies

Stéphane Charlot [1,2]

[1]Institut d'Astrophysique du CNRS, 98 bis Boulevard Arago, F-75014
Paris, France
[2]Kitt Peak National Observatory, 950 N. Cherry Avenue, Tucson, AZ
85726, U.S.A.

Abstract: We review models of spectral evolution of galaxies and current constraints set by observations at high redshifts on the formation and evolution of galaxies. These lectures assume previous basic knowledge of the general properties of stars and galaxies.

1 Stellar Population Synthesis

1.1 Introduction

The integrated light from star clusters and galaxies should reflect the distribution of stellar masses, ages, and metallicities within them, providing us with important clues on the past history of star formation. Stellar population synthesis, the modeling of the spectral energy distribution emitted by specific populations of stars, is a natural approach to identifying such clues. Studies in this field have led to the development of population synthesis models, usually termed "evolutionary" (Crampin & Hoyle 1961; Tinsley 1978; Bruzual 1983; Arimoto & Yoshii 1987; Guiderdoni & Rocca-Volmerange 1987; Buzzoni 1989; Bruzual & Charlot 1993, 1996; Bressan, Chiosi, & Fagotto 1994; Fritze von Alvensleben & Gehrard 1994; Worthey 1994; Mayya 1995; Weiss, Peletier, & Matteucci 1995). In such models the detailed physical processes affecting gas properties and star formation efficiency in a galaxy are generally all reduced to few crude assumptions about the stellar birthrate (the relative efficiency of the various processes involved are extremely difficult to determine, even in nearby star-forming regions; e.g., Silk 1996). The main adjustable parameters are then usually the stellar initial mass function (IMF), the star formation rate, and in some cases the rate of chemical enrichment. For a given set of these parameters one computes the time-dependent distribution of stars in the theoretical Hertzsprung-Russell diagram (hereafter H-R diagram), from which the integrated spectral evolution of the stellar population can be obtained. These models are widely used to study

Rocca-Volmerange 1995). However, tracing back the onset of star formation in normal disk galaxies still appears to be a long way ahead.

References

Arimoto, N., Yoshii, Y. (1987), A&A, Vol. 173, p. 23.

Aragón-Salamanca, A., Ellis, R.S., Schwartzenberg, J.M., Bergeron, J. (1994), ApJ, Vol. 421, p. 27.

Baron, E., White, S.D.M. (1987), ApJ, Vol. 322, p. 585.

Belloni, P., Bruzual A., G., Thimm, G.J., Rose, H.J. (1995), A&A, Vol. 297, p. 61.

Bergeron, J. (1988), in *Large Scale Structures in the Universe*, Int. Astron. Union Symp. No. 130 (Reidel), p. 343.

Bressan, A., Fagotto, F., Bertelli, G., Chiosi, C. (1993), A&AS, Vol. 100, p. 647.

Bressan, A., Chiosi, C., Fagotto, F. (1994), ApJS, Vol. 94, p. 63.

Bruzual A., G. (1983), ApJ, Vol. 273, p. 105.

Bruzual A., G., Charlot, S. (1993), ApJ, Vol. 405, p. 538.

Bruzual A., G., Charlot, S. (1996), ApJ, in preparation.

Burstein, D., Faber, S.M., Gaskell, C.M., Krumm, N. (1984), ApJ, Vol. 287, p. 586.

Buzzoni, A. (1989), ApJS, Vol. 71, Vol. 817.

Caldwell, J.A.R., Ostriker, J.P. (1981), ApJ, Vol. 251, p. 61.

Charlot, S. (1996), in *From Stars to Galaxies*, ed. C. Leitherer, U. Fritze-van Alvensleben, & J. Huchra (ASP Conference Series), in press.

Charlot, S., Bruzual A., G. (1991), ApJ, Vol. 367, p. 126.

Charlot, S., Fall, S.M. (1993), ApJ, Vol. 415, p. 580.

Charlot, S., Silk, J. (1994), ApJ, Vol. 432, p. 453.

Charlot, A., Worthey, G., Bressan, A. (1996), ApJ, Vol. 457, in press.

Cohen, J.G. (1982), ApJ, Vol. 258, p. 143.

Colless, M., Ellis, R.S., Broadhurst, T.J., Taylor, K., Peterson, B.A. (1993), MNRAS, Vol. 261, p. 19.

Couch, W.J., Sharples, R.M. (1987), MNRAS, Vol. 229, p. 423.

Cowie, L.L., Gardner, J.P., Hu, E.M., Songaila, A., Hodapp, K.-W., Wainscoat, R.J. (1994), ApJ, Vol. 434, p. 114.

Crampin, J., Hoyle, F. (1961), MNRAS, Vol. 122, p. 27.

Davidge, T.J. (1992), ApJ, Vol. 397, p. 457.

Fagotto, F., Bressan, A., Bertelli, G., Chiosi, C. (1994), A&AS, Vol. 100, p. 647.

Fritze von Alvensleben, U.A., Gerhard, O.E. (1994), A&A, Vol. 285, p. 751.

Frogel, J.A., Persson, S.E., Aaronson, M., Mathews, K. (1978), ApJ, Vol. 220, p. 75.

Gallagher, J.S., Hunter, D.A., Tutukov, A.V. 1984, ApJ, Vol. 284, p. 544.

García-Vargas, M.L., Bressan, A., Leitherer, C. (1996), in preparation.

González, J.J. (1993), Ph.D. Thesis, Univ. of California, Santa Cruz.

Guiderdoni, B., Rocca-Volmerange, B. (1987), A&A, Vol. 186, p. 1.

Hesser, J.E. (1993), in *The Globular Cluster-Galaxy Connection*, ed. Smith & Brodie (ASP Conference Series, Vol. 48), p. 1.

Kauffmann, G., Charlot, S. (1994), ApJ, Vol. 430, p. L97.

Kauffmann, G., White, S.D.M., Guiderdoni, B. (1993), MNRAS, Vol. 274, p. 201.

Kennicutt, R.C. (1992), ApJS, Vol. 79, p. 255.

Kennicutt, R.C., Tamblyn, P., Congdon, C.W. (1994), ApJ, Vol. 435, p. 22.

Lançon, A., Rocca-Volmerange, B. (1995), A&A, preprint.

Lanzetta, K.M., Wolfe, A.M., Turnshek, D.A., Lu, L.M., McMahon, R.C., Hazard, C. (1991), ApJS, Vol. 77, p. 1.

Lanzetta, K.M., Wolfe, A.M., Turnshek, D.A. (1995), ApJ, Vol. 440, p. 435.

Larson, R.B. (1986), MNRAS, Vol. 218, p. 409.

Lilly, S.J., Tresse, L., Hammer, F., Crampton, D., Le Fevre, O. (1995), ApJ, preprint.

Macchetto, F., Lipari, S., Giavalisco, M., Turnshek, D.A., Sparks, W.B. (1993), ApJ, Vol. 404, p. 511.

Mayya, Y.D. (1995), AJ, Vol. 109, p. 2503.

Meier, D. (1976), ApJ, Vol. 207, p. 343.

Neufeld, D.A. (1991), ApJ, Vol. 370, p. L85.

O'Connell, R.W. (1980), ApJ, Vol. 236, p. 430.

Oke, J.B., Gunn, J.E. (1983), ApJ, Vol. 266, p. 713.

Partridge, R.B., Peebles, P.J.E. (1967), ApJ, Vol. 147, p. 868.

Pei, Y.C., Fall, S.M., Bechtold, J. (1991), ApJ, Vol. 378, p. 6.

Pei, Y.C., Fall, S.M. (1995), ApJ, in press.

Peletier, R.F. (1989), Ph.D. thesis, Univ. of Groningen.

Persson, S.E., Aaronson, M., Cohen, J.G., Frogel, J.A., Mathews, K. (1983), ApJ, Vol. 266, p. 105.

Petitjean, P., Webb, J.K., Rauch, M., Carswell, R.F., Lanzetta, K.M. (1993),

Pettini, M., Smith, L.J., Hunstead, R.W., King, D.L. (1994), ApJ, Vol. 426, p. 79.

Pickles, A.J. (1989), in *The Epoch of Galaxy Formation*, ed. C.S. Frenk, R.S. Ellis, T. Shanks, A.F. Heavens, J.A. Peacock (Kluwer), p. 191.

Rose, J. (1985), AJ, Vol. 90, p. 1927.

Rowan-Robinson, M., et al. (1993), MNRAS, Vol. 261, p. 513.

Salpeter, E.E. (1955), ApJ, Vol. 121, p. 161.

Scalo, J.M. (1986), Fundamentals of Cosmic Physics, Vol. 11, p. 1.

Schaller, G., Schaerer, D., Meynet, G., Maeder, A. (1992), A&AS, Vol. 96, p. 26.

Schweizer, F., Seitzer, P. (1992), AJ, Vol. 104, p. 1039.

Searle, L., Wilkinson, A., Bagnuolo, W.G. (1980), ApJ, Vol. 239, p. 803.

Shull, J.M., Silk, J. (1979), ApJ, Vol. 234, p. 427.

Silk, J. (1996), this volume.

Spinrad, H. (1989), in *The Epoch of Galaxy Formation*, ed. C.S. Frenk, R.S. Ellis, T. Shanks, A.F. Heavens, J.A. Peacock (Kluwer), p. 39.

Spitzer, L.: *Physical Processes in the Interstellar Medium* (Wiley)

Steidel, C.C., Dickinson, M. (1995), in *Wide Field Spectroscopy and the Distant Universe* (Cambridge University Press), in press.

Steidel, C.C., Dickinson, M., Persson, S.E. (1995), ApJ, Vol. 437., p. L75.

Thuan, T.X., Gunn, J.E. (1976), PASP, Vol. 88, p. 543.

Tinsley, B.M. (1978), ApJ, Vol. 222, p. 14.

Tinsley, B.M. (1980), Fundamentals of Cosmic Physics, Vol. 5, p. 287.

Vacca, W.D., Conti, P.S. (1992), ApJ, Vol. 401, p. 543.

van den Bergh, S. (1981), A&AS, Vol. 46, p. 79.

Weiss, A., Peletier, R.F., Matteucci, F. 1995, A&A, Vol. 296, p. 73.

White, S.D.M. (1989), in *The Interstellar Medium in Galaxies*, ed. H.A. Thronson & M. Shull (Kluwer), p. 371.

Wolfe, A.M. (1995), in *The Physics of the Interstellar Medium and Intergalactic Medium*, ed. A. Ferrara, C. Heiles, C. McKee, P. Shapiro (ASP Conference Series), in press

Worthey, G. (1994), ApJS, Vol. 95, p. 107.

Galaxy Formation

Joseph Silk

Departments of Astronomy and Physics, and
Center for Particle Astrophysics,
University of California, Berkeley, CA 94720

Abstract: The essential ingredient in understanding galaxy formation is star formation. Some tentative ideas are outlined that allow a simple, schematic derivation of the stellar initial mass function (IMF). The questions that I will address are: what determines the masses of stars and what determines the IMF? On a larger scale, an important clue to the nature of star formation in galactic disks arises from the observation that gas surface density in nearby spirals remains close to the critical value above which the gas is gravitationally unstable. Many of the global properties of galactic disks can be understood if star formation is self-regulated. A plausible physical argument is provided that self-regulation occurs and accounts for the global inefficiency of star formation in galactic disks. In contrast, there is no theoretical framework for developing a model for star formation in galactic spheroids. Major mergers that triggered major starbursts are thought to have played an important role, but the associated star formation physics is at best only schematically understood. One approach is to examine ultraluminous starbursts as possible prototypes for protoellipticals. The intracluster medium provides a natural reservoir for the debris from early phases of elliptical formation, and it has been inferred that the initial mass function in protoellipticals must have been top-heavy in order to explain the amount of intracluster iron. Implications include the inevitability of wind-driven mass loss, a systematic increase in mass-to-light ratio with luminosity as inferred from the fundamental plane, and Type II supernova yields both in the stellar populations of luminous ellipticals and in the intracluster gas.

To be published in *The Universe at High-z, Large-Scale Structure and the Cosmic Microwave Background*, ed. E. Martinez-Gonzalez and J. L. Sanz (Springer Verlag, Heidelberg)

1 Introduction

The topic of galaxy formation has traditionally focused on cosmological issues, perhaps primarily because non–dissipative dark matter is relatively simple to incorporate into large–scale structure simulations. The absence of any definitive, or even plausible, detailed theory of galactic star formation has however meant that no realistic galaxy formation models have hitherto been constructed. As a result there has been relatively little contact between theory and the new deep surveys and discoveries of high redshift galaxies. What theory there has been may be said to be largely empirical: e.g. using population synthesis models. A recent development has combined analytic theory, based on the Press-Schechter formulation of mass functions, with population synthesis models (Lacey and Silk 1991; Lacey *et al.* 1993; Kauffmann, White and Guiderdoni 1993). Even this ignores the guts of the problem, namely the fundamental issues of star formation. The acknowledgement of the star formation issue typically involves invoking a cooling time argument, and is sometimes combined with triggering by tidal interactions or mergers and feedback from supernovae. One can surely improve on this approach by utilizing models for star formation based more closely on what is actually observed.

Star formation however remains a topic that has proven remarkably elusive. We observe many of the ingredients that are necessary for a successful theory, yet no such theory has been developed that can predict such critical ingredients as the star formation rate and the initial mass function. Galaxy formation, arguably the most important problem in cosmology, remains as elusive as ever because of our lack of knowledge of the details of star formation. If we cannot account for star formation in the intensively observed Orion nebula and associated molecular clouds, we cannot hope to incorporate star formation into realistic predictions of the properties of forming galaxies.

The brute force approach of computational power is not going to elucidate the innards of star formation, at least in the foreseeable future. There are simply too many variables, including magnetic fields, ionization, dust, molecular chemistry, molecular cooling, turbulence and environment. Processes that play a role include accretion, fragmentation, coagulation of fragments, Alfven wave generation, momentum-driven flows, and binary formation. In a situation of such complexity, it is important to search for some overall guiding principle that may help elucidate the underlying physics and provide some predictive power. This approach has worked well in diverse areas of physics, including the theories of gravitation, of fundamental interactions and of elementary particles. Astrophysics is a far dirtier domain, but it is useful nevertheless to seek some unifying idea that can be applied to star formation physics.

A theory for the origin of stellar masses is central to the goal of developing a model for galaxy formation and evolution. One needs to know how star formation varies with environment, and in particular the dependence of the initial mass function, the star formation rate and the star formation efficiency on the various astrophysical input parameters. The distribution of stellar masses controls chemical evolution, mass within the luminous regions of galaxies, and light. Without a theory for the stellar initial mass function (IMF), one is compelled to assume that the IMF is universal. It may be, for example, in environments far more extreme than that of the solar neighborhood that star formation is modified. Hence these lectures will begin with an admittedly subjective view of star formation, followed by sections on the formation of disk galaxies and of elliptical galaxies.

2 Star Formation

One of the most important and unresolved issues in the theory of star formation concerns the origin of the initial mass function of stars (IMF). This is approximated by a nearly scale-free power law over more than two decades in mass. There are indications of a turn-over below $\sim 0.3 M_{\odot}$, and the upper mass limit is about $60 M_{\odot}$. Between these extremes, the mass function $dN/dM_* = AM_*^{-1-\gamma}$ satisfies $\gamma \approx 1.5(\pm 0.3)$. This range for γ includes the Salpeter IMF ($\gamma = 1.35$) and also allows for a slight steepening towards higher masses, as in the Miller-Scalo IMF. The IMF appears to be reasonably universal insofar as one can study it in different star-forming regions in the Milky Way and in nearby galaxies (Scalo 1986). There are many theories of the IMF, but no accepted explanations. Suggestions range from physical processes involving fragmentation and accretion (e.g. Silk 1976, Zinnecker 1984) to scale-free clustering of coalescing fragments (e.g. Nakano 1966; Silk and Takahashi 1979; Pumphrey and Scalo 1983) or of a fractal network of molecular cloud filaments (e.g. Larson 1992). A more recent view is that the origin of the stellar initial mass function as well as the characteristic mass of a protostar are intimately connected to feedback from protostellar outflows that limit accretion onto forming protostars (Silk 1995).

Fragmentation is the initial process that leads to star formation. Unfortunately its nonlinear evolution is poorly understood. The Jeans criterion for gravitational instability has been extended to such geometries as those of a cold, self-gravitating disk or sheet. Linear theory provides a critical scale for fragmentation. However the non-linear evolution of the fragmenting cloud involves such processes as fragment accretion and coalescence. Hence it is likely that the final fragment masses are considerably larger than the linear theory would suggest. For example, application of opacity-limited fragmentation using the Jean criterion yields a minimum mass of $\sim 10^{-3} M_{\odot}$, but one can increase this by an order of magnitude or more by incorporating the physics of fragment interactions.

The questions that I will address are: what determines the masses of stars and what determines the initial mass function (IMF)? Let me begin by dismissing a myth that is prevalent in discussions of star formation. The Jeans mass does not play an important role in determining the mass of a star. It determines the masses of the dense molecular gas cores within which stars are forming. Star typically form not in isolation but in gas cores that have masses of $10^2 - 10^3 M_{\odot}$ in warm molecular clouds such as in Orion. While lower mass cores are found in cold clouds such as Taurus, even here the multiplicity of star formation seems apparent.

The application of Jeans mass scaling to cloud collapse was an early motivation for fragmentation, and for evaluation of the effects of opacity in determining a lower bound on fragment masses. The result, an application of the linearized gravitational instability criterion to a collapsing cloud, is that in the initially, approximately isothermal, regime, the Jeans mass is found to decrease to a minimum value of about $10^3 M_\odot$, before the increasing opacity drives up the local temperature by inhibiting cooling, and the Jeans mass increases. It is tempting to identify this relatively robust minimum mass scale with the mass of a protostellar core, the Jeans scale giving the instantaneous scale of fragments, at least in the isothermal regime. In fact, it seems indisputable that such nonlinear processes as collisions between fragments, coalescence and fragment accretion help build up fragment masses to the protostellar mass regime of $\gtrsim 0.1 M_\odot$.

However more than aggregation and accretion must be involved, otherwise most stars would be massive, since the accretion time-scale t_{acc} is very short. In fact, $t_{acc} \propto M/\Delta V^3$, where M is the fragment mass and ΔV is the effective sound velocity, suitably generalized to include effects of magnetic fields and turbulence, so that

$$t_{acc} = 3 \times 10^5 \left(\frac{M}{M_\odot} \right) \times \left(\frac{0.2 \, \text{km s}^{-1}}{\Delta V} \right)^3 \, \text{yr.}$$

Given that even the coldest cores have linewidths of $\Delta V \sim 0.1 \text{km s}^{-1}$ and contain a solar mass or more, and that cold molecular clouds have lifetimes of several million years, how then could one form stars of typical mass $0.3 M_\odot$? If accretion were halted by tapping stellar nuclear energy, one would form only stars of $10 M_\odot$ or more.

2.1 Outflows and Feedback

Something else must intervene to limit accretion of gas and fragment coalescence. There is only one physical process that can intervene on a sufficiently short time-scale, namely outflows initiated during the protostar phase. The ultimate energy source that drives a protostellar outflow is the Kelvin-Helmholtz gravitational contraction and release of energy, over a time-scale that amounts to about 3×10^7yr for a solar mass star to arrive on the main sequence. However the ubiquitous observed outflows have much shorter time-scales, of $10^4 - 10^5$yr, that must correspond to the very earliest stages of protostellar evolution. Infrared observations have revealed that young stellar objects (YSOs) may be classified as having a sequence of ages that spans $10^4 - 10^8$yr for objects that include deeply embedded protostars, the youngest objects, and the optically visible T-Tauri stars, with bolometric temperatures ranging from 60K to 5000K (Chen et al. 1995). The observed correlation of temperature with age suggests that YSOs may be interpreted in terms of an evolutionary sequence defined by bolometric luminosity and bolometric temperature, in analogy with the Hertzsprung-Russell diagram.

Consider then the following simple model for a YSO. An accretion disk surrounds the growing protostellar core. Outside this, the effective photosphere determines the observed size, via L_{bol} and T_{bol}, of the YSO as the radius where the gravitational energy, generated in the protostellar core by Kelvin-Helmholtz contraction and in the disk by accretion, can ultimately be released. Note that in the collapsing phase, one may take the absorption coefficient over the wavelength range $5 - 30\mu m$ to be approximately wavelength-independent, equivalent to $100 \lesssim T_{bol} \lesssim 750K$, and one can ther efore express the associated time-scale as

$$t_{YSO} \sim \frac{GM^2/R}{constant \times M^3} \propto M^{-2}.$$

Comparison with t_{acc} shows that as the core mass increases, t_{YSO} inevitably becomes the controlling time-scale.

Protostars form by infall, and non-homologous collapse results in formation of many protostellar cores within a large cloud. Protostellar outflows are indeed ubiquitous in molecular clouds. The observed bipolar flows in combination with the inferred number of old flows that stems from time-scale considerations suffice to give enough momentum input to account for the observed linewidths. These associated outflows are critical for regulating the amount of infall, since the specific angular momentum of a molecular cloud exceeds that of a protostar by some 3 orders of magnitude. This angular momentum barrier is overcome at the accretion disk shock, where the outflows are believed to originate. A young protostellar object is embedded within an accretion disk from which it is thought that gas jets are driven along the polar axis to expand and culminate in a bipolar flow.

Although the details of the angular momentum transfer that must be responsible for inhibiting the flow of low angular momentum gas to the protostar have yet to be evaluated, I will assume here that the outflow is responsible for limiting the accretion of gas onto the central protostar (Silk 1995). The bolometric luminosity of the central source exceeds the flow energy by some two orders of magnitude, and suggests that it is at least plausible to expect that a protostellar disk-driven wind halts the accretion. In the absence of a detailed theory that connects the source with the accreting gas, I will simply assume that the outflow rate is proportional to the accretion rate (Shu et al. 1987). The bolometric luminosity of the YSO includes contributions from both the Kelvin-Helmholtz luminosity of the central protostar and from the accretion luminosity. The accretion shock occurs at the outer edge of the accretion disk, where the infall energy is released.

In order for the accretion to continue, the accretion luminosity must be radiated away. Hence a necessary condition for protostellar core growth is that

$$L_{acc} \lesssim L_{bol}.$$

The accretion disk radius R_d can be estimated by assuming centrifugal balance for the inflow, which yields (Terebey, Shu and Cassen 1984)

$$R_d = \frac{G^3 \, M^3 \, \Omega^2}{16(\Delta V)^8}$$

for an accretion disk of mass M (also identified with the protostellar mass M to within a factor of order unity) and a cloud core rotating initially at rate Ω. The accretion energy release amounts to

$$L_{acc} = GM_*\dot{M}R_d^{-1},$$

where the accretion rate

$$\dot{M} = (\Delta V^3)G^{-1}.$$

The bolometric luminosity may be written

$$L_{bol} = \frac{16}{3}\frac{\sigma T^3}{\kappa\rho}\frac{dT}{dr} \times 4\pi r^2,$$

where κ is the Rosseland mean opacity, and ρ and T are the central density and temperature. Note that the scalings of the respective luminosities are

$$L_{acc} \propto (\Delta V)^{11}\Omega^{-2} \, M_*^{-2},$$

$$L_{bol} \propto \kappa^{-1} \, M_*^3.$$

This limitation on protostellar mass is inevitable if the accretion energy release is the outflow source, from the dependence of the time scales t_{acc} and t_{bol} on core mass. One may simplify matters by appealing to the observed scaling of molecular cloud rotation rate with cloud size,

$$\Omega \propto R^{-0.4} \propto (M/\Delta V^2)^{-0.4},$$

to obtain the deceptively simple result that

$$M_* \propto 1.6(\Delta V/\mathrm{km\,s}^{-1})^{13/5} \, (\kappa/20\mathrm{cm}^2\mathrm{g}^{-1})^{1/5}\mathrm{M}_\odot.$$

To compare with observations, I note that a survey (Myers and Fuller 1993) of NH_3 emission from dense cloud cores relates the luminosity of the brightest embedded IRAS sources to cloud linewidths, over the range 0.1 to $3\mathrm{km\,s}^{-1}$ spanned by cold and warm cores. Adoption of premain-sequence tracks yields the empirical result

$$M_* = 1.8(\pm 0.1)(\Delta V)^{2.4(\pm 0.2)}M_\odot$$

over the range $0.3 < M_* < 30\mathrm{M}_\odot$.

The concordance between theory and observation suggests that a self-regulating interplay between protostellar masses and molecular cloud linewidths may be responsible for determining the physics of the origin of stellar masses. It is intriguing that the observed range of stellar masses can be understood in terms of the observed range in ΔV, in a dynamical model that does not appeal to the concept of a pressure-supported Jeans mass. One corollary is that there should generally be initially no stars below $\sim 0.1\mathrm{M}_\odot$, since the minimum cloud linewidths for realistic cooling models are rarely below $\sim 0.1\mathrm{km\,s}^{-1}$. It is the distribution of linewidths that controls the distribution of stellar masses.

2.2 Outflows and Turbulence

The linewidth distribution must result from the cumulative history of outflows in a given cloud. A molecular cloud contains a network of interacting shells, driven by fossil outflows from young stellar objects, few of which are active at any given moment because of the short lifetimes ($\sim 10^5$yr) measured for the outflow phase. There will be few high velocity young flows and many low velocity old flows. The older flows will have lost any coherence as a consequence of the interaction between different outflows and ambient gas, leading to shell instabilities (both Rayleigh-Taylor and Kelvin-Helmholtz), but the momentum will be preserved as turbulent motions that are measured via cloud linewidths.

One may idealize an individual outflow by a spherical wind. An expanding cavity of low density gas develops behind the contact discontinuity that separates the low density wind gas from the dense, shocked, swept-up molecular cloud gas. The radius of the cavity is given by $R \propto t^\delta$, where $\delta \approx 0.5$ for a wind or $\delta \approx 0.3$ for a blast wave in a uniform medium. This solution may be straightforwardly generalized to the case of an ambient density gradient or a time-dependent wind (Koo and McKee 1992). For a steady generation rate of outflows, the velocity distribution of shells is given by

$$N(> \Delta V) \propto R^3 t \propto (\Delta V)^{-\left(\frac{3\delta+1}{\delta-1}\right)},$$

where the velocity of a shell fragment, ΔV, is identified with dR/dt. If self-regulation were to be maintained, one may infer stellar masses from the distribution of turbulence ΔV according to

$$M_* \propto (\Delta V)^\epsilon, \text{ where } \epsilon \approx 2.4.$$

Combination of the protostellar mass-turbulent velocity relation with this ΔV distribution then yields a prediction for the protostellar initial mass function:

$$\frac{dN}{dM_*} \propto M^{-1-x_*}; \ x_* = \frac{3\delta+1}{\epsilon(1-\delta)}.$$

For $\epsilon \approx 2.4$ and $\delta \approx 0.4$, the expected parameter range, one infers that $x_* = 1.5$, essentially the slope of the local IMF.

If the outflows are responsible for the gas turbulence within a giant molecular cloud, a similar argument gives the mass function of cloud cores, with one important difference: the mass of a cloud core is related to the turbulent velocity dispersion within the core via the condition the core be self-gravitating and the observed scaling laws that relate ΔV to clump size and/or density ρ. Self-gravity requires

$$M_{core} \approx \frac{(\Delta V)^4}{G^{3/2} \, \rho^{1/2}}$$

where the external turbulent pressure is effectively constant, the observed scaling laws yielding (Myers and Goodman 1988) $\rho \Delta V^2 \approx constant$, or $p/k \approx 4 \times 10^5 \text{cm}^{-3}$K. Hence for cloud cores, $\epsilon \approx 4$, and the corresponding mass function (with $\delta = 1.4$ as before) has $x_{core} = 0.9$. In general, because $x \propto \epsilon^{-1}$, one expects the clump mass function to be flatter than the IMF. This is consistent with what is observed (Tatematsu *et al.* 1993): $x_{core} = 0.5 - 1$, whereas $x_* \approx 1.3 - 1.8$.

2.3 Implications

The cloud linewidths must be sub-Alfvenic to avoid strong dissipation of turbulent motions. The linewidths of dense starless cores are similar to those of neighboring cores that contain embedded young stars (Caselli and Myers 1995). Alfven wave excitation provides a means of coupling outflows to nearby gas clumps which contain no embedded stars. Another means of magnetic coupling is via the ionization generated by flare activity of young stellar objects. The observed x-ray ionization is the dominant ionization source in the Rho Ophiuchus cloud. Ionization controls the ambipolar diffusion rate. Enhanced ionization will enable gas clumping to couple more strongly to the magnetic field and thereby delay or inhibit star formation. In this manner, one can perhaps understand why it is that starless cold dark cloud cores have only marginally smaller linewidths than cores containing embedded young stellar objects.

In the absence of a SPH simulation, one can make order-of-magnitude estimates that generally support the turbulence model. Suppose that at any given instant 0.1 percent of the mass of a molecular cloud was participating in high velocity bipolar outflows via driving winds at $\sim 300\,\mathrm{km\,s^{-1}}$. The turbulence persists for a few clump crossing times due to magnetohydrodynamic cushioning of clump collisions as well as geometrical effects. Hence one could generate linewidths of $\sim 1 - 3\,\mathrm{km\,s^{-1}}$ for 10 or more cloud dynamical times, when ~ 1 percent or more of the cloud would have formed stars. The characteristic stellar mass, essentially the peak in the IMF, is determined by the amount of turbulence and the efficiency of cooling. The derived initial mass function has a slope comparable to what is observed, and is insensitive to physical conditions: it should be the same for clouds of solar or of primordial composition, and for quiescent star formation or for starbursts.

One could imagine that in a situation where internal cloud turbulence is being stirred up by application of external energy input, such as with cloud collisions, the proportion of cold clouds with low ΔV would be greatly reduced. While this might happen to some extent within a spiral arm, the cloud collision rate would be far more dramatically enhanced with a galaxy merger. Since the IMF is the superposition of regions of low and of high ΔV, one might speculate that in ultraluminous starbursts, such as occur in galaxy mergers or in strong tidal interactions, and most likely during galaxy formation, the ΔV distribution is shifted to larger values. The IMF might then be top-heavy, and weighted by the expected distribution of ΔV towards more massive stars than would be characteristic of the solar neighborhood IMF.

Elliptical galaxies most likely formed by mergers that involved an ultraluminous and highly efficient phase of star formation. Perhaps when elliptical galaxies formed, the violent mergers would have resulted in strong winds that enriched the local enviro nment. Indeed, the high abundance of iron in the intracluster medium of rich clusters, as well as recent indications of Type II supernovae yields in the intracluster gas, might bear witness to such a top-heavy IMF.

One can also imagine a much more quiescent situation. Perhaps when the galaxy halo formed, the primordial clouds far from the inner galaxy were dense, cold, low mass globules, and consequently non-interacting. Stars might have formed under very low ΔV conditions in such clouds. Under these conditions, the IMF would have been bottom-heavy, dominated by low mass stars. The MACHO detections of gravitational microlensing towards the Large Magellanic Cloud are consistent with a population o f substellar mass objects that appears to be more abundant, by a considerable factor, than known stellar populations. Such a result, if confirmed by ongoing microlensing surveys, would favor a bottom-heavy IMF in the galactic halo.

One consequence is that there are unique signatures in the colors of distant galaxies, if a starburst is dominated by a truncated IMF (Charlot *et al.* 1993). Another possibility is that in very quiescent regions, possibly the dwarf galaxy mass precursor clouds to ordinary galaxies, one might have very low ΔV, leading to almost exclusively brown dwarf formation over a timescale that must have been short enough to have precluded any dynamical stirring by tidal interactions.

3 Formation of disk galaxies

A successful theory of star formation in galactic disks must account for the longevity of star formation and for its relative constancy with time. A major clue is provided by observations of $H\alpha$ emission from spiral galaxies, which indicate that the gravitational instability of disks plays an important role in the rate of star formation. The observations of Kennicutt (1989) demonstrate that the gas surface density in nearby spirals remains close to the critical value above which the gas is gravitationally unstable. The significance of these observations is twofold. Firstly, they provide a physical foundation for theories of global star formation based on gravitational instability. Previous attempts to model disk star formation have been mostly based on empirical formulations. For example, the Schmidt law, for which the star formation rate is proportional to some power of gas density, has been commonly used. Secondly, since $H\alpha$ emission can be related to massive star formation, one can compare directly with model calculations for the star formation rate.

3.1 A model for star formation in disk galaxies

The star formation theshold was identified by Kennicutt with the Toomre criterion for gravitational instability applied to the gas (HI and H_2) disk. A more sophisticated model (Wang and Silk 1994) incorporated the stellar disk and generalized theshold into an expression for the star formation rate given by

$$\mathrm{SFR} = \epsilon \Sigma_{\mathrm{gas}} \Omega \frac{(1-Q)}{Q}; \quad Q \leq 1.$$

Here, Q is the Toomre parameter, given by

$$Q = \frac{\kappa \sigma_g}{fG\Sigma_{gas}},$$

$\kappa \simeq \sqrt{2}\Omega$ is the epicyclic frequency, σ is the velocity dispersion of the gas, $f(< 1)$ corrects for the stellar component in the disk, and ϵ is the star formation efficiency. This star formation rate was applied to compute the star and gas surface density and abundances, as a function of radius. SN II yields were used to compare with galactic HII region oxygen abundances. Gas recycling was included, and gas infall was not necessary to obtain sufficient longevity of the star formation rate in the disk. The various observational properties of the Milky Way disk including the G dwarf problem were explained, provided an initial metallicity was adopted.

Some difficulties have been noted in the model of Wang and Silk (Prantzos and Aubert 1995). The small but nonzero number of low metallicity disk stars may require infall of some unenriched gas. Recent data on oxygen abundances in disks suggests that outside the central few kpc, there is no abundance gradient. An improved model is given by

$$\text{SFR} = \epsilon \Sigma_{\text{gas}}(\Omega(\text{r}) - \Omega_\text{p})$$

where Ω_p is the spiral arm pattern speed. This is the expression for the star formation rate originally proposed by Wyse and Silk (1989). The gravitational instability of the disk is now implicit in writing the SFR as proportional to $\Omega - \Omega_p$: wherever there are spiral arms, the disk is unstable to star formation. Moreover at large galactocentric radii, the star formation rate is suppressed, and at small radii it is initially large.

3.2 Star Formation Efficiency

The star formation efficiency may be estimated as follows. Assume that heating from supernova remnants (and HII regions) provides momentum input to the interstellar medium and drives the gas turbulence and cloud motions. Cooling of the interstellar medium is very effective, so that one may plausibly hypothesize that the star formation inefficiency is proportional to the net momentum input to the gas. It follows that

$$\epsilon = \sigma_g/p_{SN},$$

where

$$p_{SN} = 2E_{SN}v_{cool}^{-1}m_{SN}^{-1}$$

is the specific momentum injected, that for simplicity I have identified with supernova injection, E_{SN} is the initial supernova remnant energy, m_{SN} is the mass formed in stars per supernova, and v_{cool} is the supernova remnant velocity at the onset of the approximately momentum-conserving snowplow phase. For typical parameter values, $E_{SN} = 10^{51}$ ergs, $m_{SN} = 100\,\text{M}_\odot$, $v_{cool} \approx 300\,\text{km s}^{-1}$, and $\sigma_g \approx 10\,\text{km s}^{-1}$, $p_{SN} \approx 1000\,\text{km s}^{-1}$, and $\epsilon \approx 0.01$. This is comparable to the star formation efficiency observed in giant molecular clouds, and also not coincidentally meets the global requirement for the observed galactic star formation rate that is derived from the galactic molecular cloud mass that would otherwise collapse over a cloud free-fall time.

Self-regulation of star formation arises naturally because the star formation rate is an explicit function of gas velocity dispersion, which is itself generated both by disk gravitational instabilities on large scales and on smaller scales by gas acceleration produced by energy input from H II regions and supernovae. This results in feed-back into star formation: active star formation is both associated with, and induces, high velocity dispersion, which in turn hampers the further increase of the instability in the gas.

3.3 The Tully-Fisher Relation

One may approximate the star formation rate by

$$SFR = \epsilon \Sigma_{gas} \Omega.$$

The global star formation rate is

$$L_{SFR} = 2 \left(\frac{\epsilon}{0.01} \right) \left(\frac{4 \times 10^7 \mathrm{yr}}{\Omega^{-1}} \right) \left(\frac{M_{gas}}{6 \times 10^9 M_\odot} \right) M_\odot \, \mathrm{yr}^{-1}.$$

A more stringent test is the inferred Tully-Fisher relation, which is a consequence of the adopted star formation law. If one identifies B-band luminosity with star formation rate in spirals, one can write the luminosity of a disk galaxy as

$$L = \beta \epsilon M_{gas} \Omega = \beta \epsilon \frac{v_{rot}^3}{G} \frac{M_{gas}}{M}, \quad \beta \equiv L/SFR,$$

after applying the virial theorem, $M = R v_{rot}^2 G^{-1}$. The derived value for ϵ gives the correct normalization for the Tully-Fisher relation:

$$L = 10^{10} \left(\frac{\epsilon}{0.01} \right) \left(\frac{v_{rot}}{220 \mathrm{km \, s^{-1}}} \right)^3 \left(\frac{10 M_{gas}}{M} \right) \left(\frac{\beta}{5 \times 10^9 L_\odot \mathrm{yr} M_\odot^{-1}} \right) L_\odot.$$

Note that $L \propto v_{rot}^3$ fits the observed Tully-Fisher relation (in B and R bands), whereas the naive virial theorem plus constant surface density and mass-to-light ratio assumptions would yield the incorrect result that $L \propto v_{rot}^4$. The dependence on efficiency and gas fraction, constant within a Hubble type, means that environment should influence the normalization, but not the slope, of the Tully-Fisher relation.

3.4 Mass-to-Light Ratios

Consider the implications of the adopted global star formation law for the mass-to-light ratio of disks. The derived expression for L can be rewritten as

$$\frac{M}{L} = \beta^{-1} \epsilon^{-2/3} G^{-1/3} \Sigma_L^{-1/2} L^{1/6} \left(\frac{M}{M_{gas}} \right)^{2/3} .$$

Evidently for fixed star formation efficiency ϵ, gas fraction M_{gas}/M and surface brightness Σ_L, the mass-to-light ratio $\frac{M}{L}$ increases slowly with L. One recovers almost the identical weak $\frac{M}{L}$ dependence on L in disks at specified Hubble type or gas fraction to that found in the fundamental plane for ellipticals and for bulges. This is perhaps a hint that protodisks may have been the fundamental building blocks from which ellipticals and even bulges formed. The fact that dwarf ellipticals deviate from the fundamental plane suggests a different origin, as indeed is consistent wi th theoretical expectation given the shallowness of their potential wells. The low surface brightness and high $\frac{M}{L}$ of dwarfs is most likely a consequence of wind-driven mass loss.

Along the disk sequence, earlier Hubble types, which satisfy a trend in the mean of slowly increasing luminosity, have slightly higher $\frac{M}{L}$ than later Hubble types, an effect that can be interpreted as due to the increasing dominance of the older b ulge component. The observed slow increase of $\frac{M}{L}$ with L, together with the fact that Σ_L is approximately constant, implies that $\epsilon M_{gas}/M$ must also be more or less constant with Hubble type. Since M_{gas} decreases towards earlier Hubble types, the star formation efficiency parameter must increase. The fact that $\epsilon \propto \sigma_g$ suggests that to the extent that one may adopt the disk star formation efficiency estimate for a protobulge or protoelliptical, the efficiency should indeed increase with potential well depth. Consideration of the mean ages of the stellar populations suggest that star formation is necessarily more efficient in early-type galaxies than in late-type galaxies, and more efficient in bulges than in disks.

4 Formation of elliptical galaxies

Rich clusters are dominated by elliptical and S0 galaxies. Ellipticals formed stars rapidly (relative to a Hubble time) and efficiently. Unfortunately, there is no theory for star formation in dynamically hot systems. Did these systems form by mergers of cold disks? Or did stars form within dense molecular clouds orbiting within the spheroidal potential well? One can only try to address the issue of protospheroids phenomenologically. One has several clues. Star formation was far more efficient than in disks, and there are recent hints about chemical abundance signatures, including correlations with escape velocity, metallicity gradients, and abundances, both stellar and in the diffuse intracluster and halo gas, that bear fossil witness to the star formation history. There are in addition the clues locked into the fundamental plane for ellipticals and bulges, interpretable as a slow increase with luminosity of the mass-to-light ratio with luminosity, that may well be attributable to variations in the IMF or to systematic mass loss, or to some combination of the two effects.

Early galaxy formation occurred in a bottom-up fashion, with successively larger clumps merging together, according to the gravitational instability model for the origin of large scale structure. Elliptical formation was demarcated by major mergers, between objects of comparable mass. This occurred with significant probability in dense regions of the universe, where rich clusters are developing. Consider the universe at the epoch of cluster formation, $z \sim \Omega^{-1} - 1$. The most recent galaxy mergers occurred when the cluster underwent its initial collapse, a phase which, according to hierarchical models, is dominated by mergers of galaxy groups.

In low density regions, the typical mergers are likely to have been minor mergers between objects with mass ratios of 10:1 or more. Minor mergers are likely to trigger less efficient star formation. The preexisting disks can survive, as the shallower potential well of the merging cloud is relatively inefficient at retaining the debris of massive star formation. This provides a supply of gas that can dissipate over a much longer time-scale to form a disk.

4.1 Major merger model for elliptical formation

Major mergers or strong tidal interactions drive gas clouds into the inner kiloparsec of the evolving galaxy potential well. Non-circular motions are generated and gas clouds collide and radiate away orbital energy. Prior to the merger, gas clouds formed stars, more or less as inefficiently as they do in the local interstellar medium. The merger drives agglomeration of giant gas clouds that will be unstable to star formation and capable of forming stars more efficiently. The high pressure environment where the central gas concentration develops provides global efficiency by inhibiting disruption of molecular clouds.

A typical major merger, the event that led to the formation of an elliptical, must have occurred at modest redshift, when galaxy clusters were undergoing their initial collapse. The overdensity in a cluster and the frequency of galaxies with relatively low collision velocities mean that mergers were inevitable (Mamon 1992). Additional mergers occur for galaxies within subclusters just prior to virialization, and within groups in the field. A merger model provides a simple means of accounting for the dependence of elliptical galaxy fraction on local density and radial distance from the cluster center.

S0's would have formed by minor mergers. They also are concentrated in clusters, and a minor merger origin can account for the broader distribution, as a function of local density and distance from cluster center, of the S0 population relative to the ellipticals. The minor mergers are more frequent and on the average occurred more recently, because of the higher frequencies of minor mergers relative to major mergers in regions where one has not already depleted the dwarf galaxy population. Such depletion occurs near a centrally dominant galaxy, where dynamical friction operates. The most recent mergers in a cluster or in the field are expected to be S0 precursors. This enables one to understand the so-called $E + A$ phenomenon (Dressler & Gunn 1992).

Galaxies with a spectral distribution like that of an ordinary elliptical but containing A-type spectral features underwent a starburst some $1 - 2$Gyr ago. The present-day starburst galaxies are rare in present day rich clusters, but frequent at $z \gtrsim 0.5$. The blue galaxy fraction is also much larger in these distant clusters than in present-day clusters. The present-day counterparts of the blue galaxies as well as the $E + A$'s are plausibly the S0's. HST measurements suggest that the blue galaxies have disk-like morphologies, and similar claims have been made for nearby $E + A$'s. The population of galaxies in the outer parts of nearby clusters contains an enhanced fraction of blue and $E + A$ galaxies, relative to cluster cores, consistent with a merger history, since the outer part of a nearby cluster has been assembled relatively recently. Isolated $E's$ similarly show evidence of major mergers, in the form of shells and dust lanes that occurred several Gyr ago (Schweizer and Seitzer 1992). The E-forming mergers cannot have been too recent, since the stellar populations are $\gtrsim 10$Gyr old, although there may in some cases be an intermediate age admixture. In this latter case, an age of ~ 6Gyr is possible, for what may have been a significant ($\gtrsim 10$ percent) addition of stars in the most recent merger event (Charlot & Silk 1994). On the other hand, major mergers cannot have occurred at high redshift, when such events would have been exceedingly rare according to our best estimates of the density fluctuation power spectrum that determines the hierarchy of structure formation.

Early-type galaxies show little or no luminosity or spectral evolution in deep redshift surveys carried out to $z \sim 1$. It is likely that the typical ellipticals formed at $z \approx 2 - 3$: this estimate must be increased if $\Omega < 1$ by a factor $\sim 1/\Omega - 1$. Coincidentally this is also the epoch where the quasar population peaks. It has often been suggested that the quasar phenomenon may be identified with events that demarcate the formation of the (active) nucleus of an elliptical galaxy, such as formation of a massive black hole, and accretion of gas or disruption of stars by the central hole.

The clustering of galaxy clusters is much stronger than that of elliptical galaxies, suggesting that for a Gaussian density field, clusters formed from $\sim 3\sigma$ fluctuations. One can have a simple fit to the power spectrum, with variance approximately $\sigma\left(M\right) \propto M^{-\frac{1}{2}-\frac{n}{6}}$ and $n \approx -2$ on galaxy scales, to estimate that if galaxies represent 2σ rms density peaks, then

$$1 + z_{cluster} = [3\sigma(M_{cluster})/2\sigma(M_{galaxy})](1 + z_{galaxy}) \approx \frac{1}{2}(1 + z_{galaxy}).$$

Since the merging hierarchy on galaxy scales is going to spread over at least a factor of 2 in background expansion factor, it seems inevitable that galaxy mergers accompany cluster formation.

4.2 Evidence for winds

The injection of energy into the interstellar medium during a starburst is observed to drive galactic winds. Nearby examples such as NGC253 and M82 require a considerable number of supernovae to drive a wind. It has been argued that in the case of M82, the IMF must be top-heavy, weighted towards massive stars relative to the solar neighborhood, to account for the observed infrared luminosity and wind (Doane & Mathews 1993). The momentum input from supernovae succeeded in pressurizing the massive, dense molecular clouds that must have at least initially been present in order to have star formation proceed at the requisite efficiency. Indeed, infrared and millimeter wavelength observations of starbursts provide crude estimates of gas mass, molecular fraction, and pressure. Starbursts tend to be centrally concentrated, and simple estimates of the pressure, based on the observed luminosity converted to star formation and hence to supernova rate, immediately yield pressures that are far in excess of the local interstellar medium pressure within the central kiloparsec. This already is sufficient to overcome the Jeans mass barrier, one of the major obstacles to efficient star formation in our own galactic disk, and motivates the possible development of a top-heavy IMF in the starburst.

For a giant protoelliptical undergoing a starburst during its final gas-rich merger to develop a wind and leave it gas-poor certainly requires a top-heavy IMF during the burst. Wind-driven gas loss is the most likely explanation of the low gas content of early-type galaxies. Since ellipticals inside and outside clusters have old stellar populations with little recent star formation, one cannot appeal to ram pressure stripping by the intergalactic medium to exhaust the gas supply. A merger will concentrate much of the gas until the starburst develops.

There is more than a hint of fossil evidence that giant ellipticals once underwent a top-heavy starburst and were stripped of gas by a wind. Consider first the nucleosynthetic evidence. Giant ellipticals have an excess by about a factor of 2 in Mg/Fe relative to the solar value (Worthey, Faber & González 1992). This is precisely what one sees in old halo stars at $[Fe/H] < -1$. One understands this as the SN II contribution, from exploding massive stars, to stellar abundances, and this dominates over the SN I contribution for the first few Gyr of galactic chemical evolution. The apparent signature of a massive star-dominated IMF in giant ellipticals can be explained as a top-heavy burst, following which much of the remaining gas was ejected. This could help suppress any extended duration of the period of low mass star formation that could contribute to the SN I rate, as well as any SN I's that a starburst with a normal IMF would eventually have generated. Most of the mass loss in the form of SN I ejecta cannot be retained in the halo of the elliptical without exceeding the gas-phase Fe/H abundance inferred from X-ray observations (Forman *et al.* 1993). A top-heavy burst generates a wind that is capable of cleaning out the interstellar gas. Provided that the bursts occur after *most* of the stars in the present-day elliptical have formed, later SN I's, exploding in an almost interstellar medium-free elliptical, will still pollute the gaseous halo, but contribute only a small fraction of the Fe generated either in the burst, or prior to the burst.

Moreover in early-type galaxies, metallicity correlates with galaxy luminosity. However, the correlations of metallicity with central velocity dispersion over a wide luminosity range (Bender 1992) and local escape velocity for giant E's (Franx and Illingworth 1990) have significantly less dispersion. Such correlations are most simply understood if spherical systems, from dwarfs to giants, all underwent mass loss by galactic winds. The correlation of metallicity with galaxy mass does not have a unique explanation. It may require that galaxies of increasing mass eject systematically less gas in a wind than do less massive galaxies. However this would presumably lead to the more massive ellipticals being younger, in terms of mean age of the stellar population, and bluer, than lower mass counterparts. This prescription fails to account for the color–magnitude correlation of early type galaxies. A more exotic explanation of the metallicity–mass correlation must result in giant, old, and red, galaxies. The properties of the intracluster gas provide important constraints on the past evolution of ellipticals.

4.3 The Intracluster Medium

All stellar ejecta from stars in ellipticals, if not recycled during the starburst, are ejected, as argued above, by a wind, and collect in the intracluster medium. The intracluster gas iron abundance is well measured in clusters to be about 35% of the solar value (Ohashi 1995). Since the cluster gas fraction is about $8h^{-3/2}$ percent of the total cluster mass (White et al. 1993), one infers that there is about as much iron in the gas as in the stars. Arnaud et al. (1992) found that the iron mass in the gas is proportional to the luminosity for early-type (E + S0) galaxies, with

$$M_{Fe} \approx 2 \times 10^{-2} L_v.$$

The source of the iron must be from galaxies, and specifically from giant ellipticals and S0's. The proportionality between light and iron abundance suggests this, as does the dominance of early-type galaxies in clusters. In fact most of the stellar iron is in the giant galaxies, and most of the iron synthesized in the past and ejected from the galaxies must have originated in these galaxies. The observed intracluster iron abundance per unit mass in stars is about 5 times higher than in the Milky Way. The intracluster iron mass requires a source that is indicative of a top-heavy IMF, boosting heavy element yields by up to an order of magnitude without affecting the present epoch luminosity. The temperature of the intrastructure gas is measured to be higher than the equivalent kinetic temperature that corresponds to the galaxy velocity dispersion, $T_{gas} \approx 2T_{gal}$. This suggests that supernova-driven galactic winds indeed both heated and enriched the intracluster gas.

Galaxy groups are generally less enriched than clusters, by a factor $\sim 2 - 3$, although there is a large spread in abundance, and abundance determinations are far less certain at ASCA spectral resolution for gas at $T \lesssim 1$keV. Isolated ellipticals contain gas fractions of a few percent, and the gas iron abundance is considerably less than the stellar iron abundance (near solar in giant ellipticals). There are examples of locally enhanced gaseous iron abundances, around central

cD galaxies, e.g.. in Centaurus, M87/Virgo. These deeper potential wells are likely to retain more of the gas ejecta from supernovae than would be trapped in the halos of normal E's.

4.4 The Galaxy Connection

Elliptical galaxies, and more generally, spheroids, define a plane in the parameter space of velocity dispersion, half-light radius and absolute magnitude. This so-called fundamental plane minimizes the dispersion of the observational parameters. In combination with the virial theorem, the fundamental plane is equivalent to

$$\frac{M}{L} \propto L^{1/5},$$

for the visible light band. The fundamental plane has a dispersion within a given galaxy cluster amounting to $\lesssim 15 - 20$ percent that presumably reflects the initial conditions that determined the relations between galaxy structural and dynamical parameters.

The fact that the relation reduces to a weak dependence of M/L on L strongly suggests that an explanation must be sought within the context of star formation. One possibility is that more luminous galaxies are more metal-rich and hence would have generated a higher fraction of stellar remnants than less luminous galaxies. However this fails quantitatively to reproduce the observed M/L variation with L. Moreover such a model would predict that more luminous galaxies have recycled more gas and metals for a given gas fraction before termination of the star formation process than less luminous galaxies. This would suggest that the stellar content is systematically younger, in the mean, for the most processed galaxies. More massive, luminous galaxies form stars more recently and are younger, yielding a luminosity-color relation that would be too flat in slope, in the opposite sense to what is observed.

A starburst model accounts for the fundamental plane relation provided that the IMF in the starburst is top-heavy. There is no late-time contribution to the light, but the starburst contributes enrichment and remnants. This approach fails however if the starburst occurs when the galaxy forms, since in this case the galaxy overenriches its own stars.

One can arrange to produce the requisite metallicity required for the ICM and the remnants need to account for galactic M/L, without overproducing stellar metallicities if the starbursts are delayed, only occurring after a certain amount of star formation has already been underway. In fact, delayed starbursts are mandatory if the starburst IMF is top-heavy. Dust formation would occur as the heavy elements are formed, during the minor mergers that precede the final dramatic starburst, which involves a central concentration of interstellar matter generated by the merger and could consequently be very dusty.

The merger history anticipated in a bottom-up sequence of structure formation suggests that the duration of star formation is extended over a series of minor and major mergers. The typical mass of merging remnants increases with time. The last, most significant, merger would have been a major merger that resulted in the formation of an elliptical galaxy. Early, and most, star formation occurred with a normal IMF, culminating in a final major merger and associated top-heavy starburst. A major merger is a critical ingredient in the formation history of an elliptical galaxy. Any preexisting disks are mostly destroyed. Efficient star formation is triggered. The dynamical friction induced as a massive substructure orbits into the center of the merged galaxy effectively transfers angular momentum. The result is formation of a dense stellar core with anisotropic velocity dispersion.

The typical luminosity of a bright elliptical is $L_* \approx 10^{10} h^{-2} L_\odot$. Several empirical results suggests that similar physics governs elliptical formation down to a luminosity of about $0.03 L_*$. The more luminous ellipticals are strongly supported by the anisotropic pressure of stellar velocity dispersion, whereas the figures of low luminosity ellipticals are not anisotropically supported. The fundamental plane is sharply defined for the more luminous ellipticals, but at low luminosities, there is considerable dispersion in galaxy parameters. Finally, the surface brightness–absolute magnitude correlation for ellipticals peaks towards a brightness of about $0.1 L_*$, with surface brightness decreasing towards both lower and higher luminosities. The metallicity-luminosity relation requires sub-L_* ellipticals to have subsolar metallicity, and so they are unlikely to have played a major role in enriching the intracluster medium. I will argue that ellipticals above $\sim 0.03 L_*$ have undergone a similar history that is established as a consequence of a major merger.

4.5 A Starburst Model for Ellipticals

A major merger that resulted in a starburst with a top-heavy IMF provides a means of simultaneously enhancing the M/L ratio and of producing enriched gas that is driven out of the galaxy by the starburst induced galactic winds. Thus one can understand both the fundamental plane correlation and the enrichment of the intracluster gas. One finds that the mass fraction in remnants inferred from interpreting the $M/L \propto L^{1/5}$ relation as arising from a top-heavy starburst gives just enough iron to account for the observed abundance in the ICM. The dispersion in the fundamental plane is effectively attributed to a systematic increase of starburst fraction with galaxy mass (i.e. as $M^{1/6}$) that reflects the merging history of cluster galaxies.

The wind velocity, of order the escape velocity from the elliptical, is sufficient to drive the enriched gas out of elliptical galaxy halos, and to some extent, even from galaxy groups. Thus one would expect to find lower Fe/H in elliptical halos and in galaxy groups than in clusters of galaxies. The rich cluster ICM retains all of the ejecta in elliptical winds. The primary source of cluster Fe is from the metal-rich L_* ellipticals: low luminosity, metal-poor, dwarf ellipticals, although common and easily stripped by winds, can have synthesized only a factor f_E^{-1} more Fe than survives in their stellar component, assuming a normal IMF, with iron fraction f_E relative to the solar value. This would make only a small contribution to the ICM iron mass.

The starburst-driven enrichment hypothesis predicts that the intracluster gas should be enriched by ejecta that are dominated by remnants of Type II supernova. Outside rich clusters, there are enough ellipticals to significantly enrich the intergalactic medium. If 30 percent of ellipticals are outside rich clusters, then one might expect the typical IGM enrichment to attain a level of about 10 percent of the ICM enrichment. This assumes that two-thirds of galaxies are spheroids that do not contribute to the enrichment. The enrichment will occur late, perhaps by $z \sim 1$.

There are also consequences for ellipticals themselves. The ellipticals are wind-stripped and gas-poor, but the stars should retain the metallicity abundance anomalies characteristic of Type II supernovae. This would result in luminous ellipticals having enhanced O and Mg relative to Fe by a factor of ~ 2, as seen in old halo stars where Type II SN yields are inferred. The supernova rate per unit luminosity implied by our model of ICM enrichment is a factor of ~ 5 larger than in conventional Galactic models, the ratio of M_{Fe}/M_* in clusters to its value in the Milky Way, where there were approximately 6 SNII's for every SNIa (Tsujimoto et $al.$ 1995). Hence the rate of SNII in protoellipticals is enhanced by a factor of about 6 relative to the SNII rate requirements of the standard chemical evolution model for our Galaxy. Also, as previously remarked, cD galaxies underwent more frequent mergers than normal ellipticals, and one might expect their local environment to retain evidence of some starburst-induced enrichment that is diluted by later SNI ejecta. Such an enhanced Fe abundance is indeed observed in the vicinity of nearby cD's, and the model advocated here predicts that this gas should reveal abundance ratios intermediate between those of Type I and Type II supernova yields. Isolated ellipticals are expected to be rare, forming via late major mergers in the field. These objects should have an intermediate age population characterizing the last merger event. It is this population, rather than the underlying old stellar population, that is expected to have the characteristic SNII yield signature.

The synthesis of the intracluster metals requires a stellar energy source whose contribution to the diffuse background light is difficult to hide. In fact, given that starbursts are dusty, one would expect perhaps half of the resulting radiation to appear as a diffuse far infrared background. Puget et al (1995) report the detection of the cosmic far infrared diffuse background near 300 microns at a level of $\nu i_\nu \approx (3 - 10)\,\mathrm{nwatts\,m^{-2}sr^{-1}}$. This is comparable to the diffuse background light generated in synthesizing the metallicity in massive stars required by the preceding model of protoelliptical formation (Zepf and Silk 1995).

5 Summary

The local phenomenology of molecular clouds provides vital clues that give insights into the origin of the IMF. We know that molecular clouds form stars inefficiently and are long-lived relative to the dynamical collapse time-scale. Star formation is observed to occur in clusters within dense cloud cores. Dense cores are generally found to contain many embedded young stellar objects. All of these are inferred to undergo an outflow phase. Since the outflow lifetimes are observed to be of order 10^5 years and cloud lifetimes are $10^6 - 10^7$ years, only ~ 1 percent of the outflows are expected to be active at any given time. Nevertheless the momentum injection into the molecular gas is plausibly sufficient to account for the observed, often supersonic, linewidths, and hence for cloud support against gravitational collapse. The observed correlation between molecular linewidths and protostellar luminosity, and especially between NH_3 linewidths and proximity of the nearest young stellar clusters, argues for the causal interplay between outflows, cloud turbulence and protostellar masses.

The global star formation rate has been formulated, based on the theory of gravitational instability in a cold self-gravitating disk. In the case of a gaseous disk, it can be shown that the fastest growth mode has a timescale $t_{\text{inst}}^{-1} = Q^{-1}(1 - Q^2)^{1/2}\kappa$, where $Q = \alpha\kappa\sigma_g/\pi G\mu_g$ is the Toomre Q parameter, κ is the epicyclic frequency, σ_g is the gas velocity dispersion, G is the gravitational constant, and $\alpha \simeq 0.3$ self-consistently allows for the stabilizing effects of the stellar component of the disk. If the gas surface density decreases below some critical value, $\mu_c = \alpha\kappa\sigma_g/\pi G$, differential rotation of the disk and random motion of clouds will prevent the gas from further collapsing, halting formation of stars. As the disk becomes less gas-rich, the effective value of Q increases and helps prolong the gas supply. The disk cools as momentum input ceases from dying massive stars, and the effective value of Q decreases. In this way, self-regulation is maintained, with $Q \sim 1$.

One can now compute the star formation rate in disk galaxies at early times, obtaining the star-forming history as a function of galaxy age. This is particularly interesting because a history of star formation is essential for interpreting observations of distant galaxies. For example, the disk star formation model that is generic to reproducing the Milky way properties is proportional to disk rotation rate and hence to roughly R^{-1}. This means that disks form inside out. Thus at a lookback time of $\gtrsim 5$ Gyr, disks should appear to be substantially smaller than their nearby counterparts.

A phenomenological model has been presented for starbursts, where we may well in rare examples be witnessing the nearby counterpart of a protospheroid. There are several reasons for pursuing this analogy: the tidal interaction trigger, observed for starbursts and presumed to be a feature common to bottom-up models for galaxy formation; the star formation rate and efficiency, directly inferred for starbursts and indirectly deduced in order to account for the population content and stellar dynamical support of ellipticals; and even the surface brightness profiles.

The intracluster medium in rich clusters provides a reservoir for all ejecta from cluster galaxies over the past Hubble time. The abundances of heavy elements in the intracluster gas consequently provides a time integral, in conjunction with stellar metallicities, of the chemical history of cluster galaxies.The observed iron abundance motivates the possibility bias towards the fraction of massive stars increasing with protoelliptical galaxy mass, identified with the last major starburst event. The mass fraction of stars in this high-mass mode of star formation can be constrained by requiring the resulting stellar remnants to account for the observed increase in the mass-to-light ratio of ellipticals with increasing galaxy mass. The mass and abundance ratios of metals produced by this proposed population of massive stars are consistent with observations of the mass and abundance ratios of metals in the intracluster gas. The predicted energy density produced by this stellar population approaches current limits on the extragalactic background at both optical wavelengths, into which the ultraviolet radiation of the massive stars is likely to be redshifted, and at far-infrared wavelengths, where starlight reprocessed by dust associated with the starburst will be observed.

6 Acknowledgements

I thank my collaborators and colleagues, including S. Charlot, F. Palla, B. Wang, R. Wyse and S. Zepf, for many discussions of the topics presented here. This research has also been supported in part by grants from NASA.

References

Arnaud, M., Rothenflug, R., Boulade, O., Vigroux, L. and Vangioni-Flam, E. 1992, A&A, 254,49.

Bender, R. 1992, in *The Stellar Populations of Galaxies*, ed. B. Barbuy (Dordrecht: Kluwer), p. 267.

Caselli, P. and Myers, P. C. 1995, ApJ, in press

Charlot, S., Ferrari, F., Mathews, G.J., and Silk, J. 1993, ApJ, 419, L57.

Charlot, S. & Silk, J. 1994, ApJ, 432, 453.

Chen, H., Myers, P. C., Ladd, E. F. and Wood, D. O. S. 1995, ApJ, 445, 377.

Doane, J.S., & Mathews, W.G. 1993, ApJ, 419, 573

Dressler, A. & Gunn, J. E. 1992, ApJS, 78,1.

Forman, W., Jones, C., David, L., Franx, M., Makishima, K. & Ohashi, T. 1993, ApJ, 418, L55.

Franx, M. & Illingworth, G. 1990, ApJ, 359, L41.

Kauffmann, G., White, S. D. M. and Guiderdoni, B. 1993, MNRAS, 264, 201

Kennicutt, R. 1989, ApJ, 344, 685

Koo, B.-C. and McKee, C. F. 1992, ApJ, 388, 103

Lacey, C. and Silk, J. 1991, ApJ, 381, 14

Lacey, C., Guiderdoni, B., Silk, J. and Rocca-Volmerange, B. 1993, ApJ, 402, 15

Mamon, G. A. 1992, 1990, ApJ, 401, L3.

Myers, P. C. and Goodman, A. A. 1988, ApJ, 329, 392

Myers, P. C. and Fuller, G. 1992, ApJ, 396, 631

Ohashi, T. 1995, in Dark Matter, ed. S.S. Holt & C.L. Bennett (New York: A.I.P.), 255

Puget, J.-L. *et al.* 1995, A&A, in press.

Schweizer, F. and Seitzer, P. 1992, AJ, 104, 1039.

Shu, F. H., Adams, F. C. and Lizano, S. 1987, *ARAA*, 25, 23

Silk, J. 1995, ApJ, 438, L41

Tatematsu, K. *et al.* 1993, ApJ, 404, 643

Terebey, S., Shu, F. H. and Cassen, P. 1984, ApJ, 286, 529

Tsujimoto, T., Nomoto, K., Yoshii, Y., Hashimoto, M., Yanagida, S., & Thielemann, F.-K. 1995, MNRAS, in press.

Wang, B, & Silk, J. 1994, ApJ, 427, 759

White, S. D. M., Navarro, J. M., Evrard, A. E. and Frenk, C. S. 1993, Nature, 366,429.

Worthey, G., Faber, S.M., & González, J.J. 1992, ApJ, 398, 69

Wyse, R. F. G. and Silk, J. 1989, ApJ, 339, 700

Zepf, S. and Silk, J. 1995, ApJ, submitted.

Primordial Nucleosynthesis and Light Element Abundances

Rafael Rebolo

Instituto de Astrofísica de Canarias, 38200 La Laguna. Spain

Abstract: A brief overview of the physics involved in the primordial production of the light elements, D, ^{3,4}He and ^{7}Li and of their abundances in related astrophysical contexts is presented. Recent results on the determination of primordial abundances are discussed and confronted with predictions of standard Big Bang Nucleosynthesis. The standard model is in good agreement, within current uncertainties, with the "best" estimation of primordial abundances, providing tight constraints on the baryonic density parameter. An improvement in our knowledge of deuterium abundances in high redshift systems and lithium in extreme metal poor stars, feasible in a near future with the new large-diameter telescopes, may either definitively confirm the success of the standard model or demand an alternative one.

1 Introduction

Big Bang Nucleosynthesis provides a quantitative experimental test to standard and non-standard cosmological models. Nuclear reactions taking place in the primordial plasma during the first minutes after the initial singularity have as a consequence the production of low atomic mass elements, either copiously as ^{4}He, or in a barely measurable amount like ^{7}Li (the most abundant of the stable Li isotopes). The predicted abundances depend mainly on the universal baryonic density which may be considered the only free parameter in the nucleosynthesis calculations. Measurements of the primordial abundances of the light elements test the consistency of primordial nucleosynthesis and in turn set very valuable constraints on the baryonic density, frequently measured by the ratio of baryons to photons $\eta = n_b/n_\gamma$, a parameter that has remained constant since the positron-electron annihilation epoch to the present day.

2 Primordial Nucleosynthesis

2.1 Standard Big Bang Nucleosynthesis (SBBN)

The physics of SBBN is well understood, it involves modestly complicated reaction networks operating at temperatures where the cross sections and relevant reaction rates are known from the laboratory. During the nucleosynthesis epoch, the density of the plasma at nuclear scales is such that many body reactions are negligible. The SBBN model adopts the following assumptions:

a. Homogeneity.
b. Isotropic expansion.
c. General Theory of Relativity.
d. A radiation dominated Universe (the total energy is dominated by the contribution of relativistic particles).

At a time $t \sim 10^{-2}$s. after the initial singularity (plasma temperature $T \sim 10$ Mev) photons, leptons (neutrinos ν_i, electrons e^-) and baryons had very high interaction rates as compared with the expansion rate. The evolution of the scale factor and therefore the expansion of the Universe varied according to the total energy density of relativistic particles ρ_R. The Universe cools through a sucession of nearly thermal states as expansion proceeds. At $t \sim 0.1$ s. the light neutrino species decouple from the plasma ($T \sim 1$ MeV, $T \sim 10^{10}$K), leaving a "neutrino cosmic background" with a present temperature $T_{\nu_0} = T_{CMB_0}/1.4 \simeq 1.96$ K (where T_{CMB_0} is the present temperature of the cosmic microwave background).This is the most relevant event before nucleosynthesis processes may actually take place. At temperatures higher than 1 MeV it is not possible to form complex nuclei in the plasma. Protons and neutrons are kept in equilibrium via the charged-current weak interactions. The neutron-proton ratio is given by $n/p = \exp(-\Delta M/kT)$ where ΔM is the neutron-proton mass difference (~ 1.29 MeV). As the Universe expands and cools the n/p ratio decreases from one to a value slightly below 0.15 for a temperature 10^9 K (0.1 MeV). After the weak interaction drops out of equilibrium, the ratio n/p changes more slowly mainly due to free neutrons decaying to protons (and residual weak interactions), the neutrino capture reactions are too slow to compensate these desintegrations.

2.1.1 The onset of nucleosynthesis

When the Universe cools down to about 0.1 MeV the first nuclei start to be synthesized. Deuterium has a very low binding energy and a large photodissociation cross section. Although some reactions between neutrons and protons took place before this epoch, the resulting D nuclei were inmediately dissociated by the bath of radiation which prevented further nuclear reactions to occur (the "deuterium bottleneck"). Below 0.1 MeV the number of photons in the tail of the Bose-Einstein photon energy distribution is too low to photodissociate the deuterons formed and through the reaction n + p $\Rightarrow$ γ + D, the deuterium abundance increases. The higher the nucleon density, the higher the temperature at which the deuterium bottleneck is overcome. When the D/H ratio reaches a value D/H$\sim$ 10^{-3} heavier elements start to be built via D + p $\Rightarrow$ ^{3}He + γ; D + n $\rightarrow$^{3}H + γ. Tritium and ^{3}He interconvert by ^{3}He + n $\Rightarrow$ p + ^{3}H and ^{3}H $\Rightarrow$^{3}He + e$^-$ + ν_e. The mass-3 nuclei are also burnt into ^{4}He via ^{3}H + p $\Rightarrow$ ^{4}He + γ, ^{3}H + D $\Rightarrow$ ^{4}He + n, ^{3}He + ^{3}He $\Rightarrow$ ^{4}He + 2p, ^{3}He + n $\Rightarrow$ ^{4}He + γ and ^{3}He + D $\Rightarrow$ ^{4}He + p.

2.1.2 Primordial abundances

The absence of stable nuclei at mass 5 and 8 and the coulomb barriers provide a bottleneck to further nucleosynthesis. ^{4}He cannot be further depleted resulting in a large primordial mass fraction, Y_p= 2n/(p+n) $\simeq$ 0.25, where n/p is the ratio at the onset of nucleosynthesis. Most of the neutrons in the Universe are incorporated into ^{4}He, some traces of D and ^{3}He survive, and almost all the protons that did not capture neutrons remain as hydrogen. Only a very small fraction of ^{4}He is transformed to ^{7}Li via reactions: ^{4}He + ^{3}H $\Rightarrow$ ^{7}Li + γ and ^{4}He + ^{3}He $\Rightarrow$ ^{7}Be + γ, decaying to ^{7}Li by K-capture. Li destruction also takes place via ^{7}Li + p $\Rightarrow$ 2 ^{4}He. Primordial nucleosynthesis ends at t$\sim$ 10^3s with a definite prediction of light element abundances as a function of the baryon to photon ratio η (see e.g. Copi et al. 1995).

The predicted primordial production of ^{4}He is a smooth function of η, it also depends on N_ν (the number of light neutrino families) and on the neutron lifetime through their effect on the initial n/p ratio, larger values of N_ν or $\tau_{1/2}$ lead to higher values of the electron-neutrino decoupling temperature, therefore a higher n/p ratio and a larger ^{4}He production. Fortunately, recent experiments have established the value of $\tau_{1/2}$(10.25±0.05 mins., Mampe et al. 1989) with a sufficient accuracy to make definite predictions on N_ν if Y_p could be determined from the observations with an uncertainty on only ± 0.005. Since the work by Steigman, Schramm and Gunn (1977), it is known that SBBN imposes an upper limit on N_ν to avoid ^{4}He overproduction, the improvement in the abundance measurements and neutron lifetime determinations led to claim N_ν <4 (Pagel 1988, Olive et al. 1990, Walker et al. 1991) which has been now confirmed in accelerator experiments on the boson Z^0, the carrier of the neutral weak inter- action, where the overall average gives N_ν=2.99 ±0.03 (ALEPH et al. 1993). It should be noted that SBBN does not exclude neutrinos that were non-relativistic at a few MeV.

2.2 Non-Standard Big Bang Nucleosynthesis (NSBBN)

In the last ten years, alternative NSBBN models considering density inhomogeneities in the epoch of nucleosynthesis originated during the quark-hadron phase transition have been subject of great interest. Around a temperature of 100 MeV quarks are expected to confine into hadrons, if this quark-hadron transition were first order it would possibly develop large inhomogeneities. Preferential diffusion of neutron versus protons out of the high density regions could produce BBN under conditions of density inhomogeneities and variable n/p ratios which would give significant differences in primordial abundance predictions with respect the standard case (Applegate and Hogan 1985, Alcock et al. 1987, Applegate et al. 1988) and also very different costraints on the nucleon density parameter when compared with the observed light element abundances. However, the initial models, that conspicuously overproduced ^{7}Li, treated separately the nucleosynthesis in high and low density regions. Malaney and Fowler (1988) argued that back neutron diffusion could destroy much of the excess ^{7}Li and Applegate, Hogan and Scherrer (1988) suggested the formation of primordial r-elements was feasible in the neutron rich regions. In addition, Boyd and Kajino found considerably higher abundances of ^{9}Be and ^{10,11}B in the inhomogeneous models than in the standard one. However, Terasawa and Sato (1990) and Kurki-Suonio et al. (1990) showed through a detailed treatment of back neutron diffusion that previous estimates were too high. As Reeves (1991) states the order of the quark-hadron transition has to be settled before we have final results on BBN yields, even then, the properties of the transition have to be calculated with good accuracy and the evolution of the bubbles followed dynamically to the epoch of nucleosynthesis including proton and neutron diffusion.

The inhomogenous nucleosynthesis computations have promoted a large effort to extend the observations of primordial abundances to other elements that were produced in negligible amounts in the standard model.

3 Observations and Estimation of Primordial Abundances

The precise determination of primordial abundances involves, not only difficult observations, but extrapolation from the present objects to the pregalactic values taking into account the astrophysical processes that may have changed the chemical composition of the Big Bang emerging matter. For an updated comprehensive review on the subject the reader is referred to the book "The Light Element Abundances" (ed. Crane 1995) where an extensive discussion can be found for each relevant element. Here, a brief description of the state of the art is presented.

3.1 Helium-4

Helium is measured in many astrophysical contexts and given its high abundance a reliable statistical estimation may be expected, however as with other elements a careful treatment of the systematic errors has to be undertaken. Observations in B-type stars, subdwarfs, globular clusters, galactic nebulae and extragalactic nebulae involving a variety of techniques show helium mass fraction values between 0.2 and 0.3. Since astration of matter has the effect of a net production of helium, the chemical evolution of ^4He is easier than for other elements. It is expected a lower helium content in low processed material and different measurements agree that the value of the primeval mass fraction is about 0.24.

It has been extensively argued (see e.g. Pagel 1995, Terlevich et al. 1995) that HII regions provide the most accurate helium abundances. The main difficulties associated with the measurements are reviewed in the previous two papers. Following the pioneer work by Peimbert and Torres-Peimbert (1974, 1976), during the past twenty years, extensive work on primordial helium has been done from extragalactic HII regions with different heavy-element abundances. The Peimbert introduced the idea of deriving Y_p from $Y=Y_p + Z\ dY/dZ$ or equivalently $Y=Y_p + (O/H)\ dY/d(O/H)$ and extrapolating to $Z=0$ (where Z and O/H are overall metallicity and oxygen abundance, respectively, and $Z{\sim}25$ (O/H)). Regressions of helium against oxygen and nitrogen can be found in Pagel et al. (1992) with some updates (see Pagel 1995). According to them the extrapolated value provided by the maximum likelihood regression lines is $Y_p=0.228\pm0.005$(s.e.), which assuming a systematic error of up to 0.005 gives at 95 per cent confidence $Y_p < 0.242$ (the same conclusion is obtained independently of the adoption of oxygen or nitrogen as variable). The work by Olive and Steigman (1995) on very metal-poor extragalactic HII regions, analyzing more than 40 HII regions with oxygen abundances down to 1/50 solar, and nitrogen down to 1% yields $Y_p = 0.232\pm0.003\pm0.005$ where the statistical error is quoted first and the systematic error appears second. Finally, Izotov et al. (1994) gives a preferred value of 0.239 ± 0.007. Since the systematic errors are not perfectly understood it seems reasonable to adopt as a safe upper limit $Y_p < 0.245$. The average value of the three determinations above, assuming same weight, is $Y_p= 0.233$. Radiative transfer effects in the formation of helium lines in HII regions seem to be the main problem to reduce the systematic errors.

A remarkable result on helium in high redshift absorption systems has been obtained by Jakobsen (1995) who showed the presence of this element at very early epochs as expected from BBN predictions.

3.2 Deuterium

From ^3He measurements in meteorites and in the solar wind it is possible to infer the D abundance at the time the solar system formed $(D/H)_{solar}=2.6\pm1.0$ x 10^{-5} (Geiss 1993) The most accurate method to measure the present abundance of deuterium in the Galaxy is the determination of D to H column densities in warm interstellar gas via Lyman series lines for the lines of sight toward selected stars. Observations with the Copernicus satellite (Rogerson and York 1973), the International Ultraviolet Explorer (Londsman et al. 1984) and, more recently, the Hubble Space Telescope (Linsky et al. 1993, Lemoine et al. 1995) show an average Interstellar Medium (ISM) abundance D/H=1.6±0.1 x 10^{-5} which should be considered as a lower limit to the primordial abundance since deuterium is destroyed as interstellar matter is astrated in stars. Galactic chemical evolution models predict values for the primordial abundance a factor 2-5 higher than the present ISM abundance (Steigman and Tosi 1992, Galli et al. 1995). These predictions are affected by considerable uncertainties limiting the ability of deuterium observations to constrain the nucleon density. Fortunately, there are expectations to obtain a good approximation to truly primordial D abundances by studying the absorption spectra of high-redshift quasars. High column density quasar absorption clouds are sites where D/H can be measured accurately via observations of the Lyman series. Songaila et al. (1994) have claimed a first detection of D lines in an absorption system at z=3.3 towards the quasar Q0014+813 from which they infer a D/H=1.9-2.5 10^{-4}. There is however, a non-negligible probability that the feature seen arises from the Lyman α line of a smaller hydrogen cloud with a relative velocity of -80 km s^{-1} simulating the isotopic shift (see also Carswell et al. 1994). At present we can confidently adopt as lower limit to the primordial D abundance $D_p \geq 3$ x 10^{-5}.

3.3 Helium-3

Measurements in the solar wind and in meteorites indicate a protosolar value of ^{3}He/H= 1.5±0.3 10^{-5}. Outside the solar system Rood et al. (1984), Bania et al. (1987) and Balser et al. (1994), using observations of the 3.46 cm hyperfine line of ^{3}He$^+$ in galactic HII regions, find ^{3}He/H abundances in the range 0.9 - 5.4 10^{-5}. They conclude that the source to source variations are not due to observational error and have difficulties to explain the abundance pattern in terms of existing chemical evolution models. Since the massive stars in HII regions are very effective ^{3}He destroyers it is possible that these sites are not representative of the cosmic ^{3}He abundance. On the other hand, the galactic evolution of ^{3}He is not well determined, for instance, it is very uncertain what fraction of ^{3}He survives stellar processing, and it is also uncertain whether low mas stars, are net producers (see e.g. Galli et al. 1994). The combined D+^{3}He abundances can be used to set interesting restrictions to BBN as argued by Yang et al. (1984).

3.4 Lithium-7

Observations of a "Li-plateau" in warm halo dwarfs suggested a value of the pregalactic Li abundance of Li/H~ 10^{-10} (Spite and Spite 1982, Rebolo et al. 1988). However, Li is a fragile element easily destroyed in stellar interiors at relatively low temperatures and it was not possible to rule out some depletion during the long lifetime of these stars. Several theoretical studies (see e.g Vauclair 1988, Deliyannis et al. 1990) have suggested that indeed some depletion has taken place, and the initial value could have been as high as the present observed abundance in the interstellar medium and young stars Li/H$\sim10^{-9}$. While there is clear evidence for Li depletion in solar metallicity dwarf stars with effective temperatures above 5500 K (see for reviews Rebolo 1991, Michaud and Charbonneau 1991), most extreme metal-poor stars with such effective temperatures show a rather uniform Li abundance. It is a matter of discussion whether the scatter of the measurements is consistent within error bars with a uniform Li abundance or, alternatively, is the result of a depletion process acting in each star. Thorburn (1994) and Norris et al. (1994) claims the detection of a positive slope in the curve Li abundance versus effective temperature (also found by Rebolo et al. 1988) which might indicate that some depletion took place in the cool part of the plateau. But this slope is very sensitive to the effective temperatures adopted, and the Li abundance reanalysis of Molaro et al. (1995) (using a technique for effective temperature determination based on Balmer line profiles) do not support its existence. A very recent paper by Ryan et al. (1996) rediscussed the problem arguing that Molaro et al. included in their sample some subgiants that would mask the trends with effective temperature or metallicity, since these stars have a different history than dwarfs. They also note the existence of very similar stars that cannot have the same Li abundance.

In the literature about Li and metal poor stars it is possible to find stars with independent observations (by different authors and different telescopes) of the equivalent width of the Li λ 6708 Å line that do not agree themselves within quoted error bars. This is fortunately not very frequent but the number of stars with more than three independent measurements is not large! and we should caution on the conclusions that sophisticated statistical analysis may provide in these circumstances. A conservative position regarding the primordial Li abundance Li_p would be to adopt as a lower limit the average of the Li abundances observed in the hotter unevolved extreme metal-poor dwarfs $(Li/H)_p \geq 2 \; 10^{-10}$. The detection of the very fragile ^6Li isotope in a hot extreme metal-poor plateau dwarf (Smith et al. 1993) strongly suggests that the initial Li abundance is close to the present observed atmospheric one, and little ^7Li depletion, if any, may have taken place in the atmosphere. This also suggests that the primordial Li abundance may not be far from the value adopted above as lower limit. Another piece of evidence in this direction comes from the many astrophysical processes and sites that may produce Li in the Galaxy: cosmic ray spallation of heavy nuclei in the ISM, asymptotic giant branch and carbon stars, supernovae explosions, accretion disks around compact objects (black holes and neutron stars). According to chemical evolution models (see e.g. D'Antona and Matteucci 1991, Abia et al. 1988, Beckman and Casuso 1995) it can be easily explained a Li enrichment by a factor 10 along the lifetime of the Galaxy. A conservative upper limit to the the primordial abundance of ^7Li would then be $(Li/H)_p \leq 1x10^{-9}$.

4 Confronting theory and observations

4.1 Restrictions on Cosmological Parameters and Dark Matter Problem

Let us adopt $\eta_{10}=\eta \times 10^{10}$, the combined restrictions on η_{10} that our "observed" primordial abundances impose in the context of SBBN model predictions (for instance using those of Copi et al. 1995) are $2.5 le\ \eta_{10}\ \geq 5$. The upper limit is provided by our upper limit to Y_p since it is more restrictive than the one obtained from D_p, and the lower limit is set by the D+^{3}He argument. If we relax this last argument then a less restrictive (but more conservative) lower limit $\eta_{10}\geq 1.5$ would be imposed from both Y_p and ^7Li_p. We note that our conservative limits on ^7Li_p show additional agreement with the model predictions, although do not provide further constraints to those offered by D_p and Y_p, but ^7Li could in principle be a very powerful discriminator. If it were proved, for instance, a high Li abundance Li/H($\sim$3x10^{-10} in extreme metal poor stars ([Fe/H]$\leq$-4), the survival of the standard model would require either $Y_p \geq 0.24$ or $Y_p \leq 0.23$, and the latter would request a D_p well above 1x10^{-4} very close to the claimed value in the high redshift absorption system of Songalia. The baryon ratio would, in turn, be defined with high precision.

Using the relationship between η and the baryonic density parameter Ω_{b0} (the fraction of the critical density provided by baryons):

$$\Omega_{b0}h_0^2 = 3.7310^{-3}(T_{CMB_0}/2.75K)^3 \eta_{10}$$

where the suffix zero refers to the present epoch and h$_0$ is the Hubble constant in units of 100 km s^{-1} Mpc^{-1}, we obtain the following restrictions:

$$0.009 \leq \Omega_{b0}h_0^2 \geq 0.018$$

Although the exact range for Ω_{b0} obviously depends on the judgement employed to adopt the final limits to primordial abundances, similar results have been found along the years, showing the robustness of the standard model. Very recently some claims have been made on the possible crisis of SBBN (see Steigman 1995), but our knowledge of primordial abundances requires significant improvement before the crisis argument can hold.

Comparison of the restrictions on baryonic density and the mass density of luminous matter ($\sim$ 1 % of the critical density) appear to suggest that most baryons are dark. If the total density of the Universe were indeed the critical one, as pointed out by inflationary scenarios, then most of the mass density should be in a non-baryonic form whose nature is still unknown.

An alternative powerful method to determine the baryonic contribution to the total density will soon be able to confirm or not the results of SBBN in an fully independent way. Cosmic microwave anisotropies in angular scales of 0.5-2 degrees are very sensitive to the baryonic content of the Universe (see e.g. Scott et al. 1994) and present searches for anisotropies at those scales are reaching the level needed to constrain the baryonic density parameter with sufficient accuracy.

Acknowledgements

It is a pleasure to acknowledge R. Watson and R. J. García López for their reading of the original manuscript.

References

Abia, C., Canal, R.: Astron. Astrophys. **189** 55 (1988)

Alcock, C., Fuller, G., Mathews, G.: Astrophys. J. **320** 439 (1987)

ALEPH, L3, OPAL, DELPHI results (1993): "Lepton-Photon meeting at Ithaca, Ney York.

Applegate, J.H., Hogan, C.J.: Phys. Rev. D. **31** 3037 (1985)

Applegate, J.H., Hogan, C.J., Scherrer, R.J.: Astrophys. J. **329** 572 (1988)

Balser et al.: Astrophys. J. **430** 667 (1994)

Bania, T.M., Rood, R.T., Wilson, T.L.: Astrophys. J. **323** 30 (1987)

Beckman, J., Casuso, E.: ESO Astrophysics Symp. "The Light Element Abundances" eds. P. Crane (Springer-Verlag, Berlin Heidelberg) p. 105 (1995)

Boyd, R.N., Kajino, T.: Astrophys. J. **336** L55 (1989)

Carswell et al.: MNRAS**268** L1 (1994)

Copi, C.J., Schramm, d.N., Turner, M.S.: Science **267** 192 (1995)

Crane, P.: "The Light Element Abundances" ESO Astrophys. Symp. Springer-Verlag (1995)

D'Antona, F., Matteucci, F.: Astron. Astrophys. **248** 62 (1991)

Deliyannis, C. Demarque, P, Kawaler S.D.: AStrophys. J. Supp. **73** 21 (1990)

Galli, D., Palla, F., Straniero, O., Ferrini, F.: Astrophys. J. **432** L101 (1994)

Geiss, J.: in "Origin and Evolution of the Elements" eds. Prantzos,N., Vangioni-Flam, E., Cassé M. Cambridge U.P. 89 (1993)

Izotov, Y.I., Thuan, T.X., Lipovetsky, V.A.: Astrophys. J. **435** 647 (1994)

Jakobsen, P.: ESO Astrophysics Symp. "The Light Element Abundances" eds. P. Crane (Springer-Verlag, Berlin Heidelberg) p. 75 (1995)

Kurki-Suonio, H., Matzner, R., Olive, K., Schramm, D.N.: Astrophys. J. **353** 406 (1990)

Linsky et al.: Astrophys. J. **402** 694 (1993)

Malaney, R.A., Fowler, W.A.: Astrophys. J. **333** 14 (1988)

Mampe, W., Ageron, P., Bates, C., Pendlebury, J.M., Steyerl, A.: Phys. Rev. Lett. **63** 593 (1989)

Michaud, G., Charbonneau, P.: Space Sci. Rev. **57** 1. (1991)

Molaro, P., Primas, F., Bonifacio, P.: Astron. Astrophys. **245** L47 (1995)

Norris, J.E., Ryan, S.G., Stringfellow, G.S.: Astrophys. J. **423** 386 (1994)

Olive, K., Schramm, D.N., Steigman ,G., Walker, T.: Phys. Lett. B **236** 454 (1990)

Olive, K., Steigman, G.: Astrophys. Suppl. Ser. **97** 49 (1995)

Pagel, B.E.J.: Gemini, Royal Greenwich Obs., January issue p.1 (1988)

Pagel, B.E.J.: ESO Astrophysics Symp. "The Light Element Abundances" eds. P. Crane (Springer-Verlag, Berlin Heidelberg) p. 155 (1995)

Peimbert, M., Torres-Peimbert, S.: Astrophys. J. **193** 327 (1974)

Peimbert, M., Torres-Peimbert, S.: Astrophys. J. **203** 581 (1976)

Rebolo, R.: IAU Symp. 145 eds. Michaud, G., Tutukov, A. Kluwer p 85. (1991)

Rebolo, R., Molaro, P., Beckman, J.E.:Astron. Astrophys. **192** 192 (1988)

Reeves, H.: Phys. Rep. **201** 335 (1991)

Rood, R.T., Bania, T.M., Wilson, T.L.: Astrophys. J. **280** 629 (1984)

Rogerson, J.B., York, D.G.: Astrophys. J. **186** L95 (1973)

Ryan, S.G., Beers, T.C., Deliyannis, C.P., Thorburn, J. A.: Astrophys. J. in press. (1996)

Smith, V.V., Lambert, D.L., Nissen, P.E.: Astrophys. J. **408** 262 (1993)

Songalia, A., Cowie, L.L, Hogan, C.J., Rugers, M.: Nature **368** 599 (1994)

Spite, F. Spite, M: Astron. Astrophys. **115** 357 (1982)

Steigman, G.: ESO Astrophysics Symp. "The Light Element Abundances" eds. P. Crane (Springer-Verlag, Berlin Heidelberg) p. 10 (1995)

Steigman, G., Schramm, D.N., Gunn, J.E. Phys. Lett. B. **66** 202 (1977)

Steigman, G., Tosi, M.: Astrophys. J. **401** 150 (1992)

Terasawa, N., sato, K.: Astropys J. **362** L47 (1990)

Terlevich, E., Skillman, E.D., Terlevich, R.: ESO Astrophysics Symp. "The Light Element Abundances" ed. P. Crane (Springer-Verlag, Berlin Heidelberg) p. 175 (1995)

Thorburn, J.A.: Astrohpys. J. **421** 318 (1994)

Vauclair, S.: Astrophys. J. **335** 971 (1988)

Walker, T., Steigman, G., Schramm, D.N., Olive, K., Kang, H.S.: Astrophys. J. **376** 51 (1991)

White, M., Scott, D., Silk, J.: Ann. Rev. Astron. Astrophys. **32** 319 (1994)

Yang, J., Turner, M.S., Steigman, G., Schramm, D.N., Olive, K.A.: Astrophys. J. **281** 493 (1984)

II

Large-scale Structure

Peculiar Motions in the Universe

Riccardo Giovanelli

Dept. of Astronomy and National Astronomy and Ionosphere Center[1],
Cornell University, Ithaca, NY 14853, USA

Abstract: A summary of the observational status of measurements of peculiar velocities in the local universe is given. Special emphasis is devoted to the discussion of
applications of the Tully–Fisher technique and its limitations. Preliminary results are
presented regarding the motions of two samples of galaxies with measured peculiar
velocities: one including 24 clusters of galaxies and a second of field objects, the latter
being used to test for large amplitude bulk flow in the peculiar velocity field.

1 Introduction

Deviations in the motion of galaxies from smooth Hubble flow are generally
thought to arise from gravitational forces originating in large scale, low amplitude
mass perturbations. Their study provides kinematical information that can be
used to make inferences on cosmological parameters, which cannot be obtained
from a simple census of the luminous matter alone. Historically, the study of
such motions, generally referred to as "peculiar", emerged together with the
concept that the Milky Way and the Local Group (LG) are part of a much
larger complex of galaxies having the Virgo cluster near its center; thus early
efforts were directed towards measuring the perturbation induced by the cluster
on the motion of the LG ("Virgo infall"). In the late 1970's, the discovery of
a dipole moment in the sky brightness distribution of the Cosmic Microwave
Background (CMB) radiation field was interpreted as arising from the Doppler
shift due to the peculiar motion of the LG. Since the apex of the dipole is
significantly removed from the direction to the center of the Virgo cluster, the
motion that would produce it could not be fully ascribed as originating in Virgo.
    Several peculiar velocity surveys have been carried out in the last decade.
Following the early work of Aaronson et al. (1982), the influential contribution
of Lynden–Bell et al. (1988) established the framework with respect to which

---

[1] The Nationals Astronomy and Ionosphere Center is operated by Cornell University
under a cooperative agreement with the National Science Foundation

much of the later work was to measure: they verified that the Virgo cluster is not the dominant gravitational influence in the local universe, and proposed that the center of the mass concentration largely shaping the local peculiar velocity field is at a redshift distance of about 4300 km s^{-1} . Such structure was dubbed the "Great Attractor" (GA). Due to its low galactic latitude, the light emission of the GA should be largely dimmed by galactic absorption. Later work, however, based both on peculiar motion surveys and on the measurement of the light dipoles of various populations of extragalactic objects, raised the possibility that the principal component of the LG motion would have a coherence length a few times larger than the distance to the GA. In the first part of these lectures, we will briefly review the observational evidence collected by these studies.

While the earlier surveys spurred important theoretical efforts, which produced increasingly sophisticated techniques for the three–dimensional recovery of the peculiar velocity and of the underlying mass density field, the observational arena remains somewhat unsettled. Important uncertainties still exist at the basic level of the derivation of raw peculiar velocities of individual galaxies from the observations, as different techniques sometimes yield conflicting results, and even from the same raw data different groups obtain discrepant peculiar velocities. In this set of lectures, we will concentrate on one of such techniques, and on the procedures followed in its application, namely the Tully–Fisher (TF) technique, introduced in 1977 by R.B. Tully and J.R. Fisher.

After a modicum of theoretical background in section 2, the observational situation will be reviewed in section 3, and the TF relation will be introduced in section 4, with an evaluation of its predictive power for the estimate of extragalactic distances. The application of the TF relation in the estimate of motions of nearby clusters and of galaxies in the field will follow in sections 5 and 6. We will refrain from extending our discussion to the recovery of the density field via Potent (Bertschinger and Dekel 1989) and analogous techniques, a topic which is extensively reviewed by Dekel (1994). The reader is also referred to the reviews of Burstein (1990) and Strauss and Willick (1995).

Throughout this work, we parametrize distance dependent parameters by assuming a distance scale $H_0 = 100h$ km s^{-1} Mpc^{-1}. Whenever explicit dependence on h is not made, a value of $h = 1$ is assumed.

2 Some Useful Equations

2.1 Peculiar Velocity and the Density Field

We start with a few rudiments of the Newtonian theory of small perturbations, which is developed in detail in the textbooks of Weinberg (1972), Peebles (1993) and Padmanabhan (1993). In an expanding universe, the development of irregularities in the matter distribution can be approximated by classical mechanics, provided that the size of the region considered is small in comparison with the Hubble length cH^{-1}, larger than the Schwarzschild radius of collapsed objects — say $10^{14} \ll r \ll 10^{28}$ cm — and provided that velocities are non–relativistic. Thus it is said that the treatment is possible in the Newtonian approximation.

Let the background solution be characterized by a mean density $\bar{\rho}$. If deviations $\delta = (\rho - \bar{\rho})/\bar{\rho}$ from the background solution can be assumed at all points to be small, i.e. $\delta \sim 1$ or less, then linear perturbation theory applies, and it can be shown that the relationship between the peculiar velocity and the peculiar acceleration produced by the perturbation field is given by

$$\mathbf{u} = \frac{2}{3} \frac{f}{\Omega_0 H_0} \mathbf{g} \tag{1}$$

where $f \simeq \Omega_0^{0.6}$ (in the following equations we'll replace f with $\Omega_0 0.6$), $\Omega_0 = \bar{\rho}/\rho_{crit}$ is the density parameter, $\rho_{crit} = 3H_0^2/8\pi G$ and H_0 is the Hubble parameter. In terms of the density perturbation, the peculiar velocity can be written as

$$\mathbf{V}(\mathbf{r}) = \frac{H_0 \Omega_0^{0.6}}{4\pi} \int \frac{\delta(\mathbf{r}')(\hat{\mathbf{r}}' - \hat{\mathbf{r}})d\mathbf{r}'}{|\mathbf{r}' - \mathbf{r}|^2} \tag{2}$$

Direct measurements of the peculiar velocity can only yield the radial component of the velocity. The measured radial velocity of a galaxy is then the combination of three quantities:

$$V_{radial} = H_0 d + V_{pec,galaxy}^{\parallel} - V_{pec,LG}^{\parallel} \tag{3}$$

where $H_0 d$ is the Hubble velocity of the object, expressed in km s^{-1}, and the other two terms are the components along the line of sight of the peculiar velocities of, respectively, the target galaxy and the LG. V_{radial} is obtained directly from the galaxy's redshift. $V_{pec,LG}^{\parallel}$ can be inferred from the CMB dipole moment. One of several distance determination techniques, such as the TF method, can produce an estimate of $H_0 d$; such techniques can generally be calibrated independently on the value of H_0, in which case all distances are expressed in km s^{-1}. The application of eqn. (3) then allows the estimate of the radial component of the peculiar velocity of the observed galaxy. By mapping the peculiar velocity field, inversion of eqn. (2) allows recovery of the density field $\delta(\mathbf{r})$. To the extent that it can be assumed that the galaxy distribution mimics the mass distribution, density perturbations can be expected to be proportional to excess galaxy counts, so that $\delta_N = (N - \bar{N})/\bar{N} = b(\rho - \bar{\rho})/\bar{\rho}$, where b is usually referred to as the *bias* parameter. Then, if light traces mass, measurements of the peculiar velocity field and galaxy counts can be used to obtain an estimate of $\beta = \Omega_0^{0.6}/b$.

2.2 Convergence Depth

Since the local number density field of galaxies $\delta_N(\mathbf{r})$ is relatively well approximated by the results of redshift surveys, it can be used to estimate the peculiar velocity of the LG, on the assumption of a constant bias parameter. We define a window function $W(r, R) = exp(-r^2/2R^2)$, and approximate the peculiar velocity of the LG produced by the mass distribution out to $\sim R$ as[2]

---

[2] The author acknowledges useful discussions with Peter Schneider on the nature of the integral in eqn. (4).

$$V_{pec,LG}(R) = \frac{H_o \Omega_o^{0.6}}{4\pi b} \int \delta_N(\mathbf{r}) \frac{\hat{\mathbf{r}}}{r^2} W(r,R) d\mathbf{r} \tag{4}$$

The asymptotic value of $V_{pec,LG}(R)$, for $R \to \infty$ can then be matched to the velocity inferred from the CMB dipole moment. Note that the gravitational effect of a galaxy is $\propto M_{gal} r^{-2}$; if $M_{gal} \propto L_{gal}$, then the gravitational contribution of that galaxy to $V_{pec,LG}$ is proportional to its flux. In principle, $V_{pec,LG}(\infty)$ can then be gauged from a complete catalog of positions and fluxes of galaxies, or any other widely distributed extragalactic population, such as clusters. Also note that the value of the bias parameter b, and therefore β, will in general depend on the type of catalog chosen, i.e. on the clustering properties of the population: for example, clusters will yield smaller values of β than galaxies, and IRAS galaxies yield larger values than optically selected galaxies. In practical terms, eqn. (4) is replaced by a summation over the sampled objects. The calculation is usually expressed in terms of the monopole and dipole of the distribution of objects:

$$\mathcal{M}(R) = \sum \frac{w_i}{r_i^2} \qquad \mathcal{D}(R) = 3\sum \frac{w_i \hat{r}_i}{r_i^2}, \tag{5}$$

where the summation is over all sampled objects within the distance R, and w_i is a suitable weight, such as the object's luminosity, corrected for sample selection bias. With increasing R, the monopole term tends to $4\pi \bar{N} R$, where $\bar{N}$ is the average number density of objects and the peculiar velocity can be rewritten as

$$V_{pec,LG}(R) = 3 \; \beta \; H_o R \; \frac{\mathcal{D}(R)}{\mathcal{M}(R)} \tag{5}$$

Fig. 1 displays calculations of $V_{pec,LG}(R)$, collated from different sources: a catalog of clusters of galaxies (Tini Brunozzi et al. 1995), an optical catalog of galaxies and a sample of IRAS galaxies (both after Scaramella et al. 1994). Each curve has been "scaled" by a different value of β, namely: $\beta_c = 0.30$ for the cluster sample, $\beta_{opt} = 0.50$ for the optical galaxies and $\beta_{ir} = 0.85$ for the IRAS galaxies. The scaling was chosen (somewhat arbitrarily) so that $V_{pec,LG}(R)$ would match a value near 630 km s^{-1} for the two outermost points for which a calculation is available. The three trends differ somewhat. More than half the amplitude of $V_{pec,LG}(\infty)$ is contributed by sources within $h^{-1}R \sim 5000$ km s^{-1}, in all the curves. However, while the IRAS galaxies suggest that convergence is reached below $h^{-1}R \sim 10^4$ km s^{-1}, the optical galaxies appear to not reach convergence even beyond that distance, and clusters may require integration to about $h^{-1}R \sim 150$ Mpc for convergence to be achieved.

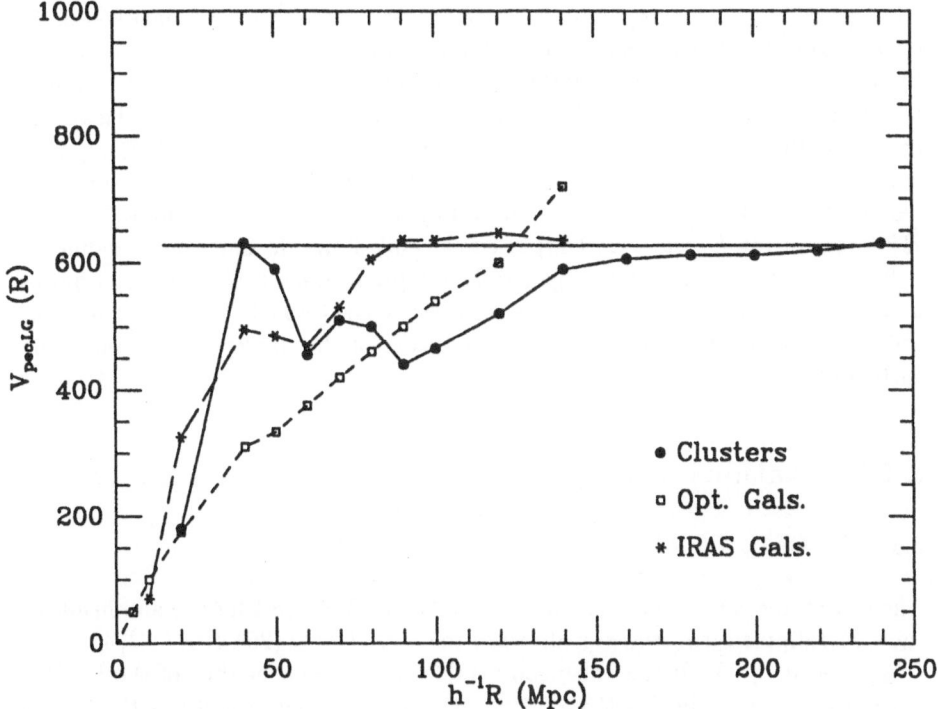

Fig. 1. Illustration of convergence depth. Plots illustrate calculations via eqn. (5) of $V_{pec,LG}(R)$ for: clusters of galaxies (solid circles; Tini Brunozzi et al. 1995), galaxies from optical an optically selected catalog (unfilled squares; Scaramella et al. 1994) and IRAS galaxies (asterisks; Scaramella et al. 1994). All curves have been scaled by the author, corresponding to values of β as given in the text.

2.3 Peculiar Velocities and the Power Spectrum

It is also useful to relate the statistical properties of the peculiar velocity field to the power spectrum of perturbations $P(k)$. As above, let $W(r)$ be a Gaussian window function and $V(r)$ the peculiar velocity field; the r.m.s peculiar velocity within a volume of radius R can be written as

$$\langle V^2 \rangle = (2\pi)^{-3} \int v^2(k) |\widetilde{W}(\mathbf{k})|^2 d\mathbf{k} \qquad (6)$$

where $v(k)$ is the square of the Fourier transform of $V(r)$ and $\widetilde{W}(\mathbf{k})$ is the Fourier transform of the window function. Using the relationship between the velocity field and the density field (2), and their respective Fourier transform, it can be shown that the r.m.s. peculiar velocity can be written in terms of the power spectrum as

$$\langle V^2 \rangle = \frac{H_o^2 \Omega^{1.2}}{(2\pi)^3} \int |\widetilde{W}(\mathbf{k})|^2 P(k) k^{-2} d\mathbf{k} \qquad (7)$$

The COBE results suggest that $P(k) \propto k$ at small k; at some value near $k \sim 0.01$–0.03 h Mpc^{-1}, $P(k)$ turns over and decreases for large values of k, as indicated by the correlation function of galaxies and clusters. CDM models require a power law drop at high k, while HDM models would require an exponential decay. We can compute $\langle V^2 \rangle^{1/2}$ using eqn. (7) and some approximation of $P(k)$, e.g. that suggested by Bond and Efstathiou (1984), normalized via the COBE and APM results; the resulting values decrease rapidly as the size of the window function R increases: for $h^{-1}R \sim 100$ Mpc, r.m.s. bulk flows do not exceed values of ~ 200 km s^{-1}, for the range of generally accepted values of Ω_o. If the LG moves at about 600 km s^{-1}, then either standard CDM models are inadequate, or the origin of our motion is local, or our corner of the universe is an unusually turbulent one.

3 Observations of Large-Scale Motions: a Brief Survey

3.1 The CMB Dipole

The COBE measurements yield a value of 3.365 ± 0.027 mK for the amplitude of the CMB dipole term, which is directed towards $(l, b) = (264.4° \pm 0.3°, 48° \pm 2°$ (Kogut et al. 1993). If the anisotropy is due to the Doppler shift of the Earth's motion relative to the CMB radiation field, the inferred velocity of the Sun is 369.5 ± 3 km s^{-1}. Accounting for solar motion with respect to the LSR [20 km s^{-1} towards $(l, b) = (57°, 23°)$] and the galactic rotation of the LSR of 220 km s^{-1}, the motion of the Galaxy with respect to the CMB is 552.2 ± 5.5 km s^{-1} towards $(l, b) = (266.5° \pm 0.3°, 29.1° \pm 0.4°)$. The motion of the Galaxy with respect to the LG is a relatively uncertain quantity; when account for it is made, using for example the Yahil et al. (1977) solution, the LG moves with respect to the CMB at 627 ± 22 km s^{-1} in the direction $(l_{cmb}, b_{cmb}) = (276° \pm 3°, 30° \pm 3°)$. This direction is about 45° away from the Virgo cluster.

3.2 Motions at the Local Supercluster Scale

The Yahil et al. (1977) solution for the solar motion with respect to the LG has Cartesian components of (-79,295,-37) km s^{-1}, directed respectively toward the galactic center, $l = 90°$ and the North galactic pole. Because the accuracy of this determination is poor, the de Vaucouleurs et al. (1976) simpler expression:

$$V_{gal} = V_{gal}^\odot + 300 \sin l \cos b \qquad (8)$$

is also frequently used.

While, as we shall see, there is substantial evidence for LG infall into the Virgo cluster, there appears to be little shear in the local peculiar velocity field (Faber and Burstein 1988). This characteristic has been referred to as the *Local Anomaly*.

De Vaucouleurs[3] (1958) pioneered the study of large scale structure in the local universe, promoting the idea of the existence of a Local Supercluster (LS). He also reasoned that the mass concentration associated with the LS should be expected to locally slow down the Hubble expansion, and because the galaxy distribution in the LS appears flattened, he invoked global rotation of the system. His earlier estimates of both infall and rotational speeds were not comforted by reliable distance determination techniques, and have been superseded by more recent work. After an earlier study by Silk (1977), in 1980 Schechter produced a model of the Virgocentric flow using the linear approach briefly described in section 2 and assuming cylindrical symmetry for the supercluster. He showed the flow to be unstable against shear only in the inner regions of the supercluster: for a mass density profile $\delta\rho \sim r^{-2}$, departures from radial motions would start occurring only inside the distance from Virgo subtended by the LG.

In 1982, Aaronson et al. obtained an estimate of the Virgocentric infall velocity of the LG of $V_{r_0} = 480 \pm 75$ km s^{-1}, using distance measurements of a sample of spiral galaxies obtained via the TF relation. A later determination by the same group, based on 306 spirals extending to a redshift distance of ~ 3000 km s^{-1}, led to a revised value of the LG Virgocentric infall of 250 ± 64 km s^{-1}. A re-analysis of the Aaronson et al. data by Kraan–Korteweg (1985) favored lower values of the Virgocentric infall, of about 200 km s^{-1}. Using the relative distance moduli of Virgo and nearby clusters, Jerjen and Tammann (1993) obtained a Hubble velocity of the LG relative to the center of the Virgo cluster of 1179 ± 17 km s^{-1}, and a Virgocentric infall velocity of 233 km s^{-1}.

3.3 The Great Attractor

The Local Supercluster alone cannot account for the motion implied by the CMB dipole moment: simple Virgo infall does not agree with either its magnitude or direction. Pure Virgocentric infall of 150 to 400 km s^{-1} would leave an unexplained residual component of 400 to 500 km s^{-1} in the LG motion with respect to the CMB. As the complexity of the large–scale structure of the galaxian distribution started to emerge, Shaya (1984), Tammann and Sandage (1985) and Lilje, Yahil and Jones (1986) suggested that the Hydra–Centaurus supercluster at a redshift of about 3000 km s^{-1} or supercluster structures hidden in the "zone of avoidance" in the general direction of (l_{cmb}, b_{cmb}) could play an important role in determining the gravitational field at the LG.

Aaronson et al. (1986,1989) expanded their observations to include clusters at distances between 4000 and 11000 km s^{-1}, beyond the Local Supercluster and Hydra–Centaurus. In their later work, they detected motions which they interpreted as produced by two main attractors, the Virgo cluster and a center just beyond the Hydra–Centaurus region. Constraining the LG velocity with respect to the CMB to be fully explained by this geometry, they obtained components of the LG motion of 240 and 486 km s^{-1}, respectively in the direction of Virgo and of Hydra–Centaurus.

---

[3] We miss you, Gerard.

With an all–sky sample of about 400 elliptical galaxies, extended in redshift to about 6000 km s^{-1} , Dressler et al. (1987) claimed that Hydra–Centaurus could not be the gravitational attractor mainly responsible for the motion that causes the CMB dipole, because the supercluster itself appears to participate in the flow. In a more detailed study, Lynden–Bell et al. (1988) explored mainly two models : (a) one with a Virgo infall velocity at the LG of 250 km s^{-1} and *bulk motion of velocity V_b of the entire sample towards direction (l_b, b_b)*, with respect to the CMB; (b) a second model, which in addition to Virgocentric and bulk flows includes that produced by a mass concentration at a distance R_m, towards which the infall field at the LG has a velocity V_m. Option (b) fit better, required a smaller Virgo infall velocity and made the bulk flow component superfluous. Their best fit yielded a Virgo infall component at the LG of only 100 km s^{-1} , $V_m = 570$ km s^{-1} and $R_m = 4350$ km s^{-1} . The large mass concentration at the distance R_m appeared to be farther than the Hydra–Centaurus supercluster, which lies between 3000 and 4000 km s^{-1} . The name of "Great Attractor" (GA) was then tagged to the structure.

The study of Lynden–Bell et al. (1988) suggested a much reduced velocity of Virgo infall for the LG. Faber and Burstein (1988) corroborated that result, suggesting that the magnitude of the Virgo infall at the LG might be as small as 85 km s^{-1} . On the other hand, re–analyzing the Aaronson et al. cluster sample within the framework of a bi–infall model (Virgo plus GA), Han and Mould (1990) found optimal solutions with Virgo infall velocities on the order of 200 km s^{-1} , and GA infall solutions of the order of 400 km s^{-1} . The question of the true magnitude of the LG Virgo infall should be considered still open.

The results of Lynden–Bell et al. placed the GA near $(l_{ga}, b_{ga}) = (309°, 18°)$, at $R_m = 4350$ km s^{-1} , and recent applications of the Potent method yield $(l_{ga}, b_{ga}) = (320°, 0°)$, at $R_m = 4000$ km s^{-1} (Dekel 1994). In this region, the surface density of galaxies is greatly reduced by galactic extinction. In order to find a counterpart to the dynamical GA, Dressler (1988) conducted a redshift survey over a wide region of sky centered roughly on (l_{ga}, b_{ga}), i.e. $-35° < b < +45°$, $290° < l < 350°$, which yielded a large overdensity between 2500 and 5500 km s^{-1} . More recently, Kraan–Korteweg and Woudt (1995) have reported that the cluster A3627, located at $l \sim 325°$, $b = -7°$ and $cz = 4300$ km s^{-1} (Abell, Corwin and Olowin 1989) is quite rich and may be an optical counterpart to the GA. It should be however clear that, while the search for massive structures in the GA region is extremely useful, the dynamical entity we refer to as the GA is likely to be the result of a complex superposition, reflecting the richness of the large–scale structure in the local universe.

3.4 Bulk Flows

In 1989 Scaramella et al. noted the presence of a large concentration of clusters of galaxies, near $cz \sim 14,000$ km s^{-1} (the "Shapley Supercluster"), located in the general direction of the GA. They postulated that infall into that overdense region may play an important role in the determination of the local velocity field. While Dressler and Faber (1990) reported back flow into the GA beyond 4500 km s^{-1}, Mathewson, Ford and Buckhorn (1992a) questioned their finding, as their independent data set indicated continued outflow of galaxies beyond the location of the GA, out to the limits of their survey near 7,000 km s^{-1}, a result supportive of the of Scaramella et al. (1989) idea.

The report of Willick (1990) that the Pisces Perseus Supercluster (PPS), at $cz \sim 5000$ km s^{-1} and antipodal to the GA, has a peculiar velocity of about 400 km s^{-1} in the general direction of the GA, appeared to corroborate the idea that the whole volume of local universe engulfing the PPS, the LS and the Hydra–Centaurus–GA region is partaking of a *bulk flow*, roughly in the direction of the GA (or the Shapley supercluster), at a velocity of 400–600 km s^{-1}. The lack of a significant gradient in the motion appeared puzzling as was the recent report of Mathewson (1995), that the flow seemed to cease near 8,000 km s^{-1}, roughly half way between the GA and the Shapley supercluster, spurring him to suggest that the source of the motion is not gravitational.

An additional controversial piece of evidence was contributed in 1994 by Lauer and Postman. Using first ranked elliptical galaxies in 119 clusters as standard candles, they reported a motion of 561 ± 284, km s^{-1} toward $l = 220°$, $b = -28°$ ($\pm 27°$), of the Local Group with respect to the reference frame of the sample of clusters, which has an effective depth of about 10,000 km s^{-1}. This is largely discrepant from the direction of the apex of the CMB dipole, and would indicate that, for a Doppler origin of the CMB dipole, the cluster sample as a whole is moving at 689 ± 178 km s^{-1} toward ($l = 343°$, $b = +53°$) (± 23). The dynamical implication of this result is that the LG motion and that of the cluster sample are caused *largely* by mass concentrations beyond 10,000 km s^{-1}. The data of Lauer and Postman have been re-analyzed independently by Colless (1995), who confirmed their results.

Courteau et al. (1993) used a sample of northern spirals to confirm (a) the Willick detection of a 400 km s^{-1} motion of the PPS with respect to the CMB, directed toward the LG, and (b) outflow in the opposite direction of the sky. They interpret the result as that of a bulk flow at 360 ± 40 km s^{-1} in the direction $l = 294° \pm 5°$, $b = 0° \pm 4°$, averaged over a top–hat sphere of radius 6,000 km s^{-1} centered on the LG. A sample of clusters of galaxies analyzed by Mould et al. (1993) yields very different results, i.e. a mostly quiescent velocity field, except for large infall in the region surrounding the GA.

4 Redshift–Independent Distance Measurement Techniques

The techniques most extensively used in peculiar velocity surveys are the TF technique, applicable to spirals, and the D_n-σ and Fundamental Plane (FP) techniques (Dressler et al. 1987b; Djorgovski and Davis 1987), applicable to spheroidal systems. Both the D_n-σ and FP methods are more accurate variants of the earlier Faber–Jackson (1976) relation between the luminosity and velocity dispersion of elliptical galaxies. Lauer and Postman (1994) used a different, purely photometric technique, applicable to brightest cluster E galaxies pioneered by Hoessel (1980). The average predictive accuracy of these methods is generally reported between 0.25 and 0.4 mag, i.e. between 12% and 20% on the distance, for an individual galaxy. Other techniques for the measurement of distances are also available, such as that pioneered by Tonry and Schneider (1988) based on surface brightness fluctuations in ellipticals, direct measurement of variable star periods, supernovae, globular cluster luminosity functions, etc. (see the review by Jacoby et al. 1992). The application of these techniques does however not lend itself to extensive survey work, or distances in excess of a 10–20 Mpc, yet. While some of them hold substantial promise, at this time the workhorses of peculiar velocity field surveys are the TF and FP or D_n-σ. In the following, we shall concentrate on the former.

4.1 The TF Relation

In 1977, R.B. Tully and J.R. Fisher proposed the relationship between luminosity and rotational speed of spiral galaxies, as a means of determining distances. Since then, the Tully–Fisher relation has been extensively used, both to map the peculiar velocity field and as a secondary method to measure the Hubble constant. TF distances have been estimated for several thousand galaxies, in their vast majority located within $cz < 7000$ km s^{-1} .

The technique relies on the combination of photometric and spectroscopic data, which jointly provide some or all of the following parameters : a spiral galaxy's redshift, rotational velocity, disk inclination to the line of sight, disk size and optical, infrared or HI line flux. After a set of corrections reduces a target galaxy's Doppler–broadened spectral line width to a rotational velocity, the latter is used to infer the expected luminosity or linear size of the object; flux or angular size are then used to obtain an estimate of its distance modulus. The essential tool of this exercise is a TF template relation: if it is assumed that a TF relation of universal validity for the spiral galaxy population exists, obtaining the best possible approximation of such a relation is generally the most delicate aspect of a TF program.

The justification for the reliability of the TF relation as a distance indicator is based on a few simple scaling arguments. Consider a pure exponential disk of central disk surface brightness $I(0)$ and scale length r_d; its total luminosity is

$$L_d \propto r_d^2 I(0) \tag{9}$$

On the other hand, the mass internal to radius R is $M(R) \propto RV^2$, and if the rotation curve flattens in the outer regions of the disk as is usually the case for spiral galaxies, the total mass is

$$M_{tot} \propto r_d V_{max}^2 \tag{10}$$

Combining eqns. (9) and (10), we can write

$$L_d \propto (M_{tot}/L_d)^{-2} V_{max}^4 / I(0) \tag{11}$$

If a dark matter halo is present so that the disk mass is $M_d = \Gamma M_{tot}$,

$$L_d \propto (M_d/L_d)^{-2} \Gamma^2 V_{max}^4 / I(0) \tag{12}$$

When the "standard" assumptions are made, i.e. that $\Gamma \sim$ const, $M_d/L_d \sim$ const and $I(0) \sim$ const (Freeman's law, 1970), $L_d \propto V_{max}^4$, which resembles the TF relation. In practice, none of the assumptions of constancy for M_d/L_d, Γ and $I(0)$ apply; all those parameters exhibit mild dependencies on V_{max} (or L_d), so that if $L_d \propto V_{max}^n$, the actual value of the exponent n in empirical TF relations is between 3 and 3.5.

Early on, the TF relation was studied and applied using blue band photometry. In the early 1980's, the introduction of H band magnitudes was hailed as an important step forward, in reducing the scatter produced by internal extinction corrections and differences among different spiral morphological types. More recently, it has become common practice to carry out surface photometry for TF work on I and R band CCD images. Sky background at those wavelengths is low, detectors have high efficiency and large fields and data acquisition is relatively fast even with small aperture telescopes. The population dominating the light at I band is comprised of stars several Gyr old. Thus disks are well outlined but of smoother appearance than farther in the blue, and their apparent inclinations to the line of sight can be more reliably determined. In addition, processes operating in clusters, that may alter the star formation rate in galaxies, will have little effect on the red and infrared light of disks; thus, smaller — if any — systematic differences are expected between the I and R TF relation of cluster and field galaxies (Pierce and Tully 1992).

The TF relation is commonly used in the form: $M_{tf} = a + b \log V_{max}$, where the coefficients (a, b) constitute the "template" relation. For an observed V_{max}, one can then estimate the TF magnitude M_{tf} obtained from the template. Its comparison with the magnitude M_{obs}, which is obtained from the observed flux by placing the galaxy at the distance indicated by its cz (e.g. as measured in the CMB reference frame), yields the difference $\delta m = M_{obs} - M_{tf}$. The radial peculiar velocity is defined by

$$V_{pec} = cz_{cmb}[1 - 10^{0.2\delta m}]. \tag{13}$$

In this definition, positive peculiar velocities are directed away from us.

4.2 The TF Relation: Corrections and Error Budget

In its most widely used form, the TF relation correlates magnitudes and velocity widths. Important corrections need to be applied to those parameters before they are ready for use. The most important correction to the magnitudes is that due to extinction, which has two components: that produced in the disk of our Galaxy, and that which arises in the disk of the target object. The first part of that correction is relatively straightforward and uncontroversial. All-sky extinction maps produced by Burstein and Heiles (1978) are widely used. The corrections are generally small, except for very low galactic latitudes. The second component, internal extinction, is generally larger in amplitude and more a subject of controversy (see for example the proceedings of a recent conference on the subject, edited by Davies and Burstein 1995). A recent set of correction recipes is discussed in Giovanelli et al. (1994, 1995).

The internal extinction correction is one that depends on the inclination, size, thickness and metallicity of the disk, and on the relative scale heights of the dust and stellar components. For use in TF work, it is customary to convert the observed magnitude to one in the face–on perspective, i.e., it is not required to solve for the total opacity of the disk, but only for the differential opacity between the observed and face–on aspect. The parametrization of the correction is simply

$$m(0) = m - \gamma \log(a/b) \tag{14}$$

where γ is a function of the luminosity of the galaxy, varying between 0.5 for the less luminous objects used in TF work, to about 1.2 for the brightest. This is due to the fact that both metallicity and path length of the line of sight through the disk tend to increase with luminosity.

Failure to apply the correct internal extinction relation will produce a biased peculiar velocity field: inferred peculiar velocities will depend on the axial ratio of the sampled object; in addition, since intrinsically bright objects require larger values of γ, a luminosity–dependent bias (which also translates in a distance–dependent bias, since samples are seldom volume–limited) is introduced if γ is assumed to be constant.

Velocity widths are generally obtained from single–dish 21 cm line spectra, in which case the galaxy is generally not resolved by the telescope beam, or from emission line (e.g. Hα) optical long–slit spectra. In some studies, Fabry–Perot imaging spectroscopy has been adopted (e.g. Schommer et al. 1993; Amram et al. 1993); while Fabry–Perot spectra contain valuable two–dimensional spatial information, the simpler, faster to obtain single slit spectra are generally favored when optical spectroscopy is sought. Synthesis instruments such as the VLA, WSRT or AT, which can deliver the full velocity field in the 21 cm line, are not extensively used for TF work due to a combination of sensitivity and instantaneous bandwidth limitations. This situation may change as currently planned upgrades materialize.

Because the HI material generally samples farther out in the disks of spirals, the 21cm velocity widths tend to provide closer approximations to V_{max} than optical spectra, which terminate at the outermost star forming regions of disks, at which radius they are often still seen to be rising. On the other hand, 21 cm single–dish spectra do not provide information on the shape of the rotation curve and on possible disturbances that might alter the spectral shape.

The impact of velocity width accuracy on the TF error budget depends on the widths themselves. As measurement errors alone tend to be relatively independent on the amplitude of the measurement, the relative error increases with decreasing width size. In addition, the relative importance of turbulent motions, vis–a–vis that of ordered rotational motions, increases as the width decreases. The correction for this contribution is quite uncertain. The relative error on V_{opt} then, rises from values which can be as low as 1–2 %, for the larger measured widths, to as high as > 30% for the smaller widths. When the width error exceeds 10–15%, the data becomes effectively worthless for TF use.

The third important ingredient of the TF corrections menu is the disk inclination, i, which enters in the determination of the internal extinction correction, as well as in the correction of the observed width necessary to infer V_{max}. Only objects with values of $\sin i$ larger than $30°–40°$ are useful for TF work. At the lower inclinations, the uncertainty introduced by errors in i in the estimate of V_{max} can be quite substantial.

Errors in the measured parameters and uncertainties in the correction recipes contribute between 0.15 and 0.30 mag to the scater of a good quality TF relation. In addition, "cosmic" sources of scatter need to be considered, such as warps and other velocity field distortions, disk and gravitational potential asymmetries, variations in the stellar population make–up and disk–to–bulge ratios of different galaxies, etc. The cosmic scatter of the TF relation is estimated to be about 0.2 mag. The addition in quadrature of the various sources of scatter produces an r.m.s. uncertainty in the TF relation that varies between 0.22 and 0.4 mag (10% to 20% on the distance), this variation being roughly monotonic between the fastest and the slowest rotators. The use of a single figure of scatter for the TF relation is a handy, but imprecise manner of referring to the quality of the technique. These estimates of scatter assume the adoption of a good TF template, of course. We consider the problems associated with obtaining such a tool, next.

4.3 TF Template Relation

Suppose the true TF relation of a population of galaxies is given by $M_{tf} = a_{true} + b_{true} \log V_{max}$, and that the cosmic scatter about that relation can be expressed in terms of a single figure ϵ_{true}. We derive estimates of a_{true} and b_{true} from incomplete, limited samples. The quality of the resulting inferences on the peculiar velocity field will depend on how well our estimates approach the true value of the TF relation coefficients. Alternatively, suppose that our estimate of the offset is too low, i.e. $a_{est} < a_{true}$; then all peculiar velocities will be biased negative, and the universe will appear to flow towards the observer from all directions. The opposite will be true if $a_{est} > a_{true}$. Suppose the slope is incorrect, so that for example the estimated TF relation lies below the true one at small widths and above it at large ones. The resulting bias in the peculiar velocity flow will produce apparent infall for objects of small width and apparent outflow for objects of large widths. Since most samples are closer to being flux–limited than to being volume–limited, it will appear as if the nearby volumes of space preferentially fall in, while the more distant parts of the sampled volumes flow out. All these biases have geocentric signatures, which assist the observer in their identification. However, when a sample does not cover the whole sky, but rather only part of a hemisphere or some restricted cone, the identification of a geocentric signature in the data is more arduous, and confusion of the spurious product of a biased TF relation with a bulk flow can occur.

How to construct a reliable TF template relation? The safest approach is that of utilizing a set of clusters of galaxies. For each cluster, a TF relation can be constructed, as all objects in the cluster can be assumed to be at the same distance. After applying bias corrections that are different for each cluster, the optimal slope is selected for the global sample via a maximum likelihood technique. Let Δm_i be the magnitude deviation of a given cluster from a TF relation of slope as just determined and arbitrary offset. The offset of the TF template can then be obtained by imposing that the average cluster deviation from the template over the whole cluster sample, $\langle \Delta m \rangle$, be zero. This will also yield a set of individual cluster motions with respect to the template/reference frame.

The question of whether the clusters partake in bulk flow or not can be answered by measuring the dipole moment of the distribution of Δm. The usual practical limitations apply to this approach: the number of clusters with well observed TF relations is relatively small (two dozens or so); the volume they sample is not very deep (most distant ones near $cz \sim 10,000$ km s^{-1} , and the majority are at $cz < 5000$ km s^{-1}), so that the influence of shear on the velocity field may be important; sampling density per cluster is poor; cluster membership assignment is occasionally of arguable quality. The infamous case of A2634 well illustrates the pitfalls of this particular exercise (Lucey et al. 1991): TF and D_n–σ distances of the cluster have yielded values that are discrepant by more than 3000 km s^{-1} , which is more than 30% of cz and several times the size of the error estimate associated with either of the two methods. This (extreme) example advises for use of much caution in the formal evaluation of distance errors.

The exercise of obtaining a template relation is not complete without an adequate treatment of biases. Seminal work on the correction for biases in TF work in clusters has been carried out by Bottinelli, Gouguenheim, Paturel and Teerikorpi, and analytical formulations have been sought by various workers (Bottinelli et al. 1986; Teerikorpi 1984, 1990, 1993; Willick 1994; Willick et al 1995 and refs. therein). A limitation of analytical treatments is that they require data sets to adhere to relatively simple selection biases, and their statistical properties to be well described by a few parameters. But scatter, as we have discussed, is multivalued. In real life, samples tend to be haphazardly constructed, responding to whims of weather, time allocation committees, instrument performance and availability, data quality, etc. Since cluster samples tend to be constituted by fewer objects than one would desire, further stripping in order to conform to some set completeness criterion is usually not a viable alternative. The selection functions are then generally different for each cluster.

An alternative to analytical modelling of bias effects is that of constructing computer simulations of the statistical process, and numerically obtaining expectation values for the bias corrections. One can then easily accomodate variable selection functions, variable scatter parameters, and the iterative nature of the bias correction process can be perhaps more accurately accomodated.

How accurate are available TF template relations? After all corrections are applied, currently available cluster samples can produce a TF relation which approaches (a_{true}, b_{true}) to within 0.05 to 0.07 magnitudes. This translates in a systematic error of about 140 km s^{-1} at distances of 5000 km s^{-1} . Since the flaw is in the calibration of the technique itself, the velocity of a given volume of space cannot be improved beyond that limit, no matter how dense the sampling. To a moderate extent, however, this calibration error can be further reduced, if a large, all–sky sample of objects is available. This can be done by judiciously minimizing the amplitude of geocentric solutions in the resulting peculiar velocity field. Ultimately, at any rate, the best approach to a better template relation should be that of observing a larger sample of distant clusters, well distributed over the sky.

5 The Motions of Nearby Clusters

A collaboration including P. Chamaraux, L. da Costa, W. Freudling, M. Haynes, T. Herter, J. Salzer, M. Scodeggio, N. Vogt, G. Wegner, and this writer has obtained velocity widths and I band magnitudes for a sample of 600 cluster galaxies. They have been combined with data published by others, to obtain a homogeneous all sky set of distances for 24 clusters, which are listed in table 1. For each cluster, great care has been taken in ascertaining membership. When strong doubts existed on the definition of a cluster's structural parameters, as occurred with A2634, a redshift survey was carried out to improve on that definition (e.g. Scodeggio et al. 1995). The cluster set was used to construct a template TF relation, and with respect to that, peculiar velocities were estimated for each cluster.

Table 1 lists for each cluster the center coordinates (cols. 2,3); the mean radial velocity measured in the CMB reference frame (col. 4); the mean peculiar velocity computed separately for a sample of strict members (V_{pec}^{in}), and for a larger, extended sample which includes both the strict members and about an equal number of peripheral objects (V_{pec}^{in+}) (cols. 5,6); the latter two values are accompanied by the estimated errors in parentheses; the angle θ_{cmb} in degrees, between the line of sight to each cluster and the CMB dipole apex direction (col 7); and finally $V_{pec}^{in+} \cos\theta_{cmb}$ (col 8). The peculiar velocities are referred to the CMB reference frame, and their reliability depends on two factors: (a) the

Table 1. Motions of Clusters of Galaxies

Cluster	RA(1950)	Dec(1950)	cz_{cmb}	V_{pec}^{in}	V_{pec}^{in+}	θ_{cmb}	V_{proj}
	hhmmss.s	ddmmss	km s^{-1}	km s^{-1}	km s^{-1}	°	km s^{-1}
NGC 383	010430.0	+321200	4865	387(281)	-68(182)	153	61
NGC 507	012000.0	+330400	4809	206(262)	44(217)	150	-38
A262	014950.0	+355440	4660	0(187)	64(138)	144	-52
A400	025500.0	+055000	6934	-411(406)	-227(237)	121	118
Eridanus	033000.0	-213000	1530	-370(090)	-348(068)	99	57
Fornax	033634.0	-353642	1321	-235(055)	-338(056)	90	2
Cancer	081730.0	+211400	4939	0(225)	-174(190)	63	-79
Antlia	102745.0	-350411	3120	100(140)	202(095)	11	198
Hydra	103427.7	-271626	4075	-332(172)	-222(143)	7	-220
NGC 3557	110735.0	-371600	3318	250(200)	388(147)	10	382
A1367	114154.0	+200700	6735	-62(206)	-109(200)	48	-73
Ursa Major	115400.0	+485300	1087	-470(049)	-495(049)	77	-113
Cen 30	124606.0	-410200	3324	285(123)	341(090)	25	309
Coma	125724.0	+281500	7185	228(214)	228(201)	62	108
ESO 508	130954.0	-230854	3210	724(183)	540(107)	29	474
A3574	134606.0	-300900	4816	44(264)	206(171)	35	168
A2197	162630.0	+410100	9162	-42(640)	-42(314)	101	8
A2199	162654.0	+393800	8996	-42(640)	-42(314)	100	7
Pavo 2	184200.0	-632000	4443	447(297)	391(182)	76	94
Pavo	201300.0	-710000	4104	352(278)	318(199)	77	69
MDL59	220018.0	-321400	2318	-554(151)	-508(109)	119	244
Pegasus	231742.6	+075557	3519	16(168)	-174(149)	161	164
A2634	233554.9	+264419	8895	61(378)	-124(343)	173	123
A2666	234824.0	+264824	7755	-199(440)		170	196

statistical accuracy, which results from the number of objects in each cluster sample, the amplitude of the scatter in the data and the quality of the cluster parameters and membership, and (b) the degree to which the cluster template corresponds to a rest frame with respect to the CMB.

The peculiar velocities derived from the extended sample of each cluster ((V_{pec}^{in+}) are statistically more reliable, due to the significantly larger number of objects leading to their definition. In the case of A2197/9, a single extended sample is used, since the two clusters are too close, both in the sky and in redshift, to allow separation into two different extended samples. In the case of A2666, its proximity in the sky to A2634 and the presence in the A2634 region of clumps at similar redshift as A2666 make the definition of an extended sample quite arduous, so a V_{pec}^{in+} for that cluster is not given.

Few clusters appear to have large, significant peculiar velocities: among those are Eridanus, Fornax, Antlia, Centaurus 30, ESO 508 and MDL59. Most of these clusters are nearby, within 2500 km s^{-1} of the LG; the more distant ones in that group — Antlia, N3557, Cen30 and ESO 508 — are close to the apex of the LG motion, and located in the foregorund of the putative GA. The remaining objects have peculiar velocities that are, on the whole, statistically undistinguishable from rest in the CMB frame. It also appears that the objects in the Pisces–Perseus supercluster region (A262, A400, N383 and N507) are not infalling as claimed by Willick (1990). Finally, while Cen30, Antlia and ESO 508 suggest substantial infall towards the GA region, Hydra and A3574, located at $cz > 4000$ km s^{-1} , exhibit significantly milder motions. The evidence for large amplitude bulk flows is nil in this data set.

The general impression conveyed by this cluster sample is one of a relatively quiescent peculiar velocity field. Motions with respect to the CMB reference frame, with amplitude in excess of that of the LG, are not seen. While motions of comparable amplitude to that of the LG are detectable among objects near the LG, peculiar velocities become milder for more distant clusters. It should be kept in mind that several nearby objects in Table 1 exhibiting high peculiar velocities are groups of low velocity dispersion, and therefore low mass, while more distant ones tend to represent, massive, main clusters in large supercluster aggregates. The latter are then more likely to be closer to the centers of large scale density inhomogeneities, while nearby groups, peripheral structures of superclusters, are more likely to trace regions where the peculiar velocity field has large amplitude.

6 A Test of the Lauer and Postman Bulk Flow

As discussed in section 3.4, Lauer and Postman (LP) measured the dipole of the distribution of the brightest elliptical galaxies in 119 clusters, distributed over a volume of 15,000 km s^{-1} radius and with an effective depth of approximately 10,000 km s^{-1} . They found that *the reference frame defined by the group of clusters is in motion with respect to the CMB, at a velocity of* 689 ± 178 *km s^{-1} , and towards the direction of galactic coordinates* $(343°, +52°)(\pm23°)$. This motion is roughly orthogonal to the CMB apex and in large disagreement with that reported by other surveys.

In this section we use preliminary data extracted from an all–sky sample of about 2000 field galaxies, of a collaboration which includes P. Chamaraux, L. da Costa, W. Freudling, M. Haynes, J. Salzer, G. Wegner and this writer, to inspect the motions of 268 spirals enclosed within two cones directed along the apex and antapex direction of the LP bulk flow. Each cone has a semi–aperture of 0.5 rad. Peculiar velocities are measured in bins of radial velocity, each 1000 km s^{-1} wide, combining in opposition the apex and antapex data sets. For each bin, a median value of the distribution of peculiar velocities is estimated. Every median value in figure 2 lies below the LP bulk velocity amplitude; in fact, not a single point exceeds half that amount. The component of the LG motion with respect to the CMB (as obtained from the CMB dipole), along the direction of the LP apex–antapex amounts to about 300 km s^{-1} . This compares well with the amplitude of the motions of the galaxies within 4000–5000 km s^{-1} , and is consistent with the idea that those may be travelling companions of the LG with respect to the CMB. Beyond that distance, the amplitude of the motion appears more uncertain; a more detailed interpretation is hampered by the presence of large infall and backflow motions in the Perseus–Pisces supercluster at 5000 km s^{-1} , which contributes part of the data in the LP antapex cone. An average

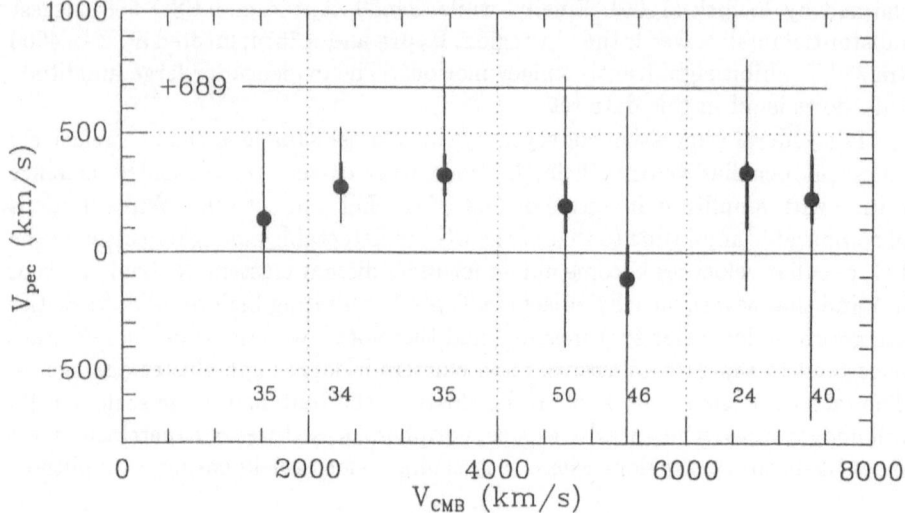

Fig. 2. Median peculiar velocities, binned by shells of radial velocity 1000 km s^{-1} wide; values refer to the total bulk velocity of the apex and antapex regions in each shell, added in opposition so that the sign is that of the net velocity in the LP apex direction. Thin error bars refer to the inner quartile range of the data around the median values, and thick error bars identify our best guess of the accuracy of the median value determination, akin to a standard error. Numbers under each symbol identify the number of galaxies used in each bin. The LP bulk flow amplitude of 689 km s^{-1} is inset as a horizontal line for comparison.

bulk flow with respect to the CMB as large as 689 km s^{-1} , of galaxies within the volume subtended by these data, which extend out to 8000 km s^{-1} , can be excluded.

In summary, the preliminary results presented in sections 5 and 6, both for clusters and field galaxies, appear consistent with a relatively quiescent picture of the peculiar velocity field. While motions of individual galaxies, or small groups, can achieve large velocities, especially in regions surrounding large mass aggregates, such as superclusters, large clusters exhibit relatively small velocities with respect to the CMB.

The author wishes to thank the organizers of the Laredo Summer School, E. Martínez–González and J. L. Sanz, for their hospitality in Cantabria, and for making a remarkable experience possible. This work was supported in part by the NSF grant AST94–20505.

References

Aaronson, M., Huchra, J., Mould, J., Schechter, P.L. and Tully, R.B. 1982, *ApJ* 258, 64

Aaronson, M., Bothun, G., Mould, J., Huchra, J., Schommer, R. and Cornell. M. 1986, *ApJ* 302, 536

Aaronson. M. et al. 1989, *ApJ* 338, 654

Abell, G.O., Corwin, H.G. and Olowin, R. 1989, *ApJSS* 70, 1

Amram, P., Le Coarer, E., Marcelin, M., Balkowski, C., Sullivan III, W.T. and Cayatte, V. 1992, *AA* 262, 384

Bertschinger, E. and Dekel, A. 1989, *ApJ* 336, L5

Bond, J.R. and Efstathiou, G. 1984, *ApJ* 285, L45

Bottinelli, L., Gouguenheim, L., Paturel, G. and Teerikorpi, P. 1986, *AA* 156, 157

Burstein, D. 1990, *Rep. Prog. Phys.* 53, 421

Burstein, D. and Heiles, C. 1978, *ApJ* 225, 40

Colless, M. 1995, *AJ* 109, 1937

Courteau, S., Faber, S.M., Dressler, A. and Willick J. 1993, *ApJ* 412, L51

Davies, J.I. and Burstein, D. 1995, eds. *The Opacity of Stellar Disks* (Dordrecht: Kluwer)

Dekel, A. 1994, *ARAA* 32, 371

de Vaucouleurs, G. 1958, *AJ* 63, 252

de Vaucouleurs, G., de Vaucouleurs, A. and Corwin, H.G. 1976, *Second Reference Catalog of Bright Galaxies* (Austin: Univ. of Texas Press)

Djorgovski, S. and Davis, M. 1987, *ApJ* 313, 59

Dressler, A. 1988, *ApJ* 329, 519

Dressler, A., Faber, S.M., Burstein, D., Davies, R., Lynden–Bell, D., Terlevich, R., Wegner, G. 1987, *ApJ* 313, L37

Dressler, A. et al. 1987b, *ApJ* 313, 43

Dressler, A. and Faber, S.M. 1990, *ApJ* 354, L45

Faber, S.M. and Burstein, D. 1988, in *Large Scale Motions in the Universe*, ed. by V.C. Rubin and G.C. Coyne (Princeton: Princeton Univ. Press)

Faber, S.M. and Fackson, R.E. 1976, *ApJ* 204,668

Freeman, K.C. 1970, *ApJ* 160, 411

Freudling, W., da Costa, L.N., Wegner, G., Giovanelli, R., Haynes, M.P. and Salzer, J.J. 1995, *AJ* in press

Giovanelli, R., Haynes, M.P., Salzer, J.J., Wegner, G., da Costa, L.N. and Freudling, W. 1995, *AJ* 107, 2036

Giovanelli, R., Haynes, M.P., Salzer, J.J., Wegner, G., da Costa, L.N. and Freudling, W. 1995, *AJ* 110, 1059

Giovanelli, R., Haynes, Chamaraux, P., da Costa, L., Freudling, W., Salzer, J. and Wegner, G. 1995, in *Examining the Big Bang and Diffuse Background Radiations*, proc. of IAU Symp. 168, M. Kafatos, ed. (Dordrecht:Kluwer)

Han, M. and Mould, J.R. 1992, *ApJ* 396, 453

Hoessel, JG 1980, *ApJ* 241, 493

Jacoby, G.H. et al. 1992, *PASP* 104, 599

Jerjen, H. and Tammann 1993, *AA* 273, 354

Kogut, A. et al. 1993, *ApJ* 307, 91

Kraan–Korteweg, R. 1985, in *ESO Worskshop on the Virgo Cluster*, ed. by O.–G. Richter and B. Binggeli, (Garching: ESO)

Kraan–Korteweg, R. and Woudt, H. 1995, preprint

Lauer, T.R. and Postman, M. 1994, *ApJ* 425, 418

Lilje, P.B., Yahil, A, and Jones, B.T. 1986, *ApJ* 307, 91

Lucey, etc. 1991, *MNRAS* 248, 804

Lynden–Bell, D., Faber, S.M., Burstein, D., Davies, R., Dressler, A., Terlevich, R.J. and Wegner, G. 1988, *ApJ* 326, 19

Mathewson, D.S., Ford, V. and Buckhorn, M. 1992a, *ApJ* 389, L5

Mathewson, D.S., Ford, V. and Buckhorn, M. 1992b, *ApJSS* 81, 413

Mould, J. et al. 1993, *ApJ* 409, 14

Padmanabhan, T. 1993, *Structure Formation in the Universe* (Cambridge: Cambridge Univ. Press)

Peebles, P.J.E. 1993, *Principles of Physical Cosmology* (Princeton: Princeton Univ. Press)

Persic, M. and Salucci, P. 1995, *ApJSS* in press

Pierce, M.J. and Tully, R.B. 1988, *ApJ* 330, 579

Pierce, M.J. and Tully, R.B. 1992, *ApJ* 387, 47

Rubin, V.C., Ford, W.K., Thonnard, N., Roberts, M.S. and Graham, J.A. 1976, *AJ* 81, 687

Scaramella, R., Baiesi–Pilastrini, G., Chincarini, G., Vettolani, G. and Zamorani, G. 1989, *Nature* 338, 562

Scaramella, R., Vettolani, G. and Zamorani, G. 1994, *ApJ* 422, 1

Schechter, P.L. 1980, *AJ* 85, 801

Schommer, R.A., Bothun, G.D., Williams, T.B., and Mould, J.R. 1993, *AJ* 105, 97

Scodeggio, M., Solanes, J.M., Giovanelli, R. and Haynes, M.P. 1995, *ApJ* 444, 61

Shaya, E.J. 1984, *ApJ* 280, 470

Silk, J. 1977, *AA* 59, 53

Strauss, M. and Willick, J.A. 1995, preprint

Tammann, G.A. and Sandage, A. 1985, *ApJ* 294, 81

Teerikorpi, P. 1984, *AA* 141, 407

Teerikorpi, P. 1990, *AA* 234, 1

Teerikorpi, (. 1993, *AA* 280, 443

Tini Brunozzi, P., Borgani, S., Plionis, L., Moscardini, L. and Coles, P. 1995, preprint

Tonry, J.L. and Schneider, D.P. 1988, *AJ* 96, 807

Tully, R.B. and Fisher, J.R. 1977, *AA* 54, 661

Tully, R.B. and Fouque', P. 1985, *ApJSS* 58, 67

Weinberg, S. 1972, *Gravitation and Cosmology*, (New York: Wiley)

Willick, J.A. 1990, *ApJ* 351, L5

Willick, J.A. 1994, *ApJSS* 92, 1

Willick, J.A., Courteau, S., Faber, S.M., Burstein, D. and Dekel, A. 1995, *ApJ* 446, 12

Yahil, A., Tammann, G.A. and Sandage, A. 1977, *ApJ* 217, 903

Non-linear Evolution of Cosmological Perturbations

Sabino Matarrese

Dipartimento di Fisica 'Galileo Galilei', Università di Padova
via Marzolo 8, I-35131, Padova, Italy

Abstract: In these lecture notes I review the theory of the non–linear evolution of cosmological perturbations in a self–gravitating collisionless medium, with vanishing vorticity. The problem is first analyzed in the context of the Newtonian approximation, where the basic properties of the Zel'dovich, frozen–flow and adhesion algorithms are introduced. An exact general relativistic formalism is then presented and it is shown how the Newtonian limit, both in Lagrangian and Eulerian coordinates, can be recovered. A general discussion on the possible role of possible relativistic effects in the cosmological structure formation context is finally given.

1 Introduction

An important theoretical issue in cosmology is to understand the physical processes that occurred during the gravitational collapse of the matter which gave rise to the observed large–scale structure of the universe. A complementary issue is to reconstruct the initial conditions of the clustering process, e.g. the value of the cosmological parameters, the type of dark matter, the statistics of the primordial perturbations, starting from observational data such as the spatial distribution of galaxies or their peculiar velocities. Much work has been recently done in the latter direction, since more and more data on peculiar velocities of galaxies, as well as very large and complete galaxy redshift surveys have become available.

A widely applied and well–motivated approximation when dealing with the dynamics of dark matter, either cold or hot, is to treat it as a system of particles having negligible non–gravitational interactions, a self–gravitating collisionless system. The dynamics of such a system is usually approached by different techniques, depending on the specific application. For instance, the evolution of small perturbations on a Friedmann–Robertson–Walker (FRW) background is followed by analytical methods. The non–linear evolution in cases where some symmetries are present can also sometimes be followed analytically: typical examples being the spherical top–hat model in the frame of the Newtonian approximation and the Tolman–Bondi solution in General Relativity (GR). There are also useful approximations valid in the mildly non–linear regime, such as the Zel'dovich approximation (Zel'dovich 1970). Besides this classical approach, a number of variants have been proposed, all trying to overcome the inability to follow the development of structures beyond caustic formation. Different approximations apply in the highly non–linear regime, such as the hierarchical closure ansatz for the BBGKY equations (e.g. Peebles 1980) . The most general problem of studying the fully non–linear dynamics of a collisionless system in Newtonian theory can only be followed by numerical techniques, such as N–body codes (e.g. Hockney & Eastwood 1981; Efstathiou et al. 1985). Finally, the non–linear relativistic evolution of a pressureless fluid has been recently studied in a number of papers (e.g. Matarrese & Terranova 1995, and references therein).

Here we review various methods, trying to show their possible interconnections. Section 2 presents the Zel'dovich, frozen–flow and adhesion approximations. Section 3 deals with the relativistic dynamics of a self–gravitating collisionless and irrotational fluid within GR. A final general discussion is given in Section 4.

2 Newtonian dynamics of self–gravitating collisionless matter

2.1 General formalism

Let us start by writing the Newtonian equations for the evolution of collisionless particles in the expanding universe (e.g. Peebles 1980). These can be written using suitably rescaled variables and in comoving coordinates. We shall assume that the universe is spatially flat and matter dominated, so that the scale factor reads $a(t) = a_0(t/t_0)^{2/3}$ (a subscript 0 will be used to define quantities at some "initial time" t_0). A generalization of these formulae to the open (and closed) universe case is given in (Catelan et al. 1995). The Euler equations read

$$\frac{d\mathbf{u}}{da} + \frac{3}{2a}\mathbf{u} = -\frac{3}{2a}\nabla\varphi, \qquad (1)$$

where $\mathbf{u} \equiv d\mathbf{x}/da$ is a rescaled comoving peculiar velocity field. The symbol $\frac{d}{da}$ stands for the total (convective) derivative

$$\frac{d}{da} = \frac{\partial}{\partial a} + \mathbf{u} \cdot \nabla. \qquad (2)$$

The continuity equation can be written in terms of the comoving matter density $\eta(\mathbf{x}, t) \equiv \varrho(\mathbf{x}, t)\, a^3(t)/\varrho_0 a_0^3$ (where ϱ_0 is the mean mass density at t_0)

$$\frac{d\eta}{da} + \eta\nabla \cdot \mathbf{u} = 0, \qquad (3)$$

while the rescaled local (or peculiar) gravitational potential $\varphi \equiv (3t_0^2/2a_0^3)\varphi_g(\mathbf{x}, t)$ is determined by local density inhomogeneities $\delta(\mathbf{x}, t) \equiv \eta(\mathbf{x}, t) - 1$ through the Poisson equation

$$\nabla^2 \varphi = \frac{\delta}{a}. \tag{4}$$

We can restrict the analysis to initially irrotational flow. According to Kelvin's circulation theorem, in the absence of dissipation, vorticity is conserved along each fluid trajectory; in particular, a fluid with vanishing initial vorticity will forever remain irrotational. However, for a collisionless fluid such a property breaks down after caustic formation: a vorticity component is created in multi-stream regions, simply because the local Eulerian velocity field takes contributions from different Lagrangian fluid elements at the same position. Outside the regions of orbit mixing and/or after applying a suitable low-pass filter one can define a velocity potential by $\mathbf{u}(\mathbf{x}, a) = \nabla \Phi(\mathbf{x}, a)$. It is then easy to derive from the Euler equation a Bernoulli equation relating the velocity potential Φ to the gravitational one φ,

$$\frac{\partial \Phi}{\partial a} + \frac{1}{2}(\nabla \Phi)^2 = -\frac{3}{2a}(\Phi + \varphi). \tag{5}$$

By integrating along the particle trajectory $\mathbf{x}(a)$ one finds a formal solution of the continuity equation,

$$\eta(\mathbf{x}, a) = \eta_0(\mathbf{q}) \exp\left\{ -\int_{a_0}^{a} da' \nabla \cdot \mathbf{u}[\mathbf{x}(\mathbf{q}, a'), a'] \right\}. \tag{6}$$

This can be compared with the formula obtained from mass conservation, $\eta(\mathbf{x}, a)d^3x = \eta_0(\mathbf{q})d^3q$, where $\mathbf{q}$ is the initial (Lagrangian) position of the particle which has reached the (Eulerian) position $\mathbf{x}$ by the time $a(t)$. One has either the well-known Lagrangian form

$$\eta[\mathbf{q}(\mathbf{x}, a), a] = \eta_0(\mathbf{q}) \| \partial x / \partial q \|^{-1}, \tag{7}$$

where $\|\partial x/\partial y\|$ is the Jacobian determinant of the transformation $x \to y$, or the Eulerian one

$$\eta(\mathbf{x}, a) = \eta_0(\mathbf{q}) \| \partial q / \partial x \|, \tag{8}$$

which however requires inverting the trajectory to find $\mathbf{q}(\mathbf{x}, a)$. These solutions are only valid as long as no shell-crossing (caustic) has occurred, so that there is a one-to-one correspondence between Lagrangian and Eulerian positions. Before caustic formation all these forms are identical.

2.2 Zel'dovich approximation

The simplest approximation is of course the linear one, which consists in neglecting the terms $\mathbf{u} \cdot \nabla \mathbf{u}$ in the Euler equations and $\nabla \cdot (\delta \mathbf{u})$ in the continuity equation. The resulting expressions read $\mathbf{u}_{LIN}(\mathbf{x}, a) = -\nabla \varphi_0(\mathbf{x})$, $\varphi_{LIN}(\mathbf{x}, a) = \varphi_0(\mathbf{x})$ and $\eta_{LIN}(\mathbf{x}, a) = 1 + a\nabla^2 \varphi_0(\mathbf{x})$, having neglected the contribution of decaying modes.

The next step is the Zel'dovich approximation (ZEL), based on the ansatz of extrapolating the equation $\mathbf{u} = -\nabla\varphi$ (i.e. $\Phi = -\varphi$) beyond linear theory; replacing this ansatz into the Euler equations gives

$$\frac{d\mathbf{u}}{da} = 0, \tag{9}$$

which has to be solved together with Eq.(3). The resulting system can be taken as the definition of ZEL. In this approximation a particle initially placed in $\mathbf{q}$ moves along a straight–line with constant "speed" $\mathbf{u}$ determined by the value of the initial peculiar gravitational potential in $\mathbf{q}$, namely

$$\mathbf{x}(\mathbf{q}, \tau_a) = \mathbf{q} - \tau_a \nabla_{\mathbf{q}}\varphi_0(\mathbf{q}), \tag{10}$$

with $\tau_a \equiv a - a_0$. The velocity field is conserved along each particle trajectory:

$$\mathbf{u}(\mathbf{x}, \tau_a) = \mathbf{u}_0(\mathbf{q}) = -\nabla_{\mathbf{q}}\varphi_0(\mathbf{q}). \tag{11}$$

The velocity potential, defined by $\mathbf{u}(\mathbf{x}, \tau_a) = \nabla_{\mathbf{x}}\Phi(\mathbf{x}, \tau_a) = \nabla_{\mathbf{q}}\Phi_0(\mathbf{q})$, which obeys the Hamilton–Jacobi equation

$$\frac{\partial\Phi}{\partial\tau_a} + \frac{1}{2}(\nabla_{\mathbf{x}}\Phi)^2 = 0. \tag{12}$$

The solution of Eq.(12) is

$$\Phi(\mathbf{x}, \tau_a) = \Phi_0(\mathbf{q}) + (\mathbf{x} - \mathbf{q})^2 / 2\tau_a. \tag{13}$$

The density field is usually represented in the Lagrangian form $\eta(\mathbf{q}, \tau_a) = \eta_0(\mathbf{q})\|\mathbf{1} + \tau_a\mathbf{D}_0(\mathbf{q})\|^{-1}$, where $\mathbf{1}$ is the unit matrix and $\mathbf{D}_0$ the deformation tensor, with components $D_{0,ij}(\mathbf{q}) = \partial^2\Phi_0(\mathbf{q})/\partial q_i\partial q_j$. The deformation tensor can be locally diagonalized, by going to principal axes X_1, X_2, X_3, with eigenvalues $\alpha_1, \alpha_2, \alpha_3$. We thus have

$$\eta(\mathbf{q}, \tau_a) = \frac{\eta_0(\mathbf{q})}{(1 + \tau_a\alpha_1(\mathbf{q}))(1 + \tau_a\alpha_2(\mathbf{q}))(1 + \tau_a\alpha_3(\mathbf{q}))}. \tag{14}$$

According to the latter expression a singularity (caustic) in Lagrangian space would form at every point $\mathbf{q}$ where at least one eigenvalue, say α_1, is negative.

Besides being, by construction, consistent with the growing mode of linear perturbations at early times, ZEL provides a good approximation up to the time of first shell crossing. The inconsistency of ZEL can be seen as follows. Inserting the ansatz $\Phi = -\varphi$ into the Poisson equation one gets an expression for the density fluctuation, $\delta_{DYN} = -a\nabla \cdot \mathbf{u}$, which is nothing but the linear theory relation between peculiar velocity and density fluctuation. This point has been discussed by Nusser et al. (1991), who refer to this determination of the density, $\eta_{DYN} = 1 + \delta_{DYN}$, as dynamical density, to distinguish it from the continuity density obtained from Eq.(14). It is possible to go further this way by replacing this expression into Eq.(3): one gets the mass density in Lagrangian form $\eta_{DYN}(\mathbf{q}, \tau_a) = \eta_0(\mathbf{q})/(1 - \tau_a\delta_+(\mathbf{q}))$, where $\delta_+(\mathbf{q}) = -(\alpha_1(\mathbf{q}) + \alpha_2(\mathbf{q}) +$

$\alpha_3(\mathbf{q})$), and the Poisson equation was used to relate φ_0 to δ_0 and we defined the (scaled) initial growing mode $\delta_+ \equiv \delta_0/a_0$.

It is then clear that the Zel'dovich ansatz is only exact for one–dimensional perturbations, where the two above expressions for the density coincide; in the general three–dimensional case it fails. An alternative understanding of the Zel'dovich approximation might be obtained in the frame of perturbation theory in Lagrangian coordinates. This approach has been recently reviewed by Buchert (1995).

2.3 Frozen–flow approximation

The frozen–flow approximation (FFA) (Matarrese et al. 1992) can be defined as the exact solution of the linearized Euler equations, where in the r.h.s. the growing mode of the linear gravitational potential is assumed. Such an equation is solved by $\mathbf{u}(\mathbf{x}, \tau_a) = \mathbf{u}_0(\mathbf{x}) = -\nabla_\mathbf{x}\varphi_0(\mathbf{x})$, plus a negligible decaying mode. In this approximation the peculiar velocity field $\mathbf{u}(\mathbf{x}, a)$ is frozen at each point to its initial value, that is

$$\frac{\partial \mathbf{u}}{\partial \tau_a} = 0, \qquad (15)$$

which is the condition for steady flow. Such an equation can be used, together with the continuity equation to define FFA. Particle trajectories in FFA are described by the integral equation

$$\mathbf{x}(\mathbf{q}, \tau_a) = \mathbf{q} - \int_0^{\tau_a} d\tau_a' \nabla_\mathbf{x}\varphi_0[\mathbf{x}(\mathbf{q}, \tau_a')] : \qquad (16)$$

particles during their motion update at each infinitesimal step their velocity to the local value of the linear velocity field, without memory of their previous motion, i.e. without inertia. This would be the case of a particle moving under the influence of a force in a medium with very large viscosity: the damping here is determined by the Hubble drag while the force is the gravitational one.

As we shall see shortly, no caustics are formed at finite time in FFA, so that all expressions for the density can be used interchangeably. It is nevertheless interesting to write the comoving mass density as given by Eq.(6), namely

$$\eta(\mathbf{x}, \tau_a) = \exp \int_0^{\tau_a} d\tau_a' \delta_+[\mathbf{x}(\mathbf{q}, \tau_a')], \qquad (17)$$

having assumed $\eta_0(\mathbf{x}_*) \approx 1$. The logarithm of the density in $\mathbf{x}$ is given by the integral of the linear density field over the trajectory of the particle which has arrived to $\mathbf{x}$ at time τ_a, starting from the Lagrangian position $\mathbf{q}$.

FFA is, by construction, consistent with linear theory, so it follows correctly the evolution at early times (precisely as it happens for ZEL). The assumption of keeping the linear approximation for the velocity potential beyond the linear regime is justified by the fact that this quantity is more sensitive to large wave-length modes than the density, it is therefore less affected by strongly non–linear evolution. Stream–lines are frozen to their initial shape, so multi–stream regions cannot form, unless they were already present in the initial velocity field. FFA therefore avoids the formation of caustics at finite τ_a, so one can try to extrapolate the approximation after the time at which the first shell–crossing would have appeared according to ZEL. A particle moving according to FFA has zero component of the velocity in a place where the same component of the initial gravitational force is zero, it will then slow down its motion in that direction approaching such a position: particles in FFA need infinite time to reach those places where a pancake, a filament or a knot will occur. Moreover, since, unlike ZEL, these particles move along curved paths, once they come close to pancake configurations they curve their trajectories, moving almost parallel to them, trying to reach the position of filaments. Again they cannot cross it, so they modify their motion, asymptotically approaching them, to finally fall, for $\tau_a \rightarrow \infty$, into the knots corresponding to the minima of the initial gravitational potential. Altogether, this type of dynamics implies an artificial thickening of particles around pancakes, filaments and knots, which mimics the real gravitational clustering around this type of structures. The physical thickening of the particle density around pancakes, filaments and knots, caused by the damped oscillations around these structures is replaced by an approximately exponential slowing down of particle motions, which however overestimates the actual particle deceleration. In the specular process of evacuation of initially underdense regions, FFA overshoots the actual dynamics. Provided one gives up resolving the trajectories of individual particles, these effects produce a density field which looks roughly similar to the real one; in this sense the method should be considered intrinsically Eulerian.

Matarrese et al. (1992) have considered the evolution of structures on large scales as described by FFA and compared it with the results of ZEL and of a N–body simulation, assuming a standard Cold Dark Matter model. Compared to the PM results, FFA recovers all the main structures in the correct places, even though they look thicker and the voids appear more empty and conspicuous. FFA leads to an excess of sub–structure, which is left on the way during the evolution instead of being erased by the hierarchical clustering process as in the true dynamics. The structures obtained by the Zel'dovich approximation, instead, are less prominent and more fuzzy, as the particles have diffused away from the caustic positions after shell–crossing. These results suggest that FFA is able to reproduce, at the statistical level, the clustering properties of the universe even on scales reached by the non–linear action of gravity.

2.4 Adhesion approximation

In the adhesion approximation (e.g. Gurbatov, Saichev & Shandarin 1989; Kofman, Pogosyan & Shandarin 1990; Weinberg & Gunn 1990; Williams et al. 1991; Kofman et al. 1992) one modifies the Zel'dovich approach by adding an artificial viscosity term to Eq.(9), which is thus replaced by

$$\frac{d\mathbf{u}}{da} = \nu\nabla^2\mathbf{u}. \tag{18}$$

The viscosity is introduced to mimic the actual sticking of particles around pancakes, caused by the action of gravity even in a collisionless medium. The parameter ν plays the role of a coefficient of kinematical viscosity, which controls the thickness of pancakes.

The previous equation is the vector generalization of the well–known non-linear diffusion or Burgers equation of strong turbulence (e.g. Burgers 1974). One can still define a velocity potential through $\mathbf{u} = \nabla\Phi$, which can be determined through the Hopf–Cole substitution $\Phi = -2\nu \ln U$; the scalar field U satisfies the linear diffusion or Fokker–Planck equation, $\partial U/\partial \tau_a = \nu\nabla^2 U$, with the initial condition $U_0(\mathbf{x}) = \exp[-\Phi_0(\mathbf{x})/2\nu]$. The resulting velocity potential reads

$$\Phi(\mathbf{x}, \tau_a) = -2\nu \ln\left[\frac{1}{(4\pi\nu\tau_a)^{3/2}} \int d^3q \exp\left(-\frac{1}{2\nu}S(\mathbf{x}, \mathbf{q}, \tau_a)\right)\right], \qquad (19)$$

where one defines the action $S(\mathbf{x}, \mathbf{q}, \tau_a) \equiv \Phi_0(\mathbf{q}) + (\mathbf{x} - \mathbf{q})^2/2\tau_a$, satisfying the Hamilton–Jacobi Eq.(12). The corresponding velocity field is easily obtained by differentiation; the Eulerian positions of the particles are found by direct integration of the integral equation

$$\mathbf{x}(\mathbf{q}, \tau_a) = \mathbf{q} + \int_0^{\tau_a} d\tau_a' \mathbf{u}[\mathbf{x}(\mathbf{q}, \tau_a'), \tau_a'], \qquad (20)$$

while the density field can be obtained from Eq.(7).

The Burgers equation is usually considered in the limit of small (but non-vanishing) ν, which corresponds to the limit of large Reynolds numbers, $\mathcal{R}_0 = u_0\ell_0/\nu$, u_0 and ℓ_0 being the characteristic amplitude and scale of the initial velocity field. The product $u_0\ell_0$ can be estimated either from the rms initial velocity potential smoothed on some scale R, $\langle\Phi_0^2(R)\rangle^{1/2}$, if this is convergent, or from the square root of the structural function of Φ_0

$$D(r) = \langle[\Phi_0(\mathbf{x})-\Phi_0(\mathbf{x}+\mathbf{r})]^2\rangle = (1/\pi^2) \int_0^\infty dk k^2 \mathcal{P}_\varphi(k)W^2(kR)[1-j_0(kr)], \qquad (21)$$

evaluated at a suitable lag, e.g. $r \approx R$. Here $\mathcal{P}_\varphi$ is the power–spectrum of the initial gravitational potential and $W(kR)$ a suitable low–pass filter.

In the small ν case the solution takes a simplified form which can be obtained from Eq.(11) through a saddle–point approximation,

$$\Phi(\mathbf{x}, \tau_a) \approx -2\nu \ln\left[\sum_\alpha \mathcal{J}(\mathbf{q}_\alpha)^{-1/2} \exp\left(-\frac{1}{2\nu}S(\mathbf{x}, \mathbf{q}_\alpha, \tau_a)\right)\right], \qquad (22)$$

where $\mathcal{J}(\mathbf{q}) = \|1 + \tau_a\mathbf{D}_0(\mathbf{q})\|$, $\mathbf{D}_0$ is the deformation tensor and $\mathbf{q}_\alpha$ are the Lagrangian points which minimize the action S at given $\mathbf{x}$ and τ_a. The Zel'dovich approximation is recovered in the limit $\nu \to 0$. This model has been applied to perform numerical simulations of the large–scale structure of the universe or to obtain some physical insight into the structure formation process in simplified cases. The model allows to obtain the skeleton of the large–scale matter distribution by a geometrical technique based on the insertion of osculating paraboloids into the hypersurface $\varphi_0(\mathbf{q})$.

Let us finally mention that a detailed statistical comparison of the various approximation schemes discussed here against full numerical simulations has been recently performed by Sathyaprakash et al. (1995).

3 Relativistic dynamics of a self–gravitating collisionless fluid

3.1 General formalism

We start by writing the Einstein's equation for a perfect fluid of irrotational dust. The formalism outlined in this section is discussed in greater detail in (Matarrese & Terranova 1995). With the purpose of studying gravitational instability in a FRW background, it is convenient to factor out the homogeneous and isotropic FRW expansion of the universe and perform a conformal rescaling of the metric with conformal factor $a(t)$, the scale–factor of FRW models, and adopt the conformal time τ, defined by $d\tau = dt/a(t)$ (τ, not to be confused with the variable τ_a of the previous section, is proportional to $t^{1/3}$ in the Einstein–de Sitter case).

The line–element is then written in the form

$$ds^2 = a^2(\tau)\left[-c^2d\tau^2 + \gamma_{\alpha\beta}(\mathbf{q},\tau)dq^\alpha dq^\beta\right] . \tag{23}$$

For later convenience let us fix the Lagrangian coordinates q^α to have physical dimension of length and the conformal time variable τ to have dimension of time. As a consequence the spatial metric $\gamma_{\alpha\beta}$ is dimensionless, as is the scale–factor $a(\tau)$ which must be determined by solving the Friedmann equations for a perfect fluid of dust

$$\left(\frac{a'}{a}\right)^2 = \frac{8\pi G}{3}\varrho_b a^2 - \kappa c^2 , \tag{24}$$

$$2\frac{a''}{a} - \left(\frac{a'}{a}\right)^2 + \kappa c^2 = 0 , \tag{25}$$

where $\varrho_b(\tau)$ is the background mean density. Here primes denote differentiation with respect to the conformal time τ and κ represents the curvature parameter of FRW models, which, because of our choice of dimensions, cannot be normalized as usual. So, for an Einstein–de Sitter universe $\kappa = 0$, but for a closed (open) model one simply has $\kappa > 0$ ($\kappa < 0$). Let us also note that the curvature parameter is related to a Newtonian squared time–scale κ_N through $\kappa_N \equiv \kappa c^2$ (e.g. Coles & Lucchin 1995); in other words κ is an intrinsically post–Newtonian quantity.

By subtracting the isotropic Hubble–flow, one introduces a peculiar velocity–gradient tensor

$$\vartheta^\alpha_{\ \beta} = \frac{1}{2}\gamma^{\alpha\gamma}\gamma_{\gamma\beta}' . \tag{26}$$

Thanks to the introduction of this tensor one can write the Einstein's equations in a more cosmologically convenient form. The energy constraint, i.e. the time–time component of the Einstein's equations, reads

$$\vartheta^2 - \vartheta^\mu_{\ \nu}\vartheta^\nu_{\ \mu} + 4\frac{a'}{a}\vartheta + c^2(\mathcal{R} - 6\kappa) = 16\pi Ga^2\varrho_b\delta , \qquad (27)$$

where $\mathcal{R}^\alpha_{\ \beta}(\gamma)$ is the conformal Ricci curvature of the three–space, i.e. that corresponding to the metric $\gamma_{\alpha\beta}$; for the background FRW solution $\gamma^{FRW}_{\alpha\beta} = (1 + \frac{\kappa}{4}q^2)^{-2}\delta_{\alpha\beta}$, one has $\mathcal{R}^\alpha_{\ \beta}(\gamma^{FRW}) = 2\kappa\delta^\alpha_{\ \beta}$. We also introduced the density contrast $\delta \equiv (\varrho - \varrho_b)/\varrho_b$.

The momentum constraint, i.e. the time–space components of the Einstein's equations, reads

$$\vartheta^\alpha_{\ \beta||\alpha} = \vartheta_{,\beta} . \qquad (28)$$

To avoid excessive proliferation of symbols, the double vertical bars are used here and in the following for covariant derivatives in the three–space with metric $\gamma_{\alpha\beta}$.

Finally, after replacing the density from the energy constraint and subtracting the background contribution, the evolution equation, coming from the space–space components of the Einstein's equations, becomes

$$\vartheta^\alpha_{\ \beta}{}' + 2\frac{a'}{a}\vartheta^\alpha_{\ \beta} + \vartheta\vartheta^\alpha_{\ \beta} + \frac{1}{4}\left(\vartheta^\mu_{\ \nu}\vartheta^\nu_{\ \mu} - \vartheta^2\right)\delta^\alpha_{\ \beta} + \frac{c^2}{4}\left[4\mathcal{R}^\alpha_{\ \beta} - (\mathcal{R} + 2\kappa)\delta^\alpha_{\ \beta}\right] = 0 . \quad (29)$$

The Raychaudhuri equation for the evolution of the peculiar volume expansion scalar ϑ becomes

$$\vartheta' + \frac{a'}{a}\vartheta + \vartheta^\mu_{\ \nu}\vartheta^\nu_{\ \mu} + 4\pi Ga^2\varrho_b\delta = 0 . \qquad (30)$$

The main advantage of this formalism is that there is only one dimensionless (tensor) variable in the equations, namely the spatial metric tensor $\gamma_{\alpha\beta}$, which is present with its partial time derivatives through $\vartheta^\alpha_{\ \beta}$, and with its spatial gradients through the spatial Ricci curvature $\mathcal{R}^\alpha_{\ \beta}$. The only remaining variable is the density contrast which can be written in the form

$$\delta(\mathbf{q},\tau) = (1 + \delta_0(\mathbf{q}))\left[\gamma(\mathbf{q},\tau)/\gamma_0(\mathbf{q})\right]^{-1/2} - 1 , \qquad (31)$$

where $\gamma \equiv \det \gamma_{\alpha\beta}$. A relevant advantage of having a single tensorial variable, for our purposes, is that there can be no extra powers of c hidden in the definition of different quantities.

Our intuitive notion of Eulerian coordinates, involving a universal absolute time and globally flat spatial coordinates is intimately Newtonian; nevertheless it is possible to construct a local coordinates system which reproduces this picture for a suitable set of observers. Local Eulerian – FRW comoving – coordinates x^A can be introduced, related to the Lagrangian ones q^α via the Jacobian matrix with elements

$$\mathcal{J}^A_{\ \alpha}(\mathbf{q},\tau) \equiv \frac{\partial x^A}{\partial q^\alpha} \equiv \delta^A_{\ \alpha} + \mathcal{D}^A_{\ \alpha}(\mathbf{q},\tau) , \qquad A = 1,2,3 , \qquad (32)$$

where $\mathcal{D}^A_{\ \alpha}(\mathbf{q},\tau)$ is called deformation tensor. Each matrix element $\mathcal{J}^A_{\ \alpha}$ labelled by the Eulerian index A can be thought as a three–vector, namely a triad, defined on the hypersurfaces of constant conformal time. They evolve according to

$$\mathcal{J}^A_{\ \alpha}{}' = \vartheta^\gamma_{\ \alpha} \mathcal{J}^A_{\ \gamma} \ , \tag{33}$$

which also follows from the condition of parallel transport of the triads relative to **q** along the world–line of the corresponding fluid element $D(a\mathcal{J}^A_{\ \alpha})/Dt = 0$.

Our local Eulerian coordinates are such that the spatial metric takes the Euclidean form δ_{AB}, i.e.

$$\gamma_{\alpha\beta}(\mathbf{q}, \tau) = \delta_{AB} \mathcal{J}^A_{\ \alpha}(\mathbf{q}, \tau) \mathcal{J}^B_{\ \beta}(\mathbf{q}, \tau) \ . \tag{34}$$

Correspondingly the matter density can be rewritten in the suggestive form

$$\varrho(\mathbf{q}, \tau) = \varrho_b(\tau)\big(1 + \delta_0(\mathbf{q})\big)\big[\mathcal{J}(\mathbf{q}, \tau)/\mathcal{J}_0(\mathbf{q})\big]^{-1} \ , \tag{35}$$

where $\mathcal{J} \equiv \det \mathcal{J}^A_{\ \alpha}$. Note that, contrary to the Newtonian case, it is generally impossible in GR to fix $\mathcal{J}_0 = 1$, as this would imply that the initial Lagrangian space is conformally flat, which is only possible if the initial perturbations vanish.

3.2 Linear approximation in Lagrangian coordinates

We are now ready to deal with the linearization of the equations obtained above. Let us then write the spatial metric tensor of the physical (i.e. perturbed) space–time in the form

$$\gamma_{\alpha\beta} = \bar{\gamma}_{\alpha\beta} + w_{\alpha\beta} \ , \tag{36}$$

with $\bar{\gamma}_{\alpha\beta}$ the spatial metric of the background space – in our case the maximally symmetric FRW one, $\bar{\gamma}_{\alpha\beta} = \gamma^{FRW}_{\alpha\beta}$ – and $w_{\alpha\beta}$ a small perturbation. The only non–geometric quantity in our equations, namely the initial density contrast δ_0, can be assumed to be much smaller than unity.

As usual, one can take advantage of the maximal symmetry of the background FRW spatial sections to classify metric perturbations as scalars, vectors and tensors. One then writes

$$w_{\alpha\beta} = \chi\bar{\gamma}_{\alpha\beta} + \zeta_{|\alpha\beta} + \frac{1}{2}\big(\xi_{\alpha|\beta} + \xi_{\beta|\alpha}\big) + \pi_{\alpha\beta} \ , \tag{37}$$

with

$$\xi^\alpha_{\ |\alpha} = \pi^\alpha_{\ \alpha} = \pi^\alpha_{\ \beta|\alpha} = 0 \ , \tag{38}$$

where a single vertical bar is used for covariant differentiation in the background three–space with metric $\bar{\gamma}_{\alpha\beta}$. In the above decomposition χ and ζ represent scalar modes, ξ^α vector modes and $\pi^\alpha_{\ \beta}$ tensor modes (indices being raised by the contravariant background three–metric).

Before entering into the discussion of the equations for these perturbation modes, let us quote a result which will be also useful later. In the $\vartheta^\alpha_{\ \beta}$ evolution equation and in the energy constraint the combination $\mathcal{P}^\alpha_{\ \beta} \equiv 4\mathcal{R}^\alpha_{\ \beta} - (\mathcal{R} + 2\kappa)\delta^\alpha_{\ \beta}$ and its trace appear. To first order in the metric perturbation one has

$$\mathcal{P}^\alpha_{\ \beta}(w) = -2\bigg[\big(\nabla^2 - 2\kappa\big)\pi^\alpha_{\ \beta} + \chi_{|}{}^\alpha_{\ \beta} + \kappa\chi\delta^\alpha_{\ \beta}\bigg] \ , \tag{39}$$

where $\nabla^2(\cdot) \equiv (\cdot)_{|\gamma}{}^{\gamma}$. Only the scalar mode χ and the tensor modes contribute to the three–dimensional Ricci curvature.

As well known, in linear theory scalar, vector and tensor modes are independent. The equation of motion for the tensor modes is obtained by linearizing the traceless part of the $\vartheta^{\alpha}{}_{\beta}$ evolution equation. One has

$$\pi_{\alpha\beta}'' + 2\frac{a'}{a}\pi_{\alpha\beta}' - c^2(\nabla^2 - 2\kappa)\pi_{\alpha\beta} = 0 , \qquad (40)$$

which is the equation for the free propagation of gravitational waves in a FRW background. The general solution of this equation is well–known (e.g. Weinberg 1972) and will not be reported here.

At the linear level, in the irrotational case, the two vector modes represent gauge modes which can be set to zero, $\xi^{\alpha} = 0$.

The two scalar modes are linked together through the momentum constraint, which leads to the relation $\chi = \chi_0 + \kappa(\zeta - \zeta_0)$. The energy constraint gives

$$(\nabla^2 + 3\kappa)\left[\frac{a'}{a}\zeta' + (4\pi Ga^2\varrho_b - \kappa c^2)(\zeta - \zeta_0) - c^2\chi_0\right] = 8\pi Ga^2\varrho_b\delta_0 , \qquad (41)$$

while the evolution equation gives

$$\zeta'' + 2\frac{a'}{a}\zeta' = c^2\chi . \qquad (42)$$

An evolution equation only for the scalar mode ζ can be obtained by combining together the evolution equation and the energy constraint; it reads

$$(\nabla^2 + 3\kappa)\left[\zeta'' + \frac{a'}{a}\zeta' - 4\pi Ga^2\varrho_b(\zeta - \zeta_0)\right] = -8\pi Ga^2\varrho_b\delta_0 . \qquad (43)$$

On the other hand, linearizing the solution of the continuity equation, gives

$$\delta = \delta_0 - \frac{1}{2}(\nabla^2 + 3\kappa)(\zeta - \zeta_0) , \qquad (44)$$

which replaced in the previous equation gives

$$\delta'' + \frac{a'}{a}\delta' - 4\pi Ga^2\varrho_b\delta = 0 . \qquad (45)$$

This is the well–known equation for linear density fluctuation, whose general solution can be found in (Peebles 1980). Once $\delta(\tau)$ is known, one can easily obtain ζ and χ, which completely solves the linear problem.

Eq.(43) above has been obtained in whole generality; one could have used instead the well–known residual gauge ambiguity of the synchronous coordinates to simplify its form. In fact, ζ is determined up to a space–dependent scalar, which would neither contribute to the spatial curvature, nor to the velocity–gradient tensor. For instance, one could fix ζ_0 so that $(\nabla^2 + 3\kappa)\zeta_0 = -2\delta_0$, so that the ζ evolution equation takes the same form as that for δ.

In order to better understand the physical meaning of the two scalar modes χ and ζ, let us consider the simplest case of an Einstein–de Sitter background ($\kappa = 0$), for which $a(\tau) \propto \tau^2$. By fixing the gauge so that $\nabla^2 \zeta_0 = -2\delta_0$ one obtains $\chi(\tau) = \chi_0$ and

$$\zeta(\tau) = \frac{c^2}{10}\chi_0\tau^2 + B_0\tau^{-3} \,, \tag{46}$$

where the amplitude B_0 of the decaying mode is an arbitrary function of the spatial coordinates. Consistency with the Newtonian limit suggests $\chi_0 \equiv -\frac{10}{3c^2}\varphi_0$, with φ_0 the initial peculiar gravitational potential, related to δ_0 through $\nabla^2\varphi_0 = 4\pi G a_0^2 \varrho_{0b}\delta_0$. One can then write

$$\zeta(\tau) = -\frac{1}{3}\varphi_0\tau^2 + B_0\tau^{-3} \,. \tag{47}$$

This result clearly shows that, at the Newtonian level, the linearized metric is $\gamma_{\alpha\beta} = \delta_{\alpha\beta} + \zeta_{|\alpha\beta}$, while the perturbation mode χ is already post–Newtonian.

These results also confirm the above conclusion that in the general GR case the initial Lagrangian spatial metric cannot be flat, i.e. $\mathcal{J}_0 \neq 1$, because of the initial "seed" post–Newtonian metric perturbation χ_0.

3.3 Recovering the Newtonian approximation in the Lagrangian picture

The Newtonian equations in Lagrangian form can be obtained from the full GR equations by an expansion in inverse powers of the speed of light; as a consequence of our gauge choice, however, no odd powers of c appear in the equations, which implies that the expansion parameter can be taken to be $1/c^2$.

Let us then expand the spatial metric in a form analogous to that used in the linear perturbation analysis above.

$$\gamma_{\alpha\beta} = \bar{\gamma}_{\alpha\beta} + \mathcal{O}\left(\frac{1}{c^2}\right) \,. \tag{48}$$

To lowest order in our expansion, the evolution equation and the energy constraint imply that $\bar{\mathcal{P}}^\alpha_{\ \beta} \equiv \mathcal{P}^\alpha_{\ \beta}(\bar{\gamma}) = 0$, and recalling that $\kappa = \kappa_N/c^2$, one gets $\bar{\mathcal{R}}^\alpha_{\ \beta} \equiv \mathcal{R}^\alpha_{\ \beta}(\bar{\gamma}) = 0$: in the Newtonian limit the spatial curvature identically vanishes. This important conclusion implies that $\bar{\gamma}_{\alpha\beta}$ can be transformed to δ_{AB} globally, i.e. that one can write $\bar{\gamma}_{\alpha\beta} = \delta_{AB}\bar{J}^A_{\ \alpha}\bar{J}^B_{\ \beta}$, with integrable Jacobian matrix coefficients. In other words, at each time τ there exist global Eulerian coordinates x^A such that

$$\mathbf{x}(\mathbf{q}, \tau) = \mathbf{q} + \mathbf{S}(\mathbf{q}, \tau) \,, \tag{49}$$

where $\mathbf{S}(\mathbf{q}, \tau)$ is called the displacement vector, and the deformation tensor becomes in this limit

$$\bar{\mathcal{D}}^A_{\ \alpha} = \frac{\partial S^A}{\partial q^\alpha} \,. \tag{50}$$

The Newtonian Lagrangian metric can therefore be written in the form

$$\bar{\gamma}_{\alpha\beta}(\mathbf{q}, \tau) = \delta_{AB} \left(\delta^A_{\ \alpha} + \frac{\partial S^A(\mathbf{q}, \tau)}{\partial q^\alpha} \right) \left(\delta^B_{\ \beta} + \frac{\partial S^B(\mathbf{q}, \tau)}{\partial q^\beta} \right) . \qquad (51)$$

One can rephrase the above result as follows: the Lagrangian spatial metric in the Newtonian limit is that of Euclidean three–space in time–dependent curvilinear coordinates q^α, defined at each time τ in terms of the Eulerian ones x^A by inversion. As a consequence, the Christoffel symbols involved in spatial covariant derivatives (which will be indicated by a single bar or by a nabla operator followed by greek indices) do not vanish, but the vanishing of the spatial curvature implies that these covariant derivatives always commute.

Contrary to the evolution equation and the energy constraint, the Raychaudhuri equation and the momentum constraint contain no explicit powers of c, and therefore preserve their form in going to the Newtonian limit. These equations therefore determine the background Newtonian metric $\bar{\gamma}_{\alpha\beta}$, i.e. they govern the evolution of the displacement vector $\mathbf{S}$.

The Raychaudhuri equation becomes the master equation for the Newtonian evolution; it takes the form

$$\bar{\vartheta}' + \frac{a'}{a}\bar{\vartheta} + \bar{\vartheta}^\mu_{\ \nu}\bar{\vartheta}^\nu_{\ \mu} + 4\pi G a^2 \varrho_b \left(\bar{\gamma}^{-1/2} - 1 \right) = 0 , \qquad (52)$$

where

$$\bar{\vartheta}^\alpha_{\ \beta} \equiv \frac{1}{2}\bar{\gamma}^{\alpha\gamma}\bar{\gamma}'_{\gamma\beta} , \qquad (53)$$

and, for simplicity, $\delta_0 = 0$ was assumed (a restriction which is, however, not at all mandatory). We also used the residual gauge freedom of our coordinate system to set $\bar{\gamma}_{\alpha\beta}(\tau_0) = \delta_{\alpha\beta}$, implying $\bar{J}_0 = 1$, i.e. to make Lagrangian and Eulerian coordinates coincide at the initial time. That this choice is indeed possible in the Newtonian limit can be understood from our previous linear analysis, where this is achieved by taking, e.g., $\zeta_0 = 0$.

The momentum constraint, $\bar{\vartheta}^\mu_{\ \nu|\mu} = \bar{\vartheta}_{,\nu}$, is actually related to the irrotationality assumption.

Let us also notice a general property of our expression for the Lagrangian metric: at each time τ it can be diagonalized by going to the local and instantaneous principal axes of the deformation tensor. Calling $\bar{\gamma}_\alpha$ the eigenvalues of the metric tensor, $\bar{J}_\alpha$ those of the Jacobian and $\bar{d}_\alpha$ those of the deformation tensor, one has

$$\bar{\gamma}_\alpha(\mathbf{q}, \tau) = \bar{J}^2_\alpha(\mathbf{q}, \tau) = \left(1 + \bar{d}_\alpha(\mathbf{q}, \tau) \right)^2 . \qquad (54)$$

From this expression it becomes evident that, at shell–crossing, where some of the Jacobian eigenvalues go to zero, the related covariant metric eigenvalues just vanish. On the other hand, other quantities, like the matter density, the peculiar volume expansion scalar and some eigenvalues of the shear and tidal tensor will generally diverge at the location of the caustics. This diverging behaviour makes the description of the system extremely involved after this event. Although dealing with this problem is far outside the aim of the present notes. let us just mention that a number of ways out are available. One can convolve the various dynamical variables by a suitable low–pass filter, either at the initial time, in order to postpone the occurrence of shell–crossing singularities, or at the time when they form, in order to smooth the singular behaviour; alternatively one can abandon the perfect fluid picture and resort to a discrete point–like particle set, which automatically eliminates the possible occurrence of caustics, at least for generic initial data. At this level, anyway, we prefer to take a conservative point of view and assume that the actual range of validity of this formalism is up to shell–crossing.

3.4 Recovering the Newtonian approximation in the Eulerian picture

As demonstrated above, it is always possible, in the frame of the Newtonian approximation, to define a global Eulerian picture. This will be the picture of the fluid evolution as given by an observer that, at the point $\mathbf{x} = \mathbf{q} + \mathbf{S}(\mathbf{q}, \tau)$ and at the time τ observes the fluid moving with physical peculiar three–velocity $\mathbf{v} = d\mathbf{S}/d\tau$. From the point of view of a Lagrangian observer, who is comoving with the fluid, the Eulerian observer, which is located at constant $\mathbf{x}$, is moving with three–velocity $d\mathbf{q}(\mathbf{x}, \tau)/d\tau = -\mathbf{v}$.

The line–element characterizing the Newtonian approximation in the Eulerian frame is well–known (e.g. Peebles 1980)

$$ds^2 = a^2(\tau)\left[-\left(1 + \frac{2\varphi_g(\mathbf{x}, \tau)}{c^2}\right) c^2 d\tau^2 + \delta_{AB} dx^A dx^B\right], \tag{55}$$

with φ_g the peculiar gravitational potential, determined by the mass distribution through the Eulerian Poisson equation,

$$\nabla_x^2 \varphi_g(\mathbf{x}, \tau) = 4\pi G a^2(\tau)\varrho_b(\tau)\delta(\mathbf{x}, \tau), \tag{56}$$

where the Laplacian ∇_x^2, as well as the nabla operator ∇, have their standard Euclidean meaning. The perturbation in the time–time component of the metric tensor here comes from the different proper time of the Eulerian and Lagrangian observers.

It is now crucial to realize that all the dynamical equations obtained so far, being entirely expressed in terms of three–tensors, keep their form in going to the Eulerian picture, only provided the convective time derivatives of tensors of any rank (scalars, vectors and tensors) are modified as follows:

$$\frac{D}{D\tau} \rightarrow \frac{\partial}{\partial \tau} + \mathbf{v} \cdot \nabla, \qquad \mathbf{v} \equiv \frac{d\mathbf{S}}{d\tau}. \tag{57}$$

This follows from the fact that, for the metric above, $\bar{\Gamma}^0_{AB} = \bar{\Gamma}^A_{0B} = \bar{\Gamma}^A_{BC} = 0$, which also obviously implies that covariant derivatives with respect to x^A reduce to partial ones.

The irrotationality assumption now has the obvious consequence that we can define an Eulerian velocity potential Φ_v through

$$\mathbf{v}(\mathbf{x}, \tau) = \nabla \Phi_v(\mathbf{x}, \tau) \,. \tag{58}$$

The Newtonian peculiar velocity–gradient tensor then becomes

$$\bar{\vartheta}_{AB} = \frac{\partial^2 \Phi_v}{\partial x^A \partial x^B} \,, \tag{59}$$

because of which the momentum constraint gets trivially satisfied and the magnetic Weyl tensor becomes identically zero in the Newtonian limit.

We can now write the Raychaudhuri equation for the Eulerian peculiar volume expansion scalar $\bar{\vartheta}$, and use the Poisson equation to get, as a first spatial integral, the Euler equation

$$\mathbf{v}' + \mathbf{v} \cdot \nabla \mathbf{v} + \frac{a'}{a} \mathbf{v} = -\nabla \varphi_g \,. \tag{60}$$

This can be further integrated to give the Bernoulli equation

$$\Phi_v' + \frac{a'}{a} \Phi_v + \frac{1}{2} \left(\nabla \Phi_v \right)^2 = -\varphi_g \,. \tag{61}$$

All these equations would of course recover the form of Section 2, if the time variable and the peculiar velocity were rescaled as described at the beginning of that section. Having shown the equivalence of this method, in the Newtonian limit, with the standard one, it would be also trivial to recover the Zel'dovich approximation in this frame. This point is further discussed by Matarrese & Terranova (1995).

4 Conclusions and discussion

Aim of these notes was to introduce the reader to the theory of the dynamics of cosmological perturbations beyond the linear approximation. Let me spend this final section to discuss what I consider an open issue in this field. The issue is whether there exists a range of scales where relativistic effects and non–linear evolution both come into play. The standard Newtonian paradigm states that the lowest scale at which the approximation can be reasonably applied is set by the amplitude of the gravitational potential and is given by the Schwarzschild radius of the collapsing body, which is negligibly small for any relevant cosmological mass scale. What is completely missing in this criterion is the role of the shear, which causes the presence of non–scalar contributions to the metric perturbations. A non–vanishing shear component is, in fact, an unavoidable feature of realistic cosmological perturbations and affects the dynamics in at least three ways, all related to non–local effects, i.e. to the interaction of a given fluid element with the environment. First, at the lowest perturbative order the shear is related to the tidal field generated by the surrounding material by a simple proportionality law. Second, it is related to a dynamical tidal induction: the modification of the environment forces the fluid element to modify its shape and density. Third, and most important here, a non–vanishing shear field leads to the generation of a traceless and divergenceless metric perturbation which can be understood as gravitational radiation emitted by non–linear perturbations. Note that the two latter effects are only detected if one allows for non–scalar perturbations in physical quantities. Truly tensor perturbations are in fact dynamically generated by the gravitational instability of initially scalar perturbations, independently of the initial presence of gravitational waves.

Using a post–Newtonian expansion in Lagrangian coordinates, Matarrese & Terranova (1995) obtained a general formula for the tensor modes π_{AB} produced by non–linear evolving perturbations. In the standard case, where the cosmological perturbations form a homogeneous and isotropic random field, they obtained a heuristic perturbative estimate of their amplitude in terms of the *rms* density contrast and of the ratio of the typical perturbation scale λ to the Hubble radius $r_H = cH^{-1}$ (where H is Hobble's constant). They found $\pi_{rms}/c^2 \sim \delta_{rms}^2(\lambda/r_H)^2$. These tensor modes give rise to a stochastic background of gravitational waves which gets a non–negligible amplitude in the so–called *extremely-low-frequency* band (e.g. Thorne 1995), around $10^{-14} - 10^{-15}$ Hz. One can roughly estimate that the present–day closure density of this gravitational–wave background would be $\Omega_{gw}(\lambda) \sim \delta_{rms}^4(\lambda/r_H)^2$. In standard scenarios for the formation of structure in the universe, the closure density on scales $1 - 10$ Mpc would be $\Omega_{gw} \sim 10^{-5} - 10^{-6}$. The amplitude of this gravitational–wave contribution, $\pi \sim \delta^2(\lambda/r_H)^2$, is an important counter–example to the standard paradigm stated above, according to which relativistic effects should be proportional to $\varphi_g/c^2 \sim \delta(\lambda/r_H)^2$.

References

Buchert, T. 1995, to appear in Proc. Enrico Fermi School, Course CXXXII, *Dark Matter in the Universe*, Varenna 1995, preprint astro-ph/9509005.

Burgers, J.M. 1974, *The Nonlinear Diffusion Equation*, Dordrecht: Reidel.

Catelan, P., Lucchin, F., Matarrese, S. & Moscardini, L. 1995, MNRAS, 276, 39.

Coles, P. & Lucchin, F. 1995, *Cosmology: The Origin and Evolution of Cosmic Structure*, Chichester: Wiley.

Efstathiou, G., Davis, M., Frenk, C. & White S.D.M. 1985, ApJS, 57, 241.

Gurbatov, S.N., Saichev, A.I. & Shandarin, S.F. 1989, MNRAS, 236, 385. ref Hockney, R.W. & Eastwood, J.W. 1981, *Computer Simulations using Particles*, New York: McGraw-Hill.

Kofman, L., Pogosyan, D. & Shandarin, S. 1990, MNRAS 242, 200.

Kofman, L., Pogosyan, D., Melott, A. & Shandarin, S. 1992, ApJ, 393, 437.

Matarrese, S., Lucchin, F., Moscardini, L., Saez, D. 1992, MNRAS, 259, 437.

Matarrese, S. & Terranova, D. 1995, preprint astro-ph/9511093.

Nusser, A., Dekel, A., Bertschinger, E. & Blumenthal, G.R. 1991, ApJ, 379, 6.

Peebles, P.J.E. 1980, *The Large-Scale Structure of the Universe*, Princeton: Princeton University Press.

Sathyaprakash, B.S., Sahni, V., Munshi, D., Pogosyan, D. & Melott, A.L. 1995, MNRAS, 275, 463.

Thorne, K.S. 1995, in Proc. *Snowmass 95 Summer Study on Particle and Nuclear Astrophysics and Cosmology*, edited by Kolb E.W. & Peccei R., preprint gr-qc/9506086.

Weinberg, S. 1972, *Gravitation and Cosmology*, New York: Wiley.

Weinberg, D.H. & Gunn, J.E. 1990, MNRAS, 247, 260. ref Williams, B.G., Heavens, A.F., Peacock, J.A. & Shandarin, S.F. 1991, MNRAS, 250, 458.

Zel'dovich, Ya.B. 1970, A&A, 5, 84.

Cosmological Applications of Gravitational Lensing

Peter Schneider

Max-Planck-Institut für Astrophysik, Postfach 1523, D-85740 Garching, Germany

Abstract: The last decade has seen an enormous increase of activity in the field of gravitational lensing, mainly driven by improvements of observational capabilities. I will review the basics of gravitational lens theory, just enough to understand the rest of this contribution, and will then concentrate on several of the main applications in cosmology. Cluster lensing, and weak lensing, will constitute the main part of this review.

1 Introduction

Gravitational light deflection has been one of the key tests of Einstein's Theory of General Relativity. Several authors in the 1920's have pointed out that this effect may give rise to spectacular effects, such as multiple images or ring-like images of distant sources, but no one expressed his vision so clearly as Zwicky in 1937, when he claimed that the observation of the gravitational lens effect will be 'a certainty'; he also estimated the probability of a distant source to be multiply imaged to be a few tenth of a percent, very close to modern estimates, and he predicted that the lens effect will allow the determination of the mass of distant cosmic objects and, due to the magnification effect, allow deeper looks into the universe (for an account of the history of this field and for references, see Chap. 1 of Schneider, Ehlers & Falco 1992, hereafter SEF). These predictions were eventually verified when Walsh, Carswell & Weymann (1979) discovered the first lensed QSO, where two QSO images with redshift $z_s = 1.41$, separated by $6''$, have nearly identical spectra from radio to X-ray frequencies, with a giant elliptical galaxy at redshift $z_d = 0.36$, situated in a cluster of galaxies, between the images. Today, the number of multiply-imaged QSOs is about 15; in addition, 6 ring-shaped radio images have been found, in some cases with a (lower-redshift) galaxy at the ring center (for a recent review of the observational situation, see Refsdal & Surdej 1994). The discovery of giant luminous arcs in 1986 by Lynds & Petrosian (1986) and Soucail et al. (1987) has shown that clusters of galaxies can act as efficient lenses; cluster lensing today is one of the most active fields of gravitational lensing (for a recent review, see Fort & Mellier 1994). Finally, the impressive demonstration (Alcock et al. 1993, Aubourg et al. 1993, Udalski et al. 1993) of the feasibility of the suggestion by Paczyński (1986) to search for compact dark objects in the halo of our Galaxy, has led to an active and successful search of Galactic microlensing events, both towards the LMC and the Galactic bulge (for a recent review, see Paczyński 1996).

These discoveries have opened up a new road towards investigating massive structures in the universe. Since gravitational light deflection is insensitive to the nature and physical state of the deflecting mass, it is ideally suited to study dark matter in the universe. In this review, only some aspects of this exciting research field can be treated; whereas strong lensing applications will be discussed in Sect. 3, I will describe cluster lensing and weak lensing in Sect. 4 in somewhat more detail. However, the necessary tools must be prepared, which will be done in Sect. 2.

2 Lensing geometry

2.1 The lens equation

The formal description of gravitational lensing is basically simple geometry. Consider a mass distribution (the deflector) at some distance D_d from us, and some source at distance D_s (see Fig. 1). Then, draw a reference line ('optical axis') through lens and observer, define planes ('lens plane' and 'source plane') perpendicular to this optical axis through lens and source, and measure the transverse separations of a light ray in the source and lens plane by η and ξ, respectively. Then from simple geometry, the relation between these two vectors is

$$\eta = \frac{D_s}{D_d}\xi - D_{ds}\hat{\alpha}(\xi) \quad , \tag{1}$$

where $\hat{\alpha}(\xi)$ is the *deflection angle*. Since all deflection angles one is interested in are very small (even in clusters of galaxies, the deflection angles are well below $1'$), and thus the gravitational fields are weak, the *linearized field equation of General Relativity* can be employed, which implies that the deflection angle is a linear functional of the mass distribution. Since the deflection angle of a light ray passing a point mass M at separation r is $4GM/(rc^2)$, the deflection angle at position ξ caused by a mass distribution descibed by the *surface mass density* $\Sigma(\xi)$ becomes

$$\hat{\alpha}(\xi) = \int_{\mathbf{R}^2} d^2\xi' \, \frac{4G\Sigma(\xi')}{c^2} \, \frac{\xi - \xi'}{|\xi - \xi'|^2} \quad , \tag{2}$$

where the integral extends over the lens plane.

The simple description of a gravitational lens situation can be justified much more thoroughly from Relativity; the reader is referred to SEF, Chap. 4, and Seitz, Schneider & Ehlers (1994) for a rigorous treatment. Here it suffices to note that for all situations encountered in this review, the gravitational lens

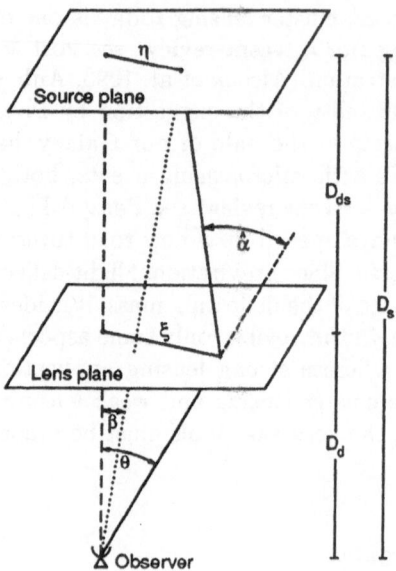

Fig. 1. The geometry of a gravitational lens

equations provide excellent approximations; in particular, the simple geometrical derivation of (1) remains valid in a Friedmann–Lemaître universe if the distances are interpreted as *angular-diameter distances*.

It is convenient to replace the physical lengths in (1) by angular variables, by defining $\beta = \eta/D_s$, $\theta = \xi/D_d$,

$$\alpha(\theta) = \frac{D_{ds}}{D_s}\hat{\alpha}(D_d\theta) = \frac{1}{\pi}\int_{\mathbf{R}^2} d^2\theta'\, \kappa(\theta')\frac{\theta - \theta'}{|\theta - \theta'|^2} \quad , \tag{3}$$

with the *dimensionless surface mass density*

$$\kappa(\theta) = \frac{\Sigma(D_d\theta)}{\Sigma_{cr}} \quad \text{with} \quad \Sigma_{cr} = \frac{c^2}{4\pi G}\frac{D_s}{D_{ds}D_d} \quad ; \tag{4}$$

then the *lens equation* simply reads

$$\beta = \theta - \alpha(\theta) \quad . \tag{5}$$

The *critical surface mass density* Σ_{cr} is a characteristic value which separates strong from weak lenses; if $\kappa \ll 1$ everywhere (i.e., $\Sigma \ll \Sigma_{cr}$), then the deflector is weak, whereas if $\kappa \sim 1$ for some θ, the lens may produce multiple images and is called strong. *Multiple images* occur if the lens equation (5) has multiple solutions θ for the same source position β.

2.2 The deflection potential, and the time-delay

Using the identity $\nabla \ln |\mathbf{x}| = \mathbf{x}/|\mathbf{x}|^2$, one sees that the deflection angle $\boldsymbol{\alpha}$ can be written as the gradient,

$$\boldsymbol{\alpha}(\boldsymbol{\theta}) = \nabla\psi(\boldsymbol{\theta}) \quad \text{with} \quad \psi(\boldsymbol{\theta}) = \frac{1}{\pi} \int_{\mathbf{R}^2} d^2\theta' \, \kappa(\boldsymbol{\theta}') \ln |\boldsymbol{\theta} - \boldsymbol{\theta}'| \quad , \tag{6}$$

of the *deflection potential* ψ. If we define the *Fermat potential*

$$\phi(\boldsymbol{\theta};\boldsymbol{\beta}) := \frac{|\boldsymbol{\theta} - \boldsymbol{\beta}|^2}{2} - \psi(\boldsymbol{\theta}) , \quad \text{then} \quad \nabla\phi(\boldsymbol{\theta};\boldsymbol{\beta}) = 0 \tag{7}$$

is equivalent to the lens equation (5). In fact, one can show that $\phi(\boldsymbol{\theta};\boldsymbol{\beta})$ is, up to an affine transformation, the *light travel time* along a light ray from the source at $\boldsymbol{\beta}$ via a point $\boldsymbol{\theta}$ in the lens plane to the observer. Hence, (7) expresses the fact that physical light rays are those for which the light travel time is stationary – which is *Fermat's principle* in gravitational lens theory.

If a source has multiple images, the light travel time along the different rays will be different. From the interpretation of ϕ it is clear that the *time delay* Δt is proportional to the difference of the Fermat potential at the image positions. One finds:

$$c\,\Delta t(\boldsymbol{\beta}) = \frac{D_{\mathrm{d}}D_{\mathrm{s}}}{D_{\mathrm{ds}}}(1 + z_{\mathrm{d}}) \left[\phi\left(\boldsymbol{\theta}^{(1)},\boldsymbol{\beta}\right) - \phi\left(\boldsymbol{\theta}^{(2)},\boldsymbol{\beta}\right) \right] \quad . \tag{8}$$

2.3 Magnification and image distortion

Light bundles are not only deflected as a whole, but differential deflection occurs. Hence, in a first approximation, a circular light bundle aquires an elliptical cross section after passing a deflector. The differential deflection changes the solid angle subtended by a source. Since the *surface brightness* (or the specific intensity) is unchanged by light deflection – this follows from Liouville's theorem, or the fact that light deflection neither creates nor destroys photons – the change in solid angle leads to a change of observed flux from a source: the flux of an infinitesimally small source with surface brightness I and solid angle $\Delta\omega$ is $S = I\,\Delta\omega$. If $\Delta\omega_0$ is the solid angle subtended by an infinitesimally small source in the absence of a deflector, then the observed flux of an image of this source at $\boldsymbol{\theta}$ is $S = \mu(\boldsymbol{\theta})\,S_0$, where the *magnification* μ of an image of an infinitesimally small source is

$$\mu(\boldsymbol{\theta}) = |\det A(\boldsymbol{\theta})|^{-1} , \quad \text{where} \quad A(\boldsymbol{\theta}) = \frac{\partial\boldsymbol{\beta}}{\partial\boldsymbol{\theta}} \tag{9}$$

is the Jacobian matrix of the lens equation[1]; in components, $A_{ij} = \partial\beta_i/\partial\theta_j \equiv \beta_{i,j}$. The matrix A describes the *locally linearized lens mapping*. Note that

---

[1] The *magnification of a source* is then the sum of the magnifications of its images; the *magnification* of an extended source is the surface-brightness averaged magnification of its source points.

$\mathrm{tr}A(\boldsymbol{\theta}) = 2[1 - \kappa(\boldsymbol{\theta})] = 2 - \nabla^2\psi(\boldsymbol{\theta})$, i.e., the deflection potential ψ satisfies a *Poisson-like equation*. The fact that the two eigenvalues of A will be different in general implies that a circular source will be imaged, to first approximation, into an ellipse. We can write the components of A as

$$A = \begin{pmatrix} 1 - \kappa - \gamma_1 & -\gamma_2 \\ -\gamma_2 & 1 - \kappa + \gamma_2 \end{pmatrix} = (1 - \kappa)\mathcal{I} - |\gamma| \begin{pmatrix} \cos(2\varphi) & \sin(2\varphi) \\ \sin(2\varphi) & -\cos(2\varphi) \end{pmatrix} ,$$

(10)

where γ is called *shear* and describes the tidal gravitational forces ($\mathcal{I}$ is the two-dimensional identity matrix). The components of the shear are given by second partial derivatives of the deflection potential,

$$\gamma_1 = \frac{1}{2}\left(\psi_{,11} - \psi_{,22}\right) , \ \gamma_2 = \psi_{,12} , \ \kappa = \frac{1}{2}\left(\psi_{,11} + \psi_{,22}\right) .$$

(11)

The eigenvalues of A are $1 - \kappa \pm |\gamma|$, where $|\gamma| = \sqrt{\gamma_1^2 + \gamma_2^2}$, the axis ratio of the elliptical image of a circular source is given by the ratio of these two eigenvalues, and the orientation of the major axis is described by the angle φ. We shall later discuss the image distortion for a general source.

Note that $\det A$ can vanish, which formally implies a diverging magnification. Of course, real magnifications remain finite. A real source is extended, and the magnification averaged over an extended source is always finite. Even if we had a point source, the magnification would remain finite: in this case, the geometrical optics approximation breaks down and light propagation had to be described by wave optics, yielding finite magnifications (see Chap. 6 of SEF). Astrophysically relevant situations involve sufficiently large sources for the geometrical optics approximation to be valid. The closed curves on which $\det A = 0$ are called *critical curves*; the corresponding curves in the source plane, obtained by inserting the critical points into the lens equation, are called *caustics*. An image close to a critical curve can have a large magnification; also, the number of images of a source changes by ± 2 if and only if the source position changes across a caustic. In this case, two images merge at the corresponding point of the critical curve, thereby brightening, and disappear once the source has crossed the caustic. The caustic is not necessarily a smooth curve, but it can develop *cusps*. A source close to, and inside a cusp has three bright images close to the corresponding point of the critical curve, whereas it has one bright image if situated just ouside the cusp.

3 Applications: Strong lensing

In this section I will discuss some of the cosmological applications of gravitational lensing which are related to galaxy-sized deflectors and those of smaller mass, keeping cluster-size lenses for the next section. The list presented here is of course non-exhaustive; I refer the interested reader to the review by Blandford & Narayan (1992) and the other reviews mentioned in the introduction.

3.1 Mass determination

The perhaps most obvious application of gravitational lensing is the determination of the mass of the deflector. The simplest situation in which a mass can be determined is that of a spherical deflector, with a source right behind the lens' center. If the lens is sufficiently strong, the source will form a ring-shaped image ('Einstein ring'), of which several examples have been found. For an axi-symmetric mass distribution, the deflection angle becomes $\hat{\alpha}(\theta) = 4GM(<\theta)/(c^2 D_d\theta)$, and so the lens equation, with the source at the origin, reads $\theta D_s = \hat{\alpha}(\theta)D_{ds}$. Combining the last two equations, one finds

$$M(<\theta) = \pi \left(D_d\theta\right)^2 \Sigma_{cr} \ . \tag{12}$$

Hence, the mean surface mass density inside the Einstein ring is the critical surface mass density, and thus the mass inside the Einstein ring can be determined once its angular diameter and the redshifts of lens and source are measured.

In fact, even if no ring-shaped image is observed, a mass estimate based on the preceding ideas is often useful and surprisingly accurate. For example, a quadruple image system allows to trace approximately the Einstein 'circle', and a mass estimate can be obtained from (12). However, more detailed modelling is warranted in such cases. It should be mentioned that the mass inside the inner $0\rlap{.}''9$ of the lensing galaxy in the quadruple QSO 2237+0305 (the so-called 'Einstein cross') has been determined with an accuracy of a few percent (Rix, Schneider & Bahcall 1992), with the largest uncertainty being due to the Hubble constant. For modelling extended images, such as radio rings, elaborate techniques have been developed (Kochanek & Narayan 1992) and successfully been applied (Kochanek et al. 1989; Kochanek 1995a; Chen, Kochanek & Hewitt 1995; Wallington, Kochanek & Narayan 1995).

Whereas the mass determination from strong lensing events is the most accurate extragalactic mass determination (again: this method does not depend on the nature or state of the matter), the limitations of this method should be kept in mind: it measures the mass inside 'cylinders', i.e., the projected mass, and it measures the mass only in the inner part of a lensing galaxy.

3.2 Measuring the Hubble constant

Refsdal (1964) pointed out that a gravitational lens system can be used to determine the Hubble constant. The basic argument is as follows: all observables in a gravitational lens system are dimensionless (angles, flux ratios – although fluxes are measured, they provide no constraint on the geometry since the intrinsic luminosity of the source is unknown –, redshifts etc.), except the time delay between any pair of images. Now consider the size of the universe to be scaled by a factor L; then, all dimensionless observables were unchanged, but the time delay would also change by a factor L. Thus, a measurement of the time delay enables one to determine the absolute size of the lensing geometry, and thus the Hubble constant.

From (8) we see that the time delay can be factorized as follows:

$$\Delta t = \frac{1}{H_0} F \,(\text{lens model}, z_\text{d}, z_\text{s}, \text{cosmology}) \quad . \tag{13}$$

The dimensionless function F depends on the cosmological parameters Ω and Λ, but this dependence is not very strong if the source and lens redshifts are smaller than ~ 2 and ~ 0.5, respectively. The redshifts of source and lens are assumed to be known. The largest uncertainty is the construction of a reliable lens model; we shall discuss this further below.

The second problem which occurs is the measurement of the time delay itself. For the double QSO 0957+561, monitoring of the two QSO images has been done in the optical (e.g., Vanderriest et al. 1989, Schild & Thomson 1995) and the radio (Roberts et al. 1991, Haarsma et al. 1996) wavebands for over 15 years. Despite this enormous observational effort, there has been no agreement on the value of Δt, with values between 410 days and 540 days occurring in the literature, because: (i) the QSO has not been very cooperative, i.e., it has not varied strongly in the last 15 years; (ii) some variability of the images must be attributed to *microlensing*[2]; (iii) the QSO is observable from the ground with optical telescopes for only 8 months a year, so that the lightcurves have gaps; this does not apply to the radio lightcurves, but due to the changing configurations of the VLA, the radio lightcurves also have gaps. A cross-correlation of the two lightcurves is thus subject to windowing effects. Furthermore, data points with underestimated errors can affect the resulting time delay and thus require the usage of robust statistical methods (for a thorough discussion of these issues, see Press, Rybicki & Hewitt 1992, Pelt et al. 1994).

Nevertheless, even if the time delay in 0957+561 is measured, its use for the determination of the Hubble constant will be limited, due to the uncertainty of the lens model. The large angular separation of this system ($\sim 6''$) implies that the image splitting is caused by a combination of the main (elliptical) galaxy at $z_\text{d} = 0.36$ and a cluster in which that galaxy is embedded; in addition, there is a second concentration of galaxies in the field, at a redshift of $z \sim 0.5$. The description of the mass distribution thus requires more parameters than available constraints from the observations, leaving a large freedom for the function F in (13) (see, e.g., Bernstein, Tyson & Kochanek 1993). In addition, if $\kappa_0(\theta)$ describes a mass distribution for the lens which is compatible with all observational constraints [image positions, relative magnification matrix $A(\theta^{(1)}) A^{-1}(\theta^{(2)})$],

---

[2] Since the matter in the lensing galaxy consists partly of stars, the mass distribution is grainy; the emitting region of the optical continuum light of QSOs is sufficiently small to be sensitive to the gravitational field of stars in the lensing galaxy, down to about Jupiter mass. Whereas the stellar gravitational field does not noticibly affect the angular position of the QSO images, it affects the magnifications, and thus the flux of the images. This effect has been clearly observed in the quadruple QSO 2237+0305 (Houde & Racine 1994, and references therein), as in this system the fluxes of the four QSO images vary independently, whereas any intrinsic variation of the QSO must show up in all four images within the expected time delay of ~ 1 day. Note that this microlensing has led to interesting upper bounds on the size of the QSO emitting region (Rauch & Blandford 1991; Jaroszyński, Wambsganss & Paczyński 1992)

then the whole family $\kappa(\theta) = \lambda\kappa_0(\theta) + (1 - \lambda)$ of mass distributions satisfies the observational constraints, but the function F in (13) scales like λ, thus affecting the resulting value of H_0 (Gorenstein, Falco & Shapiro 1988). This *mass sheet degeneracy* is always present, but is particularly severe in a case like 0957+561 where the presence of a mass sheet is in fact concluded from the presence of a cluster. The mass sheet degeneracy then implies that gravitational lensing can strictly yield only upper bounds on the Hubble constant.

Perhaps the most promising system currently known for the determination of H_0 is the Einstein ring B0218+35.7 (Patnaik et al. 1993), which contains two compact flat-spectrum image components. These compact components are expected to vary, thus enabling the measurement of the time delay, whereas the ring can be used to construct a detailed lens model. Since an extended image yields much more information about the lensing geometry than multiply imaged point-like sources, this system will be much better for constraining the function F in (13), also because the small image separation ($0\rlap{.}''35$) points towards lensing by an isolated (spiral) galaxy. Furthermore, the compact radio components are sufficiently extended (they have been resolved with VLBA observations – see Patnaik, Porcas & Browne 1995) as to not be affected by microlensing. Indeed, from the variability of the *polarized* flux, a preliminary value for the time delay ($\Delta t = 12 \pm 3$ days) has been obtained (Corbett, Browne & Wilkinson 1996).

A value of H_0 measured from lensing would be valuable for several reasons: it is a measurement which is completely independent of any local 'distance ladder', it would measure H_0 on a truly cosmic scale, and thus being independent of local peculiar velocity fields, and also because an agreement between measurements on cosmic scales with those measured locally would provide a strong support for the validity of standard Friedmann–Lemaître cosmological models.

3.3 Galactic microlensing

Among the currently most active fields of lensing research is Galactic microlensing, i.e., lensing by stars in our Galaxy. Paczyński (1986) suggested that a search for such microlensing events may lead to the discovery of, or to an upper limit on the density of compact objects in the halo of our Galaxy, which are dark matter candidates. In this case, stars in the LMC are sources which are lensed by halo objects. As a 'control experiment', he suggested (Paczyński 1991) to observe stars in the Galactic bulge; in this case, the lenses are known to exist, namely the disk stars. The signature of microlensing is a characteristic lightcurve of the lensed star which is described by only four parameters. However, the difficulty of both experiments is the incredibly small lensing probability: about 1 out of 10^{-7} stars in the LMC is lensed at any given time if the halo of our Galaxy is made of compact objects. This implies that millions of stars have to be monitored, and the microlensing events have to be extracted from these many lightcurves which include many variable stars. It therefore came as a surprise when three groups announced their detection of microlensing events in the second half of 1993. Today (Oct. 1995), more than hundred microlensing events are known, most of them towards the Galactic bulge (see Paczyński 1996 for a review). The main result of these experiments is that the microlensing rate towards the LMC is smaller than expected, by about a factor of 5, but that the event rate towards the Galactic bulge is larger by a factor of three than expected from naive Galactic mass models. The latter fact is interpreted as indicating that our Galaxy has a bar which is pointing nearly towards us, and that this bar constitutes a major fraction of the microlensing optical depth (Zhao, Spergel & Rich 1995). A variation of the optical depth to microlensing with angular position will allow detailed mass models for the Galaxy. The small microlensing event rate towards the LMC indicates that the halo of the Galaxy is not mainly composed of compact objects, its best-fitting mass fraction being about 20% (Alcock et al. 1995). However, at least part of the lensing optical depth can be provided by objects in the LMC itself or nonhalo Galactic objects.

The incredibly large frequency of publications on galactic microlensing events indicates that this research will continue to yield important results; e.g., on the Galactic mass distribution, the frequency of binary stars, on the dynamics within the Galactic bar, and can even be used to search for planetary systems. It should also be borne in mind that the results from such experiments provide an eldorado for people working on stellar variability!

3.4 Lensing statistics and compact dark matter in the universe

The fraction of all high-redshift QSOs which are multiply imaged is proportional to the number density of lenses in the universe; hence, from the observed fraction of multiply imaged QSOs it is possible to constrain the statistical properties of the lens population.

The probability that a QSO is multiply imaged depends on its redshift (the larger the redshift, the more likely is a lens in the line-of-sight), its luminosity (because of the magnification bias[3]), the number density of galaxies (and its possible cosmological evolution), the mass and mass profiles of galaxies, and the cosmological model. Furthermore, the angular separation statistics of the multiple images depends on the masses and redshifts of the lenses, as well as on the cosmological model. The observed angular separation statistics depends furthermore on the observational selection function, which takes into account the finite angular resolution of the observations and the dynamic range of flux ratios which can be observed, depending on the angular separation of the images.

---

[3] QSO samples are flux limited. If a source is magnified, it can enter the flux-limited survey, although its unlensed flux may be below the flux threshold of the sample. Since multiply imaged QSOs are always magnified – typically by a factor of ~ 4 for double QSOs, and by a factor ~ 10 or higher for quadruple QSOs – multiply imaged QSOs are overrepresented in flux-limited samples. This effect is called magnification bias, and it is larger for steep source counts: the steeper the counts, the more faint QSOs are there for any bright QSO, and thus the reservoir out of which sources can be magnified above the flux threshold is larger. QSO counts are very steep for bright QSOs with $m \lesssim 19$, and flatten considerably for fainter magnitudes (see Hartwick & Schade 1990); hence, the magnification bias is large for bright QSOs.

Several lens surveys have been completed in recent years (for references, see Kochanek 1995b), both in the optical and radio. In order to make use of the magnification bias, and thus to increase the probability that a QSO is multiply imaged, these surveys were performed for the apparently most luminous QSOs, i.e., for bright high-redshift QSOs. For these surveys, the selection function can be reasonably well determined (Kochanek 1993a).

A statistical analysis of the results of these lens surveys consists in a parametrized description of the lens population. Kochanek (1993b) modelled the lensing galaxies as singular isothermal spheres, used a Faber-Jackson relation for the dependence of velocity dispersion (the parameter characterizing the lensing properties of an isothermal sphere) on luminosity, $\sigma/\sigma_* \propto (L/L_*)^\eta$, where L_* is the characteristic luminosity which enters the (Schechter) luminosity function of galaxies. He then used a maximum-likelihood analysis to obtain the best-fitting parameter values from the lens surveys, assuming a constant comoving lens population. A similar analysis was carried out by Maoz & Rix (1993), who investigated also different mass profiles for the lensing galaxies.

The main results of these studies can be summarized as follows: the observed statistics of multiply imaged QSOs is fully compatible with the 'standard assumptions' about the galaxy population and cosmology. The best fit value of σ_* is $245 \pm 30 \, \text{km/s}$, very much in agreement with dynamically consistent models of early-type galaxies (spirals, though more numerous, contribute only little to the lensing probability), and the best-fit values for the Faber-Jackson index and the faint-end slope α of the Schechter function are $\eta \sim 4$ and $\alpha \sim -1.1$, again fully compatible with the canonical values. For flat universes with $\Omega + \Lambda = 1$, the best-fit value is $\Lambda = 0$, and a formal upper limit of $\Lambda \le 0.66$ (95% confidence) can be obtained (Kochanek 1995b). Models in which elliptical galaxies have no dark halo do not reproduce the observed statistics; they predict too few large separation systems.

There is not much room for compact 'dark' lenses with mass in excess of $10^{11} M_\odot$, given that in the majority of the multiple QSOs a (luminous) lens 'between' the images is detected. However, the constraints are less strong for lower-mass objects. For lens masses larger than about $10^6 M_\odot$, these can in principle be detected (or ruled out) with radio-interferometric observations. Kassiola, Kovner & Blandford (1991) analyzed available VLBI observations to put an upper limit of $\Omega_c \le 0.4$ on the cosmological density of compact objects in the mass range $10^7 M_\odot \lesssim M \lesssim 10^9 M_\odot$; this limit and the corresponding mass range will very soon be dramatically improved, following dedicated VLBI surveys (Augusto, Wilkinson & Browne 1996; Patnaik et al. 1996). The image splitting by lenses with $M \lesssim 10^5 M_\odot$ cannot be resolved even with VLBI; nevertheless, a significant cosmological density of lenses with $M \gtrsim 10^3 M_\odot$ can be ruled out if gamma-ray bursts are at cosmological distances; in that case, lensing of bursts would lead to multiple bursts with delay of $\sim 2 \times 10^{-5} (M/M_\odot)$ seconds (Blaes & Webster 1992), and no obvious candidates for such multiple bursts have been identified yet. Following an early idea of Canizares (1982), Schneider (1993) has obtained upper limits on the density of compact objects in the mass range

$3 \times 10^{-4} M_\odot \lesssim M \lesssim 10^{-1} M_\odot$, down to a limit of $\Omega_c \lesssim 0.1$, from constraints on the variability of high-redshift QSOs: a cosmological population of such lenses would lead to the magnification of high-redshift QSOs, and since sources and lenses are moving, the magnification will change in time, leading to lens-induced variability. If the preceding limits on Ω_c are violated, QSOs would be more variable than observed [the lower mass limit is due to the finite size of QSOs; lenses with $M \lesssim 10^{-4} M_\odot$ cannot magnify the continuum flux of QSOs significantly; the upper mass limit is due to the finite time of observations from which these constraints were obtained – the Hawkins & Véron (1993) sample of variability-selected QSOs]. Since the continuum source of QSOs is much smaller than the broad line region, lenses with $10^{-3} M_\odot \lesssim M \lesssim 10^2 M_\odot$ can magnify the continuum flux, but not the line flux; a cosmologically significant density of compact objects in this mass range would thus lead to small line-to-continuum fluxes of some high-redshift QSOs. The observed lack of this effect has led Dalcanton et al. (1994) to obtain an upper limit of $\Omega_c \lesssim 0.1$ for lenses in the above mentioned mass range.

4 Cluster lensing and weak lensing

When giant luminous arcs were first explicitly mentioned by Lynds & Petrosian (1986) and Soucail et al. (1987)[4], they came as a surprise. Whereas alternative explanations for them have been put forward, the redshift determination of the arc in the Abell cluster A370 with redshift $z_d = 0.37$, yielding $z_s = 0.724$ (Soucail et al. 1988), clearly verified the lensing hypothesis. Many giant arcs have been discovered since, and systematic surveys have been carried out (for a recent review on giant arcs and cluster lensing, see Fort & Mellier 1994). For example, Luppino et al. (1995) found giant arcs in 8 out of 40 X-ray-selected clusters with redshift ≥ 0.15, and the fraction of arc clusters increases with increasing X-ray luminosity. In Sect. 4.1 below I will discuss some selected results from the analysis of arcs in clusters. If a few background galaxies are so strongly distorted as to form these giant luminous arcs, it appears evident that many more background galaxies are more weakly distorted; Fort et al. (1988) were the first to discover so-called arclets in A370: images near the cluster center, still with a large axis ratio, and aligned in the tangential direction relative to the center of the cluster. Spectroscopy verified the lensing origin of the brightest of these arclets, situated at $z_s = 1.305$ (Mellier et al. 1991). Later, Tyson, Valdes & Wenk (1990) found several tens of aligned images of (presumably background) galaxies in the clusters A1689 and CL 1409+52. These discoveries then opened up the possibility to study the mass distribution in clusters, using giant arcs for the innermost part of the clusters, and the weakly distorted images in the outer parts. The finding of Kaiser & Squires (1993) of a parameter-free reconstruction of the surface mass density from observed image distortions has marked the beginning of a new and extremely promising field of research, of which some

---

[4] though arcs have been observed previously by several researchers

aspects and results are discussed in Sect. 4.2. The rest of this section is then devoted to other aspects of weak gravitational lensing, including the discovery of groups of galaxies through weak image distortions and magnification bias, the investigation of statistical properties of the mass distribution of galaxies, and the possibility to measure the power spectrum of density fluctuations in the universe from weak lensing.

4.1 Results from giant luminous arcs

Giant arcs are the result of very strong distortions of light bundles from background sources. Such strong distortions require that the locally linearized lens mapping, described by the matrix A (10), is nearly singular. In other words, giant arcs are formed near a critical curve of the cluster lens. Assuming for a moment that the cluster mass distribution is axially-symmetric, then the mass estimate as given by eq. (12) is valid, where now θ is the distance of the arc from the cluster center, and Σ_{cr} can be determined if the redshift of the arc is measured, or estimated from the color of the arc. This mass estimate is the most basic parameter one can infer from the observation of a single arc, and in the absence of additional information and assumptions, it is the only quantity that can be derived. Depending on the geometry of the cluster, this mass estimate is fairly robust; it loses its accuracy if the cluster is highly eccentric or has significant substructure. From a sample of numerically generated clusters, Bartelmann (1995a) has shown that this simple mass estimate typically overestimates the mass of the cluster within the arc distance by about 30%, however with a large scatter.

The discovery of arcs was a surprise, because it has been thought that clusters are not compact enough to produce critical curves. To understand this, consider a cluster mass profile; keeping the outer profile fixed, by reducing the core size (i.e., the length scale within which the cluster mass profile is roughly flat) the central surface mass density is increased. Clusters become critical (i.e., possess critical curves) only if the dimensionless surface mass density κ is of order unity at the center; this requires the core size to be sufficiently small. The core radius of clusters as estimated from X-ray observations of the intracluster gas was thought to be considerably larger than needed for critical clusters. The occurrence of arcs in clusters immediately demonstrated that the core size of clusters must be small, much smaller than estimated before.

The preceding discussion has been rather vague, since the concept of a core size of a cluster is not very well defined. Basically, it is a parameter in a parametrized profile, either of the mass or the X-ray emissivity, and different parametrizations can yield different values for the core radius. However, the differences between the core size as estimated from X-ray studies (typically in excess of $100h^{-1}$ kpc) and that estimated from lensing are larger than can be easily explained as being due to semantic problems. To wit, if the mass profile of a cluster is described by an isothermal sphere with a finite core radius, in order for the cluster to be critical, the core radius must be smaller than half the Einstein radius of the cluster. Since the arc roughly traces the Einstein radius, the core radius must be smaller than half the separation of the arc from the cluster center. Given that most arcs have a separation of $\sim 20''$ from the cluster center, this argument implies core radii $\lesssim 30h^{-1}$ kpc, in marked conflict with the results from X-ray imaging. These qualitative remarks have been substantiated in detail by Miralda-Escudé & Babul (1995) who have investigated three arc clusters in detail for which X-ray observations are available. They also outlined several possible origins for the discrepancy, e.g., projection effects (which they consider unlikely), non-thermal pressure support of the intracluster gas, or a multiphase medium. When judging the seriousness of this discrepancy, one should always bear in mind the large number of assumptions entering the X-ray investigations, e.g., hydrostatic equilibrium, symmetry, isothermal gas distribution, whereas the lensing investigation is simple and purely geometrical. Recently, Waxman & Miralda-Escudé (1995) and Navarro, Frenk & White (1995) showed that the discrepancy may be reduced if the dark matter halo profile in clusters follows a universal density law, which allows an isothermal X-ray gas in hydrostatic equilibrium to develop a flat core well outside the radius where giant arcs form.

For some clusters, the observations of arcs permit a much more detailed study of their (projected) mass density. This is the case if multiple images can be identified, or if several arcs show up, or if the brightness profile of the arc permits the identification of multiply imaged components. In the cluster Cl 2137−23 ($z_d = 0.313$), two arcs have been discovered (Fort et al. 1992): a tangential arc $15''.5$ away from the central cD galaxy and $12''$ long, and a radial arc about $5''$ long and also $5''$ away from the center of the cD galaxy. The importance of this radial arc cannot be overstated, since its position clearly indicates the turnover of the mass profile; in other words, its position directly yields the core radius of this cluster, quite independent of any details of the lens model; the resulting value is $r_{core} = 25h^{-1}$ kpc. A detailed model of this arc system was performed by Mellier, Fort & Kneib (1993). Amazingly, an elliptical isothermal mass profile (with finite core) with the same ellipticity and orientation as the cD galaxy yields an acceptable model for the tangential and the radial arc. This model then *predicts* the locations of two additional images corresponding to the source of the tangential arc, and one additional image of the source of the radial arc, and these predicted locations are impressively close to observed arclets in the cluster (within $0''.6$). Hence, in this case the lens model has predictive power, and can be safely assumed to yield a realistic description of the mass distribution within the inner $\sim 15''$ of the cluster. In the cluster A370, the detailed structure of the giant arc and several multiple image candidates were used to construct a detailed mass model for this cluster (Kneib et al. 1993); also in this case, a mass model which follows closely the distribution of light yields a satisfactory fit to the observations. The giant arc in the cluster Cl 0024+16 is split up into three segments; this is caused by a clump of cluster galaxies near the arc which locally perturb the lens potential significantly. Satisfactory models of this arc system were derived by Kassiola, Kovner & Fort (1992), and a lens inversion, using techniques similar to those used for inverting radio ring images (see Sect. 3.1), has been performed by Wallington, Kochanek & Koo (1995). In this case, the mass of the perturbing galaxies can be estimated fairly accurately. As a final example, refurbished-HST images of the cluster A2218 (Kneib et al. 1995) have revealed a most amazing collection of arcs in the central parts of this cluster; together with several redshifts measured for these arcs, the most detailed mass model for the central part of a cluster currently available has been constructed.

Several heroic attempts have been made to predict the frequency of occurrence of giant luminous arcs from the observed number density of clusters, using analytical models (e.g., Wu & Hammer 1993, Bergmann & Petrosian 1993, Miralda-Escudé 1993, Grossman & Saha 1994). The results of these studies, in particular those which consider mainly spherically symmetric mass profiles for the clusters, are to be interpreted with great care, as shown by the numerical investigation by Bartelmann & Weiss (1994), and Bartelmann, Steinmetz & Weiss (1995); the probability for forming arcs in these numerically generated cluster mass profiles is substantially higher than that of more symmetric mass profiles, say with the same mass. The reason for that is that asymmetries and substructure increases the total length of the caustic curve. Another way to view this fact is that the shear is increased by substructure, such that critical curves can occur in regions where κ is considerably less than unity (Bartelmann 1995a).

4.2 Cluster mass reconstruction from weak lensing

The fact that the sky is densely covered by faint galaxy images allows the statistical study of distortions of light bundles from these high-redshift sources. The basic idea here is that the shape of a galaxy image is affected by the tidal gravitational field along its corresponding light bundle. This tidal field causes a circular galaxy to form an elliptical image. Since galaxies are not round intrinsically, this effect can not be detected in individual galaxy images (except when the distortion is so strong as to lead to the formation of arcs), but since the intrinsic orientation of galaxies can be assumed to be random, a coherent alignment of images can be detected from an ensemble of galaxies. In this and the next two subsections, we shall discuss several aspects of this general idea.

If one considers the line-of-sight towards a cluster of galaxies, one can assume that the main contribution to the tidal gravitational field along light bundles corresponding to galaxies behind the cluster comes from the cluster itself, unless there are other clusters near this line-of-sight. The tidal field, or the shear, is then related to the gravitational potential ψ of the cluster, as given in (11). Combining eqs.(11) and (6), and defining the complex shear γ by $\gamma = \gamma_1 + i\gamma_2$, one finds the relation between shear and surface mass density κ to be

$$\gamma(\theta) = \frac{1}{\pi} \int_{\mathbf{R}^2} d^2\theta' \, \mathcal{D}(\theta - \theta') \, \kappa(\theta') \quad , \tag{14a}$$

with the complex function

$$\mathcal{D}(\theta) = \frac{\theta_2^2 - \theta_1^2 - 2i\theta_1\theta_2}{|\theta|^4} \quad . \tag{14b}$$

Since the relation (14a) between shear and surface mass density is a convolution-type integral, it can be inverted, e.g., by Fourier methods, to yield (Kaiser & Squires 1993)

$$\kappa(\boldsymbol{\theta}) = \frac{1}{\pi} \int_{\mathbf{R}^2} d^2\theta' \, \mathcal{R}e\big[\mathcal{D}^*(\boldsymbol{\theta} - \boldsymbol{\theta}') \gamma(\boldsymbol{\theta}')\big] + \kappa_0 \quad , \tag{15}$$

where the asterisk denotes complex conjugation, and $\mathcal{R}e(x)$ is the real part of the complex variable x. Hence, if the tidal field γ can be measured, the surface mass density of the cluster can be obtained from (15) up to an overall constant. The reason for this constant to occur is that a homogeneous mass sheet does not cause any shear.

One can think of several methods to characterize the shape of a galaxy image. A convenient method is provided by using the matrix of second brightness moments,

$$Q_{ij} = \frac{\int d^2\theta \, I(\boldsymbol{\theta}) \, (\theta_i - \bar{\theta}_i) \, (\theta_j - \bar{\theta}_j)}{\int d^2\theta \, I(\boldsymbol{\theta})} \quad , \tag{16}$$

where $I(\boldsymbol{\theta})$ is the surface brightness distribution, and $\bar{\boldsymbol{\theta}}$ is the center of light of the galaxy image, defined such that $\int d^2\theta \, I(\boldsymbol{\theta}) \, (\boldsymbol{\theta} - \bar{\boldsymbol{\theta}}) = 0$. Defining in analogy the tensor of second brightness moments $Q_{ij}^{(s)}$ of the intrinsic brightness distribution of the galaxies, one finds from the lens equation (5) and the conservation of surface brightness, $I(\boldsymbol{\theta}) = I^{(s)}(\boldsymbol{\beta}(\boldsymbol{\theta}))$ that $Q^{(s)} = A\,Q\,A$, where A is given by (10).

In the following, we shall for simplicity restrict our attention to non-critical clusters only, i.e., we shall assume that $\det A > 0$ everywhere. The reader is referred to Schneider & Seitz (1995) and Seitz & Schneider (1995a) for the treatment of critical clusters. One then defines the complex ellipticity of an image as

$$\epsilon = \frac{Q_{11} - Q_{22} + 2iQ_{12}}{Q_{11} + Q_{22} - 2\sqrt{Q_{11}Q_{22} - Q_{12}^2}} \quad , \tag{17}$$

and correspondingly the ellipticity $\epsilon^{(s)}$ of the intrinsic brightness profile of the galaxy in terms of $Q_{ij}^{(s)}$. For example, if an image has elliptical contours of axis ratio $r \leq 1$, then $|\epsilon| = (1 - r)/(1 + r)$. From the relation $Q^{(s)} = A\,Q\,A$ one then derives the transformation between intrinsic and observed ellipticity (Schneider 1995)

$$\epsilon^{(s)} = \frac{\epsilon - g}{1 - g^*\epsilon} \quad , \tag{18}$$

where

$$g = \frac{\gamma}{1 - \kappa} \tag{19}$$

is the (complex) reduced shear. Finally, averaging over a set of galaxy images, together with the assumption that the intrinsic ellipticity distribution is isotropic, so that $\langle \epsilon^{(s)} \rangle = 0$, one finds that

$$g = \langle \epsilon \rangle \quad . \tag{20}$$

Several comments have to made at this point:

(a) The definition (16) of the quadrupole moments cannot be applied to real images, as the integration extends to infinity. In order not to be completely dominated by noise, a weighting function has to be included in the integrals. However, with an angle-dependent weight function, the relation between Q and $Q^{(s)}$ no longer has a simple form and is only approximately given by $Q^{(s)} = A Q A$; the deviations from this law depend on the intrinsic brightness profile of the source and the weighting function. Even worse is the effect of seeing and an anisotropic point-spread-function (PSF), in particular if the latter is not known very precisely. Several methods to deal with these complications have been discussed in the literature (e.g., Bonnet & Mellier 1995; Kaiser, Squires & Broadhurst 1995). In particular, a calibration of the relation between ϵ and $\epsilon^{(s)}$ is obtained from numerical simulations and from applying these methods to degraded HST images. It is clear that HST images with their unprecedented angular resolution are best suited for this kind of work, and that ground-based images are much more difficult to analyse. Future ground-based observations will make use of the calibration that can be obtained from HST images, in particular if an HST field is centered on the ground-based image.

(b) The fact that the observable g has to be obtained from averaging over an ensemble of galaxy images implies that this method has a finite resolution. I.e., the averaging process is performed over the galaxy images within a certain smoothing length from the point of interest. Several methods of smoothing have been discussed (Kaiser & Squires 1993, Seitz & Schneider 1995a); we prefer smoothing with Gaussian weights. Since the number of images over which the average is perfomed is finite, the relation $\langle \epsilon^{(s)} \rangle = 0$ is not strictly valid due to the finite width of the intrinsic ellipticity distribution; only the expectation value of $\epsilon^{(s)}$ vanishes. The smoothing length need not be kept constant, but can be adapted to the local 'strength of the signal'.

(c) It is clear from (20) that only the reduced shear is an observable, but not the shear itself as needed in the inversion equation (15). If the lens is weak in the sense $\kappa \ll 1$, then $g \approx \gamma$, and (15) can be applied directly. In general, one can replace γ in (15) by $(1 - \kappa)g$, which then yields an integral equation for $\kappa(\theta)$. As shown in Seitz & Schneider (1995a), this integral equation can be easily solved in a few iteration steps. If this nonlinear correction is taken into account, then $\kappa(\theta)$ is no longer determined up to an overall additive constant as implied by (15), but there exists a global invariance transformation (Schneider & Seitz 1995)

$$\kappa(\theta) \to \lambda\kappa(\theta) + (1 - \lambda) \quad , \tag{21}$$

which leaves all image shapes invariant. Note that this invariance transformation is the same as the mass sheet degeneracy discussed in Sect. 3.2. Of course, the allowed values of λ are restricted by the requirement that the resulting mass distribution is non-negative. Hence, this constraint always allows to obtain a lower limit on the mass. An alternative way to obtain a lower limit to the mass inside circular apertures has been discussed by Kaiser (1995a) – the so-called aperture densitometry – which also allows a rigorous estimate of the uncertainty of this lower limit. Also, if the data field is sufficiently large, one might expect that κ decreases to near zero at the boundary of the field, which then yields a plausible range for λ; this in fact is one of the arguments to demand wide-angle fields.

(d) The integral in (15) extends over the whole sky; on the other hand, data are given only on a finite data field (CCD field) $\mathcal{U}$. If the field $\mathcal{U}$ is not sufficiently large, and the contributions of the integral (15) from outside the data field are neglected, the estimate of the surface mass density is no longer unbiased, but boundary artefacts occur. Kaiser (1995a) noticed that there exists a *local* relation between the gradient of κ and certain combinations of first derivatives of the shear components (which is due to the fact that both of these quantities are third derivatives of the deflection potential ψ). Performing averages over line integrations of this local relation allows the construction of unbiased finite-field inversion formulae (Schneider 1995, Kaiser et al. 1995, Bartelmann 1995c, Seitz & Schneider 1995b). In the latter of these papers, an inversion formula has been derived which filters out a particular noise component in the data which is readily identified as such, and a quantitative comparison with other inversion formulae has been performed.

(e) The transformation (21) leaves all image shapes invariant, but affects the magnification, $\mu \to \mu/\lambda^2$. Hence, this invariance transformation can be broken if the magnification can be measured. Two possibilities have been mentioned in the literature: Broadhurst, Taylor & Peacock (1995) noticed that the magnification effect changes the local number density of galaxy images (see footnote 3), $n(S) = n_0(S/\mu)/\mu$, where $n(S)$ are the cumulative number counts, and $n_0(S)$ are the counts in the absence of lensing. Assuming a local power law, $n_0(S) \propto S^{-\alpha}$, then $n(S)/n_0(S) = \mu^{\alpha-1}$. The blue galaxy counts have $\alpha \approx 1$, and so no magnification bias effect is observable. However, counts in the red have a flatter slope, $\alpha \approx 0.75$, and a number density decrease should be seen in regions of high magnifications. The number counts of galaxies with a red color has an even flatter slope, and the magnification effects become stronger. Indeed, this effect has been clearly seen in the cluster A1689 (Broadhurst 1995). The magnification effect also changes the redshift distribution at fixed apparent magnitude. Bartelmann & Narayan (1995) noticed that individual galaxy images become apparently brighter, at fixed surface brightness. Assuming a sufficiently tight intrinsic magnitude - surface brightness relation, the magnification can be obtained locally. The additional information coming from the magnification effects cannot be incorporated easily in a direct inversion formula such as (15), and there are two possibilities to make use of it: one could obtain the surface mass distribution from a direct inversion, such as (15), and use the magnification information afterwards to fix the transformation parameter λ in (21). Or, one could use a reconstruction method which takes into account the *local* magnification information. One possibility for the latter is a maximum-likelihood approach (Bartelmann et al. 1995) for the reconstruction of the deflection potential ψ.

(f) We have implicitly assumed that all sources have the same redshift, i.e., that the critical surface mass density Σ_{cr} is the same for all sources. This assump-

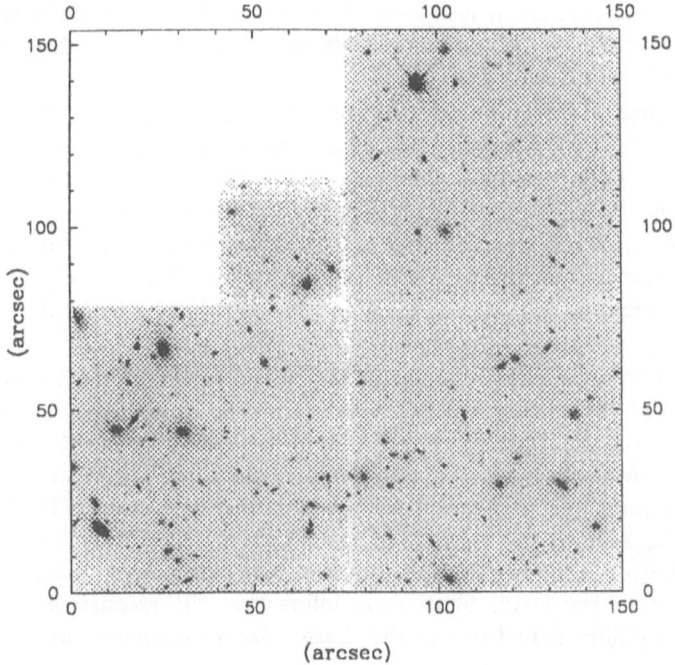

Fig. 2. The WFPC2 image of the cluster Cl0939+4713 (A851); North is at the bottom, East to the right. The coordinates are in arcseconds. The cluster center is located at about the upper left corner of the left CCD, a secondary maximum of the bright (cluster) galaxies is seen close to the interface of the two lower CCDs, and a minimum in the cluster light is at the interface between the two right CCDs. In the lensing analysis, the data from the small CCD (the Planetary Camera) were not used

tion is not too bad if the cluster is at a sufficiently low redshift, since then the ratio D_{ds}/D_s can be assumed constant for faint galaxies. In general, however, the redshift distribution of galaxies has to be taken into account. In the weak lensing regime ($\kappa \ll 1$, $|\gamma| \ll 1$), only the mean value of D_{ds}/D_s enters the reconstruction. The non-linear case is more complicated (Seitz & Schneider 1996) and requires the functional form of the redshift distribution. On the other hand, this dependence may also allow to obtain constraints on the redshift distribution of the faintest galaxies. Alternatively, Bartelmann & Narayan (1995) pointed out that the expected strong dependence of surface brightness on the redshift of galaxies, together with the dependence of the lensing strength on source redshift, may allow to determine the redshift distribution of galaxies by studying the variation of lensing strength (i.e., mean ellipticity) as a function of surface brightness. Also, the comparison of lens reconstruction of clusters at different redshifts might allow conclusions about the redshift distribution as a function of magnitude (Smail, Ellis & Fitchett 1994).

The cluster construction method described above has been applied to several clusters. Fahlman et al. (1994) analyzed the shear field of the cluster MS1224 and obtained a mass-to-light ratio of $\sim 800h$, where h is the Hubble constant in units of $100\,\text{km/s/Mpc}$; in particular, the mass derived is much larger than that obtained from a virial analysis. For the cluster A1689, an M/L-ratio of about $450h$ was found by two independent groups (Kaiser 1995b; Tyson & Fischer 1995). A similar value for the M/L-ratio was found for two clusters by Smail et al. (1995).

We (Seitz et al. 1995) have recently analyzed the 'weak' lensing effects in the cluster Cl0939+4713 (A851), using WFPC2 data (Dressler et al. 1994). Since the WFPC2 field is fairly small, we have data only in the center of the cluster, where the lensing is not weak. Also, the small field requires the use of an unbiased finite-field inversion technique, and we used the one derived in Seitz & Schneider (1995b). Fig. 2 shows the WFPC2 image of the cluster, and the reconstructed mass distribution, together with results from a bootstrapping analysis, is shown in Fig. 3. From the latter figure, one infers that the reconstruction yields basically four significant features in the mass map: a maximum close to the position where the cluster center is predicted from optical observations, a secondary maximum roughly in the lower right CCD, an overall gradient in the lower two CCDs increasing 'to the left', and a pronounced minimum at the interface between the two right CCDs. Comparing these features with the image (Fig. 2) one sees that the maximum is clearly visible in the bright (cluster) galaxies, but also the secondary maximum and the minimum in the light distribution. In addition, the two maxima may be traced by the X-ray emission, as indicated by the ROSAT PSPC-map. Hence, in this cluster we have strong evidence of significant substructure in the mass, and that the light distribution on average follows this substructure. It will be interesting to compare the mass map with a detailed HRI map which will be obtained soon (S. Schindler, private communication). The M/L-ratio of the cluster within the WFC field depends on the assumed redshift distribution of the background galaxies. Assuming that the mean redshift of galaxies with $24 \leq R \leq 25.5$ is about unity, we find that $M/L \sim 200h$, a value significantly lower than for, e.g., MS1224. However, this is not too surprising, since A851 is the highest-redshift cluster in the Abell catalog which clearly biases towards high optical luminosity. In this cluster, we also have detected the magnification effect discussed above, which has allowed us to obtain not only a strict lower limit on the mass inside the data field, but also to obtain an estimate of the mass, which led to the above value for the M/L-ratio. Note, however, that this mass calibration is uncertain due to the fact that an (unknown) fraction of the faint galaxies are cluster members which renders the estimate of the magnification effect uncertain.

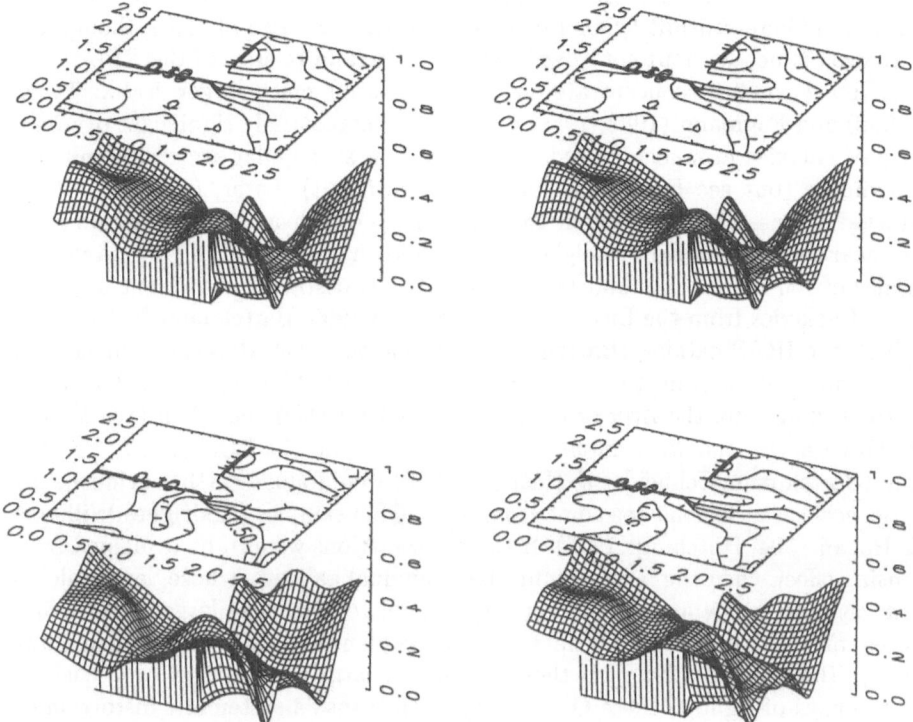

Fig. 3. The lower right panel shows the reconstructed mass distribution of A851, assuming a mean redshift of the $N = 295$ galaxies with $24 \leq R \leq 25.5$ of $\langle z \rangle = 1$. The other three panels show reconstructions obtained from the same data set via bootstrapping, i.e., selecting randomly (with replacement) $N = 295$ galaxies from the galaxy sample. The similarity of these mass distributions shows the robust features of the reconstruction, i.e., a maximum, a secondary maximum, an overall gradient, and a pronounced minimum; these features can be compared with the light distribution as shown in Fig. 2

4.3 Magnification effects in high-redshift QSO samples

The magnification bias which has been discussed in footnote 3, can affect the number counts of objects, provided the optical depth (or lensing probability) is sufficiently large, and the number counts of these sources are sufficiently steep. Whereas there has been a long debate of estimating the importance of this magnification bias on QSO counts, it now appears that the counts are not dramatically changed by lensing (for references and a detailed discussion, see Sect. 12.5 of SEF). Nevertheless, the fraction of magnified sources in a flux-limited sample can still be appreciable. A sign of a magnification bias could be found if high-redshift QSOs were associated with potential lenses along their lines-of-sight. Such associations have been found: on scales of a few arcseconds, several claims have been made of a detection of a statistically significant overdensity of galaxies around high-redshift QSOs (for references, see Sect. 12.3 of SEF), though the situation is not without controvercies (Wu 1996, and references therein).

Here I want to concentrate on associations on much larger scales: Tyson (1986) and Fugmann (1988, 1989) discovered a statistically significant overdensity of galaxies around high-redshift quasars on an angular scale of about one arcminute (but see Fried 1992 for a negative result). Later, Fugmann (1990) started a series of investigations to search for an overdensity of foreground matter near the lines-of-sight to high-redshift radio quasars from the 1-Jy catalog on scales of ten arcminutes and larger. Indeed, a statistically significant overdensity of galaxies from the Lick catalog (Fugmann 1990, Bartelmann & Schneider 1993), the IRAS catalog (Bartelmann & Schneider 1994, Bartsch, Schneider & Bartelmann 1996), and the APM catalog (Benitez & Martinez-Gonzalez 1995), with clusters from the Zwicky (Seitz & Schneider 1995c) and Abell (Wu & Han 1995) catalogs, and with diffuse X-ray emission from the ROSAT All Sky Survey (Bartelmann, Schneider & Hasinger 1995) were found. Further evidence for large-scale associations has come from other QSO samples (Rodrigues-Williams & Hogan 1994; Hutchings 1995). If these associations were to be explained by a lensing effect, then the lenses cannot be individual galaxies, whose 'lens-scale' is only at most a few arcseconds, but groups, clusters, or even larger-scale structures must be responsible; in fact, such an explanation works at least qualitatively (Bartelmann 1995b). If there are indeed large-scale matter overdensities in the lines-of-sight to these QSOs, they might cause a systematic distortion of background galaxies. This was the motivation for Fort et al. (1995) to image the faint galaxies around several high-redshift 1-Jy QSOs. For several of them, they obtained clear evidence for a coherent shear pattern around these QSOs, which can also be spatially related to local concentrations of faint galaxies. These concentrations may indicate the presence of a group or a cluster, but they are so faint optically that they would not appear in any cluster catalog. What this might suggest is that there exists a population of clusters with a much larger mass-to-light ratio than those clusters which are selected because of their high optical luminosity, i.e., which appear in optically-selected cluster catalogs. If these findings are confirmed (e.g., by HST observations), one has found a way to obtain a *mass-selected sample* of clusters and/or groups.

4.4 Galaxy-galaxy lensing

The shear field around clusters is sufficiently strong to measure their mass distributions – see Sect. 4.2. One can easily show that, assuming an isothermal mass profile, the 'detection efficiency' of a lens scales like σ^4, where σ is the velocity dispersion. This scaling then implies that individual galaxies are too weak for their presence to be detected in their shear field[5], but one should be able to

---

[5] assuming a number density of 50 galaxies/arcmin^2, the minimum velocity dispersion for which a 3-σ detection would be possible is about 350 km/s (Miralda-Escudé 1991, Schneider & Seitz 1995).

detect this effect from a large ensemble of galaxies, if the signals from the individual galaxies are added statistically. The signal one would expect is a slight tangential alignment of background galaxies relative to the direction connecting this background galaxy with a near foreground galaxy.

Tyson et al. (1984) have investigated this effect using $\sim$ 60000 galaxies; they obtained a null result. More recently, Brainerd, Blandford & Smail (1995) have analyzed a deep field; they have divided their galaxy sample into 'foreground' and 'background' galaxies, according to the optical magnitudes, and then studied the angle between the major axis of the background galaxy and the line connecting the background galaxy with the nearest foreground galaxies. The distribution of this angle shows a deficit at small angles, and an excess at large angles, indicating the expected tangential alignment. Since an accurate measurement of image ellipticities from the ground is very difficult, only galaxies brighter than $r = 24$ were used; the effect disappears for fainter galaxies, which most likely shows the effect of the PSF on small images. Brainerd et al. have then simulated data, treating galaxies as truncated isothermal spheres, and distributing them in redshift, and they showed that the effect they observe is in accordance with expectations from their modelling. Recalling that this effect was detected (at a 3-σ level) with 'only' 506 'background' galaxies, it appears that one can use galaxy-galaxy lensing as a tool to investigate statistically the mass distribution in galaxies, since larger samples will become available soon (also, ground-based images with a smaller and/or more stable PSF will allow the use of fainter galaxies). Schneider & Rix (1995) have proposed a maximum likelihood method for the analysis of galaxy-galaxy lensing, which is very sensitive to the characteristic velocity dispersion of the galaxies, and which can also yield significant lower bounds on the halo size of galaxies.

4.5 Lensing by the large-scale structure

The cosmological density fluctuations out of which the structure in the universe has formed (at least in the conventional model of gravitational instability – which has received impressive support from the detection of microwave background fluctuations by COBE) can also distort the images of high-redshift galaxies. The corresponding distortions have been calculated by Blandford et al. (1991) and Kaiser (1992 and references therein), and are expected to be small; nevertheless, depending on the cosmological model, these distortions are measureable in principle, either by averaging the ellipticity of galaxy images over large fields, or by considering the two-point correlation function of galaxy ellipticities on large scales. If such an effect can be measured, it will allow a direct measurement of the power spectrum of the density fluctuations on the appropriate scales, very much like COBE has done. What is important to note is that the power spectrum of the density fluctuations in cosmogonies is normalized either by the amplitude of fluctuations in the microwave background, or by rms variations of galaxy numbers in 'big volumes'. Both of these normalizations are such that *relative density fluctuations* $\delta\rho/\rho$ are normalized. However, the lensing effect depends of $\delta\rho$, and not on the ratio $\delta\rho/\rho$. This implies that the gravitational distortion of images of background galaxies is proportional to the mean cosmic density Ω (Villumsen 1995a).

The same data from which galaxy-galaxy lensing was detected by Brainerd et al. (1995) have been used to search for the 'cosmic shear'; keeping in mind the difficulties to measure accurate ellipticities of very faint images from the ground, it is not surprising that Mould et al. (1994) did not find a statistically significant shear signal on a field of $4\overset{\prime}{.}8$ radius. Using the same data, but a different method for analyzing the image ellipticities (basically, giving less weight to 'small' images, which are most contaminated by the PSF), Villumsen (1995b) obtained a shear signal with a formal 5-σ significance. Further observations are needed to confirm this result; as mentioned before, the observations are very difficult to carry out, and the expected effects are so small that even tiny systematical effects which escape detection can mimic a significant detection.

5 Outlook

Predicting the future is a dangerous business; however, it is easy to foresee that the current developments in observational astronomy will continue to increase the usefulness of gravitational lensing for studying the universe. Concerning strong lensing, new big lens surveys, such as the CLASS survey (see, e.g., Myers et al. 1995), will allow to set much stronger constraints on the density of galaxy-mass objects in the universe. Hopefully, some of the newly discovered multiply-imaged systems will turn out to be useful for determining the Hubble constant. MACHO-type searches for compact objects in our Galaxy will continue and expand, allowing to get stronger constraints on the density of compact objects in our halo, and to measure the mass distribution in the central part of the Galaxy. Concerning weak lensing, we have just scratched the surface. On the observational side, wide-field cameras and imaging with 8m-class telescopes will dramatically increase the rate and quality of data, allowing surveys for dark matter concentrations. The refurbishment of the HST has enabled images of faint galaxies with unprecedented image quality and resolution. These images, together with new theoretical developments, will allow us to understand better the relation between observed image shapes and the true image shapes, before degradation with a PSF. The combination of dark matter maps from weak lensing and X-ray and dynamical studies of clusters will yield fresh insight into the structure, dynamics, and history of these systems. If the systematic effects of ground-based imaging can be understood sufficiently well, we might be able to obtain the cosmic density and the power spectrum of density fluctuations directly from lensing.

I would like to thank M. Bartelmann for carefully reading this manuscript. This work was supported by the "Sonderforschungsbereich 375-95 für Astro–Teilchenphysik" der Deutschen Forschungsgemeinschaft.

References

Alcock, C. et al. 1993, Nat 365, 621.

Alcock, C. et al. 1995, Phys. Rev. Lett. 174, 2867.

Aubourg, E. et al. 1993, Nat 365, 623.

Augusto, P., Wilkinson, P.N. & Browne, I.W.A. 1996, in: *Astrophysical applications of gravitational lensing*, C.S. Kochanek & J.N. Hewitt (eds.), Kluwer: Dordrecht, p. 399.

Bartelmann, M. 1995a, A&A 299, 11.

Bartelmann, M. 1995b, A&A 298, 661.

Bartelmann, M. 1995c, A&A 303, 643.

Bartelmann, M. & Narayan, R. 1995, ApJ 451, 60.

Bartelmann, M., Narayan, R., Seitz, S. & Schneider, P. 1995, ApJ (submitted).

Bartelmann, M. & Schneider, P. 1993, A&A 271, 421.

Bartelmann, M. & Schneider, P. 1994, A&A 284, 1.

Bartelmann, M., Schneider, P. & Hasinger, G. 1994, A&A 290, 399.

Bartelmann, M., Steinmetz, M. & Weiss, A. 1995, A&A 297, 1.

Bartelmann, M. & Weiss, A. 1994, A&A 287, 1.

Bartsch, A., Schneider, P. & Bartelmann, M. 1996, A&A (submitted).

Benitez, N. & Martinez-Gonzalez, E. 1995, ApJ 448, L89.

Bergmann, A.G. & Petrosian, V. 1993, ApJ 413, 18.

Bernstein, G.M., Tyson, J.A. & Kochanek, C.S. 1993, AJ 105, 816.

Blaes, O.M. & Webster, R.L. 1992, ApJ 391, L63.

Blandford, R.D. & Narayan, R. 1992, ARA&A 30, 311.

Blandford, R.D., Saust, A.B., Brainerd, T.G. & Villumsen, J.V. 1991, MNRAS 251, 600.

Bonnet, H. & Mellier, Y. 1995, A&A 303, 331..

Brainerd, T.G., Blandford, R.D. & Smail, I. 1995, ApJ, in press.

Broadhurst, T.J. 1995, in: *Dark matter*, AIP Conf. Proc. 336, eds. S.S. Holt & C.L. Bennett (New York: AIP).

Broadhurst, T.J., Taylor, A.N. & Peacock, J.A. 1995, ApJ 438, 49.

Canizares, C.R. 1982, ApJ 263, 508.

Chen, G.H., Kochanek, C.S. & Hewitt, J.N. 1995, ApJ 447, 62.

Corbett, E.A., Browne, I.W.A. & Wilkinson, P.N. 1996, in: *Astrophysical applications of gravitational lensing*, C.S. Kochanek & J.N. Hewitt (eds.), Kluwer: Dordrecht, p. 37.

Dalcanton, J.J., Canizares, C.R., Granados, A., Steidel, C.C. & Stocke, J.T. 1994, ApJ 424, 550.

Dressler, A., Oemler, A., Butcher, H. & Gunn, J.E. 1994, ApJ 430, 107.

Fahlman, G., Kaiser, N., Squires, G. & Woods, D. 1994, ApJ 437, 56.

Fort, B., Le Fèvre, O., Hammer, F. & Cailloux, M. 1992, ApJ 399, L125.

Fort, B. & Mellier, Y. 1994, A&AR 5, 239.

Fort, B., Mellier, Y., Dantel-Fort, M., Bonnet, H. & Kneib, J.-P. 1995, A&A, in press.

Fort, B., Prieur, J.L., Mathez, G., Mellier, Y. & Soucail, G. 1988, A&A 200, L17.

Fried, J.W. 1992, A&A 254, 39.

Fugmann, W. 1988, A&A 204, 73.

Fugmann, W. 1989, A&A 222, 45.

Fugmann, W. 1990, A&A 240, 11.

Gorenstein, M.V., Falco, E.E. & Shapiro, I.I. 1988, ApJ 327, 693.

Grossman, S.A. & Saha, P. 1994, ApJ 431, 74.

Haarsma, D.B., Hewitt, J.N., Burke, B.F. & Lehár, J. 1996, in: *Astrophysical applications of gravitational lensing*, C.S. Kochanek & J.N. Hewitt (eds.), Kluwer: Dordrecht, p. 43.

Hartwick, F.D.A. & Schade, D. 1990, ARA&A 28, 437.

Hawkins, M.R.S. & Véron, P. 1993, MNRAS 260, 202.

Houde, M. & Racine, R. 1994, AJ 107, 466.

Hutchings, J.B. 1995, AJ 109, 928.

Jaroszyński, M., Wambsganss, J. & Paczyński, B. 1992, ApJ 396, L65.

Kaiser, N. 1992, ApJ 388, 272.

Kaiser, N. 1995a, ApJ 439, L1.

Kaiser, N. 1995b, preprint (astro-ph/9509019).

Kaiser, N. & Squires, G. 1993, ApJ 404, 441.

Kaiser, N., Squires, G. & Broadhurst, T. 1995, ApJ 449, 460.

Kaiser, N., Squires, G., Fahlman, G., Woods, D. & Broadhurst, T. 1995, preprint.

Kassiola, A., Kovner, I. & Blandford, R.D. 1991, ApJ 381, 6.

Kassiola, A., Kovner, I. & Fort, B. 1992, ApJ 400, 41.

Kneib, J.-P., Ellis, R.S., Smail, I., Couch, W.J. & Sharples, R.M. 1995, preprint (astro-ph/9511015).

Kneib, J.-P., Mellier, Y., Fort, B. & Mathez, G. 1993, A&A 273, 367.

Kochanek, C.S. 1993a, ApJ 417, 438.

Kochanek, C.S. 1993b, ApJ 419, 12.

Kochanek, C.S. 1995a, ApJ 445, 559.

Kochanek, C.S. 1995b, preprint (astro-ph/9510077).

Kochanek, C.S., Blandford, R.D., Lawrence, C.R. & Narayan, R. 1989, MNRAS 238, 43.

Kochanek, C.S. & Narayan, R. 1992, ApJ 401, 461.

Luppino, G.A., Gioia, I.M., Hammer, F., Le Fèvre, O. & Annis, J. 1995, ApJ (in press).

Lynds, R. & Petrosian, V. 1986, BAAS 18, 1014.

Maoz, D. & Rix, H.-W. 1993, ApJ 416, 425.

Mellier, Y., Fort, B., Soucail, G., Mathez, G. & Cailloux, M. 1991, ApJ 380, 334.

Mellier, Y., Fort, B. & Kneib, J.-P. 1993, ApJ 407, 33.

Miralda-Escudé, J. 1991, ApJ 370, 1.

Miralda-Escudé, J. 1993, ApJ 403, 497.

Miralda-Escudé, J. & Babul, A. 1995, ApJ 449, 18.

Mould, J., Blandford, R., Villumsen, J., Brainerd, T., Smail, I., Small, T. & Kells, W. 1994, MNRAS 271, 31.

Myers, S.T. et al. 1995, ApJ 447, L5.

Navarro, J.F., Frenk, C.S. & White, S.D.M. 1995, ApJ (submitted).

Paczyński, B. 1986, ApJ 304, 1.

Paczyński, B. 1991, ApJ 371, L63.

Paczyński, B. 1996, ARA&A (in press).

Patnaik, A.R., Browne, I.W.A., King, L.J., Muxlow, T.W.B., Walsh, D. & Wilkinson, P.N. 1993, MNRAS 261, 435.

Patnaik, A.R., Garrett, M.A., Polatidis, A. & Bagri, D. 1996, in: *Astrophysical applications of gravitational lensing*, C.S. Kochanek & J.N. Hewitt (eds.), Kluwer: Dordrecht, p. 405.

Patnaik, A.R., Porcas, R.W. & Browne, I.W.A. 1995, MNRAS 274, L5.

Pelt, J., Hoff, W., Kayser, R., Refsdal, S. & Schramm, T. 1994, A&A 286, 775.

Press, W.H., Rybicki, G.B. & Hewitt, J.N. 1992, ApJ 385, 404.

Rauch, K.P. & Blandford, R.D. 1991, ApJ 381, L39.

Refsdal, S. 1964, MNRAS 128, 307.

Refsdal, S. & Surdej, J. 1994, Rep. Prog. Phys. 56, 117.

Rix, H.-W., Schneider, D.P. & Bahcall, J.N. 1992, AJ 104, 959.

Roberts, D.H., Lehar, J., Hewitt, J.N. & Burke, B.F. 1991, Nat 352, 43.

Rodrigues-Williams, L.L. & Hogan, C.J. 1994, AJ 107, 451.

Schild, R. & Thomson, D.J. 1995, AJ 109, 1970.

Schneider, P. 1993, A&A 279, 1.

Schneider, P. 1995, A&A 302, 639.

Schneider, P., Ehlers, J. & Falco, E.E. 1992, *Gravitational lenses*, Springer: New York (SEF).

Schneider, P. & Rix, H.-W. 1995, preprint.

Schneider, P. & Seitz, C. 1995, A&A 294, 411.

Seitz, C., Kneib, J.-P., Schneider, P. & Seitz, S. 1995, A&A (submitted).

Seitz, C. & Schneider, P. 1995a, A&A 297, 287.

Seitz, C. & Schneider, P. 1996, A&A (submitted).

Seitz, S. & Schneider, P. 1995b, A&A, in press.

Seitz, S. & Schneider, P. 1995c, A&A 302, 9.

Seitz, S., Schneider, P. & Ehlers, J. 1994, Class. Quan. Gravity 11, 2345.

Smail, I., Ellis, R.S. & Fitchett, M.J. 1994, MNRAS 270, 245.

Smail, I., Ellis, R.S., Fitchett, M.J. & Edge, A.C. 1995, MNRAS 273, 277.

Soucail, G., Fort, B., Mellier, Y. & Picat, J.P. 1987, A&A 172, L14.

Soucail, G., Mellier, Y., Fort, B., Mathez, G. & Cailloux, M. 1988, A&A 191, L19.

Tyson, J.A. 1986, AJ 92, 691.

Tyson, J.A. & Fischer, P. 1995, ApJ 446, L55.

Tyson, J.A., Valdes, F., Jarvis, J.F. & Mills Jr., A.P. 1984, ApJ 281, L59.

Tyson, J.A., Valdes, F. & Wenk, R.A. 1990, ApJ 349, L1.

Udalski, A. et al. 1993, Acta Astro. 43, 289.

Vanderriest, C., Schneider, J., Herpe, G., Chevreton, M., Moles, M. & Wlérick, G. 1989, A&A 215, 1.

Villumsen, J.V. 1995a, MNRAS, in press.

Villumsen, J.V. 1995b, MNRAS, submitted.

Wallington, S., Kochanek, C.S. & Koo, D.C. 1995, ApJ 441, 58.

Wallington, S., Kochanek, C.S. & Narayan, R. 1995, preprint (astro-ph/9511137).

Walsh, D., Carswell, R.F. & Weymann, R.J. 1979, Nat 279, 381.

Waxman, E. & Miralda-Escudé, J. 1995, ApJ (in press).

Wu, X.-P. 1996, Fund. Cosm. Phys. (in press).

Wu, X.-P. & Hammer, F. 1993, MNRAS 262, 187.

Wu, X.-P. & Han, J. 1995, MNRAS 272, 705.

Zhao, H.S., Spergel, D.N. & Rich, R.M. 1995, ApJ 440, L13.

III

The Cosmic Microwave Background

10

The Cosmic Microwave Background

Detections of Cosmic Microwave Background Anisotropies at Large and Intermediate Angular Scales: Data Analysis and Experimental Results

Laura Cayón

Lawrence Berkeley Laboratory and Center for Particle Astrophysics 1 Cyclotron Road, build. 50, room 205 Berkeley, CA 94720 USA

Abstract: The first detection of Cosmic Microwave Background (CMB hereinafter) anisotropies was claimed by the COBE/DMR team in April 1992. Since then, eight groups have published detections: Saskatoon, Tenerife, ACME/SP94, IAB, Python, ACME/MAX3-4, MSAM and Argo. It is the aim of this contribution to review the experimental setups, data analyses and results of these experiments. The COBE/DMR experiment and the implications of its observations are considered in detail.

1 Designing an Experiment to Measure CMB Anisotropies

An experiment built to measure CMB anisotropies is basically a radiotelescope (radiometer) consisting of an antenna, a receiver and a data collecting system. The power reaching the receiver is directly related to the antenna temperature T_A. The temperature detected by the antenna is determined by its effective area A (the effective area of an antenna is usually defined by the Full Width at Half Maximum (FWHM) of its gaussian fit) and the intensity of the source (I_ν):

$$T_A(\nu) \propto \int \int d\Omega I_\nu A(\Omega) , \qquad (1)$$

where ν refers to the working frequency and the integral is evaluated over the solid angle viewed by the antenna. The CMB radiation detected by us is emitted in different directions from the last scatering surface, and not from a point source. It is therefore very important to control all the possible contaminants that could affect the CMB data. Lets start this section by describing all the possible sources that can contribute to the observations (Sect. 1.1). These contaminants are taken into account in the final design of an experiment. The experimental setups commonly used by the existing CMB anisotropy experiments are described in Sect. 1.2.

1.1 Contaminants

Radiation coming from the **ground** and from **astronomical sources** (moon, sun, etc) can contaminate data taken at microwave wavelengths. In order to avoid this, first of all, it is important to design antennas with very low sidelobes so that no radiation outside the effective area can enter the telescope. Moreover, a shield is normally build up around the experiment in order to protect it from any radiation coming from these sources.

Ground-based and balloon experiments collect data that are affected by **atmospheric emission.** As can be seen in Fig. 1, although the atmospheric contribution is reduced at high balloon altitudes ($\approx$ 30 km), it is at best approximately two orders of magnitude larger than CMB anisotropy level ($\sim 10^{-5}$). Working frequencies are chosen among those that are less affected by the atmospheric emission. Double beam switching or four beam techniques (see below) have been developed to reduce this contamination.

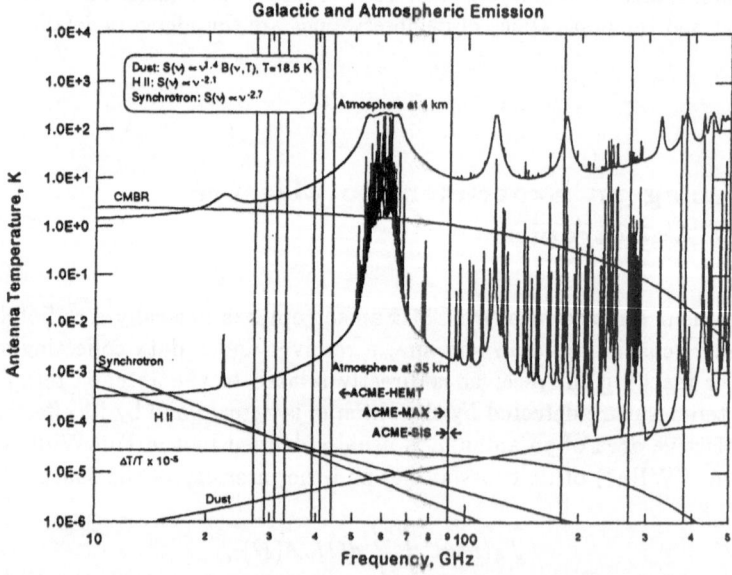

Fig. 1. Contribution of different contaminants to the antenna temperature as a function of the working frequency. As an example, working frequencies of some of the ACME experiments are indicated in the figure.

The area sampled by an experiment could include some **extragalactic foreground sources** that may produce observable signal. Catalogs are used to check the possible existence of this type of sources, as well as to extrapolate the amplitude that a certain experiment could be detecting in case they are in the field of view.

Among all the possible contaminants that can be affecting the CMB anisotropy data, the radiation coming from the **Galaxy** is the main concern for anyone involved in the design of these experiments or in the data analysis process. The sources of this contribution are synchrotron and free-free emissions and dust. Synchrotron radiation is produced by relativistic electrons moving in the magnetic field of the Galaxy. It is characterized by an intensity that varies with frequency as ν^β with $\beta \sim -0.7$ (spectral index). Free-free emission is produced by interactions of electrons with ions. The spectral index that characterizes this is $\beta \sim -0.15$. The dust present in the Galaxy is found to be at temperatures ~ 20.4 K (warm) and ~ 4.77 K (cold) and it is characterized by an spectral index $\beta \sim 2$. Contributions from these processes at different wavelengths are presented in Fig. 1. As the amplitudes of these three sources change with wavelength, a way to determine the amount of contamination in the observed data is to take measurements at different frequencies. Unfortunately, there are not many observations of these processes at several wavelengths. To establish the possible amplitude of synchrotron and free-free emissions, in a certain set of CMB anisotropy data, extrapolations are taken from sky surveys at 408, 820 and 1420 MHz. Free-free contributions are also extrapolated from H_α observations. Possible dust contribution have been extrapolated from the IRAS map at 100 μm.

1.2 Experimental Setups

In this subsection we briefly describe the different types of telescopes and receivers used by the CMB anisotropy experiments. For detailed technical information one can see Marton 1976, Christiansen and Hogbom 1985 or any of the references indicated in Sect. 2 for the different experiments. An explanation of techniques used to measure CMB anisotropies is included at the end.

.-**Telescopes** CMB anisotropy experiments require measurements of fluctuations that are $\leq 10^{-5}$. Absolute measurements of the temperature of the CMB radiation would require a very high sensitivity. The technique normally used is based on differences of the temperature observed in two or more positions of the sky (as we will see at the end of this subsection). Therefore, telescopes are designed so chopping can be performed without changing the orientation.

A very common design consists if an aluminium chopping flat that reflects the CMB photons to a parabolic mirror at who's focus the receiver is located. This is the type of telescope used, for example, in the Python experiment. A very similar design is used by the Tenerife experiment where the mirror is substituted by horn antennas with the receiver located behind them. Experiments such as Argo or MSAM use Cassegrain telescopes with a chopping secondary mirror. Gregorian telescopes, also including a chopping secondary mirror, are used by the ACME experiments. In both types of telescopes the receiver is located behind the primary (avoiding direct pointing to possible sources of contamination like the earth). In Cassegrain telescopes the secondary is a convex hyperbolic mirror situated in front of the primary's focus. The secondary of a Gregorian telescope is a concave mirror located behind the focus of the primary.

.-Receivers The main elements that comprise a receiver are a filter (or filters) used to select bandwidth at which the observations are performed, a detector, amplifiers and a data acquisition system. The detector used is a function of the working frequency. Observations at high frequency (≥ 100 GHz) use bolometers (MSAM, Python, etc) while low frequency observations are performed using other types of detectors as square-law detectors, diodes, etc. In the case of systems operating at low frequencies two types of configurations are considered. In the case of the ACME/SP and Saskatoon experiments, the receiver is designed including the following elements: Inmediately after the filters, a mixer (SIS-Superconductor-Insulator-Superconductor) is used to reduce the input frequency. HEMT (High Electron Mobility Transistor) amplifiers sensitive to low frequencies are used to amplify the signal that would be sent into the detector. In the Tenerife and COBE/DMR experiments Dicke switch elements are introduced at the begining to perform differences with a reference signal. After the difference is detected a synchronous demodulator would give the output value of the signal.

.-Observational Strategies As indicated above CMB anisotropy measurements are normally performed by taking differences of the CMB temperature at two or more positions on the sky. Different strategies together with the effective area of the antenna, determine the scales at which an experiment would be sensitive. To understand this, lets consider the expansion of the CMB temperature in spherical harmonics:

$$T(\boldsymbol{n}, \sigma_b) = \sum_l \sum_{m=-l}^{l} a_{lm} Y_{lm} e^{-\frac{\sigma_b^2}{2}(l+1/2)^2} , \qquad (2)$$

where $\boldsymbol{n}$ is the unit vector in the direction of observation, σ_b is the dispersion of the beam related to the $FWHM$ by $\sigma_b \simeq 0.425 FWHM$, the a_{lm} refer to multipole coefficients and the Y_{lm} are spherical harmonics. The RMS value of the temperature fluctuations averaged over the sky is given by

$$\Delta T_{RMS} = < \Delta T(\boldsymbol{n}, \sigma_b)^2 >_{sky} = \sum_{l=1} \frac{(2l+1)}{4\pi} < |a_{lm}|^2 >_{sky} W_l , \qquad (3)$$

where the window function $W_l = e^{-\sigma_b^2(l+1/2)^2}$ determines the range of multipoles (l values) at which an experiment is sensitive. However, this would be the case when the experiment is taking absolute measurements of temperature of CMB photons (this is only the case of the COBE/DMR data in which after taking data as differences between two directions separated in the sky sixty degrees, a sky map of absolute temperatures is constructed, see Sect. 4).

Single beam switching techniques perform differences in between two directions $\boldsymbol{n}_1, \boldsymbol{n}_2$ separated a certain angle $\alpha = cos^{-1}(\boldsymbol{n}_1\boldsymbol{n}_2)$. In this case, the RMS value would be:

$$\Delta T_{RMS} = < (\Delta T(\boldsymbol{n}_1, \sigma_b) - \Delta T(\boldsymbol{n}_2, \sigma_b))^2 > =$$

$$\sum_{l=1} \frac{(2l+1)}{4\pi} < |a_{lm}|^2 >_{sky} 2(1 - P_l(\cos\alpha)) e^{-\sigma_b^2(l+1/2)^2}, \qquad (4)$$

where the window function is given by $W_l = 2(1 - P_l(\cos\alpha)) e^{-\sigma_b^2(l+1/2)^2}$ and P_l are the Legendre polinomials.

In order to reduce atmospheric and terrestial contributions, double beam switching and four beam difference techniques are used by some experiments (Tenerife and Python respectively). In the former case, the central temperature is compared with the temperatures of two positions symmetrically located on the right and the left:

$$\Delta T_{RMS} = < (0.5[(\Delta T(\boldsymbol{n}_1, \sigma_b) - \Delta T(\boldsymbol{n}_2, \sigma_b)) - (\Delta T(\boldsymbol{n}_3, \sigma_b) - \Delta T(\boldsymbol{n}_1, \sigma_b))])^2 > =$$

$$\sum_{l=1} \frac{(2l+1)}{4\pi} < |a_{lm}|^2 >_{sky} 0.5(3 - 4P_l(\cos\alpha) + P_l(\cos2\alpha)) e^{-\sigma_b^2(l+1/2)^2}, \qquad (5)$$

where $\boldsymbol{n}_1\boldsymbol{n}_2 = \boldsymbol{n}_1\boldsymbol{n}_3 = \cos\alpha$. Four beam differences are considered by calculating $-0.25T(\boldsymbol{n}_1) + 0.75T(\boldsymbol{n}_2) - 0.75T(\boldsymbol{n}_3) + 0.25T(\boldsymbol{n}_4)$ where $\boldsymbol{n}_1\boldsymbol{n}_2 = \boldsymbol{n}_2\boldsymbol{n}_3 = \boldsymbol{n}_3\boldsymbol{n}_4 = \cos\alpha$. The window function introduced by this technique would be

$$W_l = [5/4 - 15/8 P_l(\cos\alpha) + 3/4 P_l(\cos2\alpha) - 1/8 P_l(\cos3\alpha)] e^{-\sigma_b^2(l+1/2)^2}. \qquad (6)$$

The window functions of the CMB anisotropy experiments, considered in this contribution, are plotted in Fig. 2. As can be seen from this figure, the only experiment sensitive to very low multipoles is the COBE/DMR experiment. Difference techniques filter out low and very high multipoles as it occurs for the rest of the experiments. The angular scale θ subtended by a proper length λ_o at present depends on the density parameter of the Universe Ω_o as well as on the Hubble parameter $H_o = 100h^{-1}$ km/sec/Mpc, $\theta = 34''.4(\Omega_o h)(\lambda_o/1\text{Mpc})$. The Hubble radius at decoupling $z \sim 1000$ subtends an angle of $\sim 1°$. This angular scale corresponds to $\lambda_o \sim 100\text{Mpc}$ and multipoles of the order of 60. Experiments working at angular scales above or approximately 1 degree would be observing CMB temperature anisotropies that are directly connected to the primordial matter density fluctuations. Temperature anisotropies at smaller scales would be affected by physical processes that took place during recombination.

2 Data Analysis

The aim of this section to explain how the results of the CMB anisotropy experiments are presented and discuss the main assumptions in the data analyses performed. Up to approximately two years ago or so, most of the experimental results were given in terms of the RMS value of CMB temperature fluctuations observed, applying maximum likelihood techniques (Lasenby and Davies 1988, Martin 1971) (see Sect. 2.1). The need for comparing detections obtained by

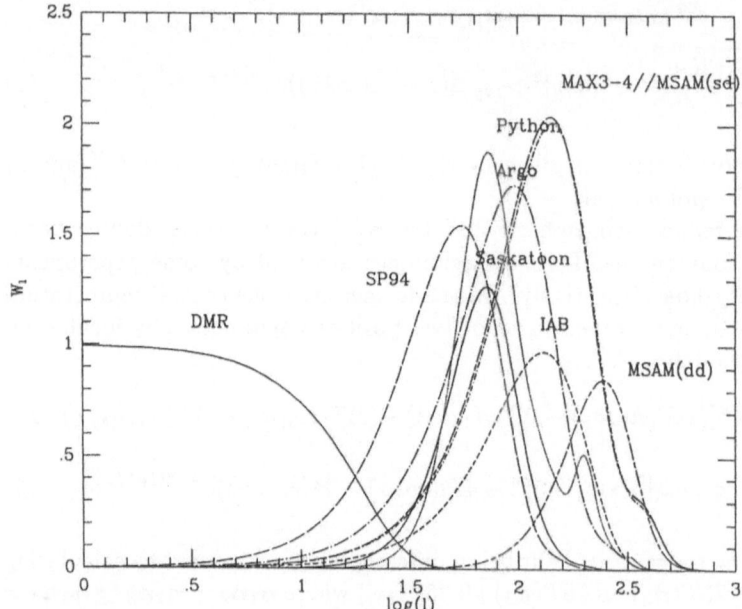

Fig. 2. Window functions of the experiments that have claimed the detections of CMB anisotropies

different experiments working at different angular scales initiated the practice where in experimental results are quoted by their band powers (Bond 1995) (see Sect. 2.2).

2.1 Maximum Likelihood

Given a set of data $\boldsymbol{\Delta T} \equiv (\Delta T_1, \Delta T_2, ..., \Delta T_{nd})$, where ΔT_i refers to the observed CMB temperature fluctuations at nd different directions $\boldsymbol{n_i}$. Assuming that CMB temperature fluctuations are gaussian distributed, the probability of measuring such data set if produced by a cosmological signal of dispersion σ_s is given by the likelihood function

$$L(\boldsymbol{\Delta T}/\sigma_s) \propto |V|^{-0.5} exp(-0.5 X^T V^{-1} X) , \qquad (7)$$

where $X \equiv \boldsymbol{\Delta T}$ and V is the covariance matrix. The elements of the covariance matrix are built taking into account the experimental noise that would appear in the observations. Modeling the noise by a Gaussian of dispersion σ_i and assuming that it is independent from one position to another, V would be given by $V \equiv N + M$. N is the noise covariance matrix of elements $(N)_{ij} = \sigma_i^2$ if $i = j$ and 0 otherwise. The elements of the "model" matrix M are given by the theoretical correlation of CMB temperature fluctuations $(M)_{ij} = C(\theta, \sigma_b, \sigma_s)$, where $\boldsymbol{n_i n_j} = cos\theta$ and σ_b refers to the beam dispersion (as defined above). This theoretical correlation function is very often assumed to be gaussian, parametrized by the amplitude C_o (related to σ_s) and the coherence angle $\theta_c = (-C(0,\sigma_b)/C''(0,\sigma_b))^{0.5}$:

$$C(\theta,\sigma_b,C_o) = \frac{C_o\theta_c^2}{2\sigma_b^2+\theta_c^2}exp(-\theta/2(2\sigma_b^2+\theta_c^2)) \ . \tag{8}$$

Another functional form assumed for the theoretical correlation depends on the amplitude A and the spectral index n of the matter power spectrum. At large angular scales the temperature fluctuations are directly related to the matter density fluctuations at last scattering surface (Sachs and Wolfe 1967). If these are assumed to be gaussian with mean value zero (as predicted by Inflation), it is enough to define the second order moment to characterize them. If δ_k is the Fourier transform of the matter density fluctuation δ, the second order moment is given by the matter power spectrum $P(k) = < |\delta - k| > = Ak^n$. In this case, the theoretical correlation is given by:

$$C(\theta,\sigma_b) = \sum_{l=2}(2l+1)f(A,n)P_l(cos\theta)W_l \ , \tag{9}$$

where P_l are the Legendre polinomials, W_l is the window function introduced in the previous section and the dependence on the matter power spectrum goes as:

$$f(A,n) = 2^{n-1}A\frac{\Gamma(3-n)\Gamma(l+(n-1)/2)}{\Gamma(2-n/2)^2\Gamma(l+(5-n)/2)} \ . \tag{10}$$

One can also refer to the quadrupole of the radiation power spectrum Q_{rms-PS} as related to the amplitude of the matter power spectrum by:

$$A = \frac{4\pi}{2^{n-1}5}Q_{rms-PS}^2\frac{\Gamma((9-n)/2)\Gamma(2-n/2)^2}{\Gamma((3+n)/2)\Gamma(3-n)} \ . \tag{11}$$

Therefore, the result obtained from a certain data set, would be quoted by the values of (C_o,θ_c) or (Q_{rms-PS},n), depending on the assumed theoretical correlation function, at which the likelihood function has a maximum. A detection occurs when the likelihood function presents a significant maximum. Otherwise, the values of the parameters are just upper limits of the cosmological signal. The confidence interval of the detected signal is estimated from the probability of having the detected value of the cosmological signal considering the observational data set $P(\sigma_s/\Delta T) \propto L(\Delta T/\sigma_s)P(\sigma_s)$, where $P(\sigma_s)$ is a prior probability (Baye's formula). Assuming a uniform prior probability, the confidence interval is directly obtained from the likelihood function.

2.2 Band Power

The RMS value of the CMB temperature fluctuations is given by (3), where the dispersion of the multipole coefficients $< |a_{lm}|^2 >$, averaged over the whole statistical ensemble, can be theoretically calculated for different cosmological models. The radiation power spectrum is given by $l(l+1) < |a_{lm}|^2 > /2\pi = l(l+1)C_l/2\pi$. In order to compare different experiental results with a certain theory one can assume $< C_l >_{bp} \equiv l(l+1)C_l/2\pi$ to be constant over the range of multipoles at which a certain experiment is more sensitive. This is called the band power and it is related to the experimental measured RMS value ΔT_{RMS} by

$$< C_l >_{bp} = \frac{(\Delta T_{RMS})^2}{\sum_{l=l_{min}}^{l_{max}} (2l+1)W_l/2l(2l+1)} , \qquad (12)$$

where (l_{mim}, l_{max}) indicate the range of multipoles viewed by a considered experiment. To compare with theory it would be enough to substitute the RMS in (12) by the expression given in (3).

3 Detections: Characteristics and Results of CMB Anisotropy Experiments

As previously mentioned, the detection of CMB anisotropies has been claimed by eight groups up to now.

Five experiments are mounted on the **ground**: Tenerife, Saskatoon, Python, IAB and ACME/SP94. For details about the Tenerife experiment see the contribution by R. Rebolo in this proceedings. **Saskatoon** is an experiment built by a group from Princeton University that has taken data twice. It takes its name from the observing site, Saskatoon in Canada. The first set of observations were presented in Wollack et al 1993. The analysis and observations of the data taken in 1994 can be found in Netterfield et al 1995. **Python** does not have any other meaning than "a large, nonvenomous Old World snake that coils around and suffocates its prey", as defined by its designers. This experiment has taken data twice from the Amudsen-Scott South Pole Station in Antarctica. Both set of observations are presented in Dragovan et al 1994 and Ruhl et al 1995. **IAB** is a ground based experiment that collected data from 1991 December 30 to 1992 January 16, from the Italian Antartic Base. These observations were presented in Piccirillo and Calisse 1993. The **ACME** (Advanced Cosmic Microwave Explorer) experiment was build by P. Lubin's group at the University of California, Santa Barbara, in collaboration with R. Wilson's group at Bell Laboratories. The **ACME/SP94** experiment refers to the observations taken by this experiment from the South Pole during the 1993-1994 austral summer (Gundersen et al 1995).

Four detections have been claimed from data taken from **balloons**: ACME/MAX3, ACME/MAX4, Argo and MSAM. The **ACME/MAX** (MAX stands for Millimeter-wave Anisotropy eXperiment) experiment is a collaboration between P. Lubin's group and the Berkeley group (P. Richards/A. Lange). Unlike the ACME/SP experiment, this experiment uses bolometric detectors. This experiment has been flown already five times. Detections have been claimed from the data taken in January 1991, **MAX3** (Gundersen et al 1993) and in January 1993, **MAX4** (Devlin et al 1994, Clapp et al 1994). At the time this proceedings were written, no public announcement has been made in relation to MAX5 flight. **Argo** is an Italian experiment developed by F. Melchiorri's group in Rome (De Bernardis et al 1994). The analysis of the observations taken by the **MSAM** (Medium Scale Anisotropy Measurement) experiment in June 1992 were presented by Cheng et al 1994. A confirmation of the detection of anisotropies has come through the analysis of the data taken during the second flight of this experiment (Cheng et al 1995).

The **COBE/DMR** (COsmic Background Explorer/Differencial Microwave Radiometer) experiment is a NASA experiment that was taking data from the COBE **satellite** from November 1989 to January 1994. A detailed discussion of this experiment is given in Sect. 4.

Brief descriptions of the experimental characteristics and of the data of these experiments are given in the following subsections.

3.1 Setups

– **Saskatoon:** Off-axis parabola coupled to an aluminium flat oscillating around a vertical axis ($FWHM \approx 1°.4$). Observational strategy: Double beam switching ($\alpha = 2°.5$). Receiver: Radiometer using HEMT amplifiers. Working frequencies: 1993 observations were taken in three channels working at 26-29, 29-32 and 32-35 GHz. 1994 observations taken in six channels working at $26 - 29, 29 - 32, 32 - 36$ (Ka band), $36 - 39.5, 39.5 - 43$ and $43 - 46$ (Q band) GHz.
– **Python:** Off-axis primary coupled to a vertical switching flat ($FWHM = 0°.75$). Observational strategy: Four differences ($\alpha = 2°.75$). Receiver: Bolometer. Working frequencies: Four channels working at 90 GHz.
– **IAB:** Off-axis chopping parabolic primary coupled to an off-axis hyperbolic secondary ($FWHM = 50'$). Observational strategy: Single beam switching ($\alpha = 1°$). Receiver: Bolometer. Working frequencies: 150 GHz.
– **ACME/SP94:** Off-axis Gregorian telescope with a chopping secondary ($FWHM = (1/0.425)(0°.7 \pm 0°.04(27.7/\nu GHz))$ for Ka band and $FWHM = (1/0.425)(0°.47 \pm 0°.04(41.5/\nu GHz))$ for Q band. Observational strategy: Single beam switching ($\alpha = 3°$). Receiver: Radiometer with HEMT amplifier. Working frequencies: $27.25, 29.75, 32.25, 34.75$ (Ka band), $39.15, 41.45, 43.75$ (Q band) GHz.
– **ACME/MAX3:** Off-axis Gregorian telescope with a chopping secondary ($FWHM = 0°.5$). Observational strategy: Single beam switching ($\alpha = 1°.3$). Receiver: Bolometer. Working frequencies: 180, 270 and 360 GHz.
– **ACME/MAX4:** Off-axis Gregorian telescope with a chopping secondary ($FWHM = 0°.55 \pm 0°.05$ at 105 GHz and $FWHM = 0°.75 \pm 0°.05$ for the

other frequencies). Observational strategy: Single beam switching ($\alpha = 1°.3$). Receiver: Bolometer. Working frequencies: 105, 180, 270 and 360 GHz.

– **Argo:** On-axis Gregorian telescope with a chopping secondary ($FWHM = 52'$). Observational strategy: Single beam switching ($\alpha = 1°.8$). Receiver: Bolometer. Working frequencies: 150, 250, 375 and 600 GHz.

– **MSAM:** Off-axis Cassegrain telescope with a chopping secondary ($FWHM = 30'$). Observational strategy: Single beam switching ($\alpha = 80'$) and double beam switching ($\alpha = 40'$). Receiver: Bolometer. Working frequencies: 168, 270, 495 and 675 GHz.

– **COBE/DMR:** Corrugated horn antennas ($FWHM = 7°$) separated $60°$ on the sky. Observational strategy: Differencies projected into a temperature sky map. Receiver: Radiometer with HEMT amplifier. Working frequencies: 31.5, 53 and 90 GHz.

3.2 Sampled Area of the Sky

– **Saskatoon:** Fixed declination $\delta = 85°.1$, right ascension 0-25 hours.

– **Python:** Two scans separated $2°.75$ in elevation and with seven positions separated $2°.75$ in azimuth, each. Survey centered at right ascension 23.32 hours, declination $-49°.5$.

– **IAB:** Fixed declination $-85°$ and right ascensions $0.25, 2.25, 4.25, ..., 22.25$ hours. Field at 11.25 hours is not considered in the analysis (coinciding with ^3He recycling process).

– **ACME/SP94:** Fixed declination $-62°$, right ascension $23 - 67$ degrees.

– **ACME/MAX3:** Fixed declination $72°24'$ (epoch 1991), right ascension $224 - 242$ degrees. Sampled field near Gamma Ursae Minoris.

– **ACME/MAX4:** Three fields sampled centered at Near Gamma Ursae Minoris (declination $71°50'$, right ascension $15^h20^m.7$), Sigma Hercules (declination $42°46'$, right ascension 16^h30^m and Iota Draconis (declination $59°36'$, right ascension 15^h25^m). Sampling $\pm 3°$ in azimuth from the center.

– **Argo:** Centered at Hercules region smpling 63 positions separated from each other $1°.4$ in azimuth.

– **MSAM:** Two strips at fixed declination $82°$, from right ascension $14^h.44$ to $16^h.89$ and from $17^h.18$ to $20^h.33$.

– **COBE/DMR:** Full sky coverage.

3.3 Results

– **Saskatoon:** Maximum likelihood (bayesian) analysis assuming Gaussian correlation: Data taken in 1993 $C_o^{1/2} = 41^{+16}_{-13}\mu K$ at $\theta_c = 1°.2$ and spectral index $\beta = -0.3^{+0.7}_{-1.2}$, data taken in 1994 $C_o^{1/2} = 60^{+14}_{-11}\mu K$ at $\theta_c = 1°$.

– **Python:** Maximum likelihood (bayesian) analysis assuming Gaussian correlation: $C_o^{1/2} \approx 95\mu K$ at $\theta_c = 1°$.

– **IAB:** Maximum likelihood (bayesian) analysis assuming Gaussian correlation: $C_o^{1/2} \approx 122\mu K$ at $\theta_c = 40'$.

– **ACME/SP94:** Maximum likelihood analysis assuming flat power spectrum: $RMS = 42.0^{+15.8}_{-6.8}\mu$K.

– **ACME/MAX3:** Maximum likelihood (bayesian) analysis assuming Gaussian correlation: $C_o^{1/2} \approx 115^{+46}_{-30}\mu$K (95%) at $\theta_c = 25'$.

– **ACME/MAX4:** Maximum likelihood (bayesian) analysis assuming Gaussian correlation at $\theta_c = 25'$ (95%) for the three scans centered at: Gamma Ursae Minoris $87^{+98}_{-38}\mu$K, Sigma Hercules $52^{+82}_{-37}\mu$K and Iota Draconis $83^{+94}_{-56}\mu$K.

– **Argo:** Maximum likelihood (bayesian) analysis assuming Gaussian correlation: $38 < C_o^{1/2} < 82\mu$K (90%) at $\theta_c = 30'$.

– **MSAM:** Maximum likelihood (bayesian) analysis assuming Gaussian correlation: $16 < C_o^{1/2} < 60\mu$K (90%) at $\theta_c = 30'$ for single differences and $30 < C_o^{1/2} < 84\mu$K (90%) at $\theta_c = 0°.3$ for single differences.

– **COBE/DMR:** $RMS = 44.4 \pm 7.5\mu$K analysing data taken during the first two years of observations. Other results are presented in the section dedicated to the COBE/DMR experiment.

As indicated above, since the final aim of measuring CMB anisotropies is to determine the cosmological model that gives the best fit to the data, one way to present the experimental results is by their band powers. In Fig. 3 one can see the band power of the experiments that have claimed the detection of anisotropies. The situation is not very clear at present and new measurements at around $l \sim 200$ would tell us much more (see the contribution by W. Hu) in the (near ?) future.

4 COBE/DMR

April 1992 was a very exciting month for anyone interested in cosmology, specially for all the groups involved in CMB research. This was the time when the first detection of CMB anisotropies was claimed by the COBE/DMR group (Smoot et al 1992). The COBE (Cosmic Background Explorer) satellite was build at the NASA's Goddard Space Flight Center and it was launched by a Delta rocket in November 18, 1989. Three different instruments are on board of this satellite: DIRBE (Diffuse Infrared Background Experiment), FIRAS (Far InfraRed Absolute Spectrophotometer) and DMR (Differential Microwave Radiometer). They cover the wavelength range in between 1 μm and 1 cm. The satellite is orbiting around the earth at an altitude of 900 km and spinning around its axis at 0.8 r.p.m. The sky coverage is complete in approximately five months.

The COBE/DIRBE experiment was dedicated to study the infrared radiation coming from stars, interestellar dust and the background. It was working at wavelengths in between 1 μm and 0.03 cm. Characteristics of the Galactic dust are presented in Berriman et al 1994. Analysis of COBE/DIRBE data has also provided information about the stellar population and the main properties of the disk and the bulge of the Galaxy (Arendt et al 1994, Weiland et al 1994).

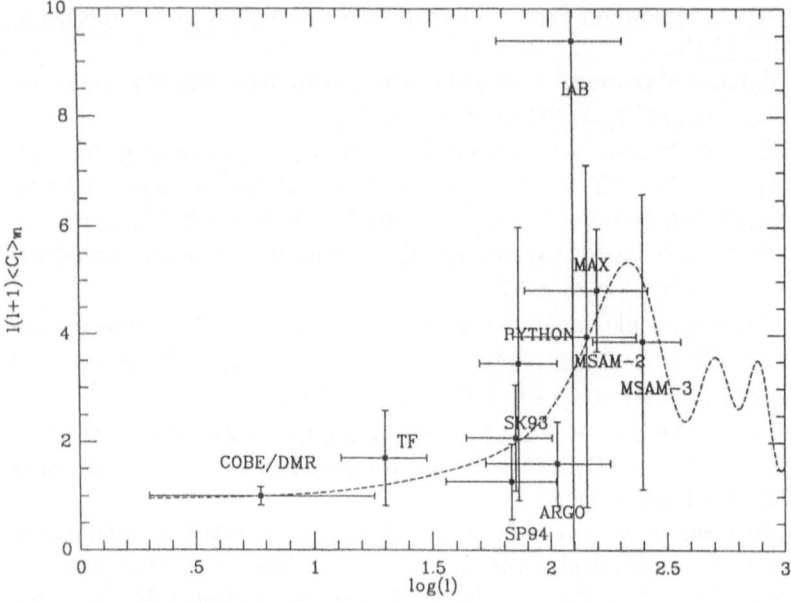

Fig. 3. Band powers of the experiments that have claimed the detection of CMB anisotropies. For comparison, the predictions of the Cold Dark Matter model are indicated by the solid line

The results from the COBE/FIRAS experiment have confirmed the black body nature of the CMB radiation spectrum. The value of the temperature of the CMB radiation at the 95% confidence level is 2.726 ± 0.01 K (Fixen et al 94, Mather et al 94). Moreover, distortions of the CMB spectrum provide information about energy release processes that could have happenned above redshifts of $\sim 10^3$. Any energy release taking place between $z = 3 \times 10^6$ and $z = 10^5$ would produce a Bose-Einstein distortion characterized by the μ parameter. If the energy release occurs at $z < 10^5$ there would be no conservation of the kinetic equilibrium and a Compton distortion would appear in the spectrum. This distortion is characterized by the y parameter. Analysis of the COBE/FIRAS data has provided strong upper limits for these two distortions: $\mu < 3.3 \times 10^{-4}$ and $y < 2.5 \times 10^{-5}$ at the 95% confidence level. These results have been compared with the ones obtained by other experiments (Wright et al 1994a) finding a good agreement among all of them. The range of wavelengths for this experiment was from 0.01 to 1 cm. Interesting results have been obtained about the type of dust present in our Galaxy (Reach et al 1995) and for the first time a detection of the Infrared background has been claimed (Puget et al. 1995).

The rest of this section focusses on the COBE/DMR experiment and what we have learned from its observations. Although the COBE satellite is still orbiting around the earth, the COBE/DMR experiment stopped working in January 1994. At present, the data analysis of the four years of observations is being performed by the COBE group. Results are expected to be released together with the data by the end of 1995. Analyses performed on the first two years of observations are presented in the next subsections. Sect. 4.1 presents some experimental characteristics and how the construction of sky maps is performed. The results obtained from a direct analysis of the COBE/DMR sky maps are shown in Sect. 4.2. What the COBE/DMR data have shown about different cosmological models is presented in Sect. 4.3. Sect. 4.4. and Sect. 4.5. are dedicated to the works that have been done on identifying individual features in the sky maps and on comparing with other large scale experimental results, respectively.

4.1 COBE/DMR Sky Maps

The COBE/DMR experiment consists of three differential microwave radiometers working at 31, 53 and 90 GHz. Each of them has two channels that provide data of CMB temperature differences taken in two directions of the sky (averaged over the antenna beam width $FWHM = 7°$) separated 60 degrees. The receiver consists of basically a Dicke switch, a square law detector, a synchronous filter and an integrator that provides a differential power each 0.5 seconds. Therefore, the data aquiered from the radiometers are temperature differences. The construction of the absolute temperature sky maps is a complicated process that, very briefly explained, consists of the following (for a detailed explanation see Lineweaver 1994): Each differential measurement $S(m)$ can be written as (Kogut tel al 1992, Bennett et al 1994)

$$S(m) = (1/G(m))(\Delta T(m) + \Delta(m) + O(m) + \sum_k W_k(m)) , \qquad (13)$$

where m refers to a certain measurement taken by one of the channels of one of the radiometers. G is the instrumental gain. Its value is determined during the calibration process. The radiometers are calibrated each two hours by reference to an internal known noise source. Other calibrations are performed by pointing to the Moon and by comparing with the dipole produced by the movement of the satellite (Earth) around the Sun. ΔT refers to the real cosmological temperature difference. The instrumental noise is represented by Δ. O is the offset that is removed by a fit to the baseline appearing in the data. The sum $\sum_k W_k$ accounts for all the known systematic effects as internal effects produced by thermal and voltage changes, magnetic changes in the ferrite switch or the effects produced by the presence of the FIRAS and DIRBE instruments. There are also external known systematic effects caused by microwave emission from the Sun and the Earth (possible contaminations entering the shield that sorrounds the three instruments on board of the COBE satellite), interferences caused by other satellites as well as emission from the Moon and planets. Therefore, after calibration, baseline subtraction and correction of systematic effects, one is left with the "true" array of temperature differences and the instrumental noise. Taking into account the directions at which each of the temperature differences were observed, on can minimize the following Chi-square to determine the value of the absolute temperature at different directions on the sky T:

$$\chi^2 = \sum_{m=1}^{m_{tot}} \frac{(D(m) - \Delta T(m))^2}{\sigma_m^2} , \tag{14}$$

where m_{tot} would be the total number of measurements taken by one of the channels (denotted by A or B) of a certain radiometer (31, 53 or 90 GHz). The array D refers to the data differences after calibration, offset and systematic errors corrections. This m_{tot} array can be written as a function of the variables $\boldsymbol{T}$ that one wants to determine, $\boldsymbol{D} = [V]\boldsymbol{T} + \boldsymbol{\Delta}$. The matrix $[V]$ consists of zeros and ones which are ordered in a manner, so as to take into account the corresponding positions of the observed measurements. If the celestial sphere is pixelized in N_{pix} with an associated direction each of them, $\boldsymbol{T}$ is a N_{pix} dimension vector and $[V]$ is a $m_{tot} \times N_{pix}$ matrix. Finally, σ_m is the disperison of the instrumental noise corresponding to the m observation. Solving (14) encounters certain problems that can affect the final solutions $\boldsymbol{T}$. Inversion of large matrices is solved by iteractive methods inducing problems in the convergence of the solution. Noise affecting different pixels is considered independent from each other. This fact has been probed to the first order in Lineweaver et al 1995. The antenna pointing adds error through the whole process and it is also known that the amplifier introduces correlations between consequtive outputs.

The next step is to get a two dimensional projection of the celestial sphere. If a cube is located inside a sphere, one can project each point of the sphere on a pixel on one of the six faces of the cube. Deformated pixels on the sphere would correspond to equal area pixels on the surface of the cube. Therefore, if we want to have a projection with 1024 pixels/cubic face, taking into account that a sphere covers 4π strad, one would have 6144 pixels of area $\sim 2°.6 \times 2°.6$ per pixel.

Solving (14) one gets the best fit absolute temperature at each pixel. Experimental noise should also appear at each pixel. The fact that each radiometer observes two independent channels is used to get the value of the instrumental noise. The RMS value of the differences between data taken by two channels at the same frequency (dispersion of the $(A - B)/2$ map) corresponds to the dispersion of the noise observed in the $(A + B)/2$ map. Considering the number of observations N_{obs} that have been performed at each pixel i, the dispersion of the noise associated to this pixel would be $RMS^0_{(A-B)/2}.5/N_{obs}(i)^0.5$.

4.2 Getting Information from the COBE/DMR Sky Maps

The COBE/DMR experiment works at angular scales of $\sim 7°$ that correpond to proper lengths larger than the Hubble radius at decoupling time. At these scales, CMB temperature anisotropies are given by

$$\Delta T(\boldsymbol{n})/T_o \equiv \frac{T(\boldsymbol{n}) - T_o}{T_o} = \frac{\boldsymbol{nv}}{c} - \delta\phi_{LS} + \frac{1}{4}\delta_{\gamma LS} , \tag{15}$$

where T_o is the absolute temperature of the CMB radiation at present. The first term on the right hand side of (15) is the Doppler effect produced by the movement of the observer $\boldsymbol{v}$ relative to the CMB. The second and third terms constitute the Sachs-Wolfe effect. In the case of adiabatic fluctuations, photon density fluctuations are related to matter density fluctuations and this effect can be written as $-(1/3)\delta\phi_{LS}$ with ϕ_{LS} being the gravitational potential at the last scattering surface. Notice that in this case, overdensities in the matter distribution ($\delta\phi > 0$) would produce cold temperature fluctuations ($\Delta T < 0$). How do all these effects appear in the COBE/DMR maps?.

A COBE/DMR map of absolute temperatures, right after the construction processes explained above, is presented in Fig. 4a (this map, as well as any other appearing in this contribution, is presented in Galactic coordinates with the Galactic plane occupaying the central part. Positive latitudes would be in the upper part, longitudes from 0 to 180° cover the band from the center to the left and from 0 to −180° from the center to the right). The observed field of temperatures is perfectly homogeneous with a mean value of $T_o = 2.726$K. If the fitted monopole is removed we are left with a map like the one presented in Fig. 4b. A clear dipole can be observed corresponding to the kinematic Doppler effect appearing as the first term in (15). The absolute value of the kinematic dipole is 3.365 ± 0.027mK (Smoot et al 1992) from the analysis of the first year of observations and 3.363 ± 0.024mK (Bennett et al 1994) from two years of observations. This dipole is pointing to the direction (in Galactic coordinates) $(l, b) = (264°.4 \pm 0.2, 48°.1 \pm 0.4)$ (Bennett et al 1994). The kinematic dipole can be removed from the map and one would observe the features shown in Fig. 4c. The main feature in this map is the central hot spot covering the Galactic plane. This is the contribution coming from the Galaxy. At present, removing the Galactic contribution as well as any other foreground sources, in the experimental CMB anisotropy data is one of the most challenging problems. Bennett et al 1992 discuss three different ways of removing the Galactic emission from the maps. None of these methods subtract the Galactic contribution completely and the way usual to proceed, before any data anlysis is performed on the maps, is to cut the Galactic band removing the pixels with latitudes $|b| < 20 - 30°$. Other foreground contributions to the COBE/DMR maps have been estimated to be very small (Bennett et al 1993, Kogut et al 1994). The rest of the features observed in the map correspond to intrinsic CMB anisotropies produced by the Sachs-Wolfe effect. Expanding the temperature fluctuations in spherical harmonics (see (2)), the intrinsic dipole corresponding to $l = 1$ is much smaller than the kinematic quadrupole. However, one can fit the temperature fluctuations observed in the map up to $l = 2$ and obtain the value of the intrinsic quadrupole (larger than the kinematic one). Analysis of the first year of data provided a value of the quadrupole $Q_{rms} = 13 \pm 4\mu$K (Smoot et al. 1992). Subtraction of the Galaxy was performed in the first plus second year analyzed maps and a value of $Q_{rms} = 6 \pm 3\mu$K was estimated (Bennett et al. 1994). The large scale Galactic signal away from the plane is predominantely quadrupolar and subtraction could also remove part of the intrinsic cosmological quadrupole.

Direct analysis of the COBE/DMR maps would also provide us with the RMS value of the observed temperature fluctuations. The estimated value of the

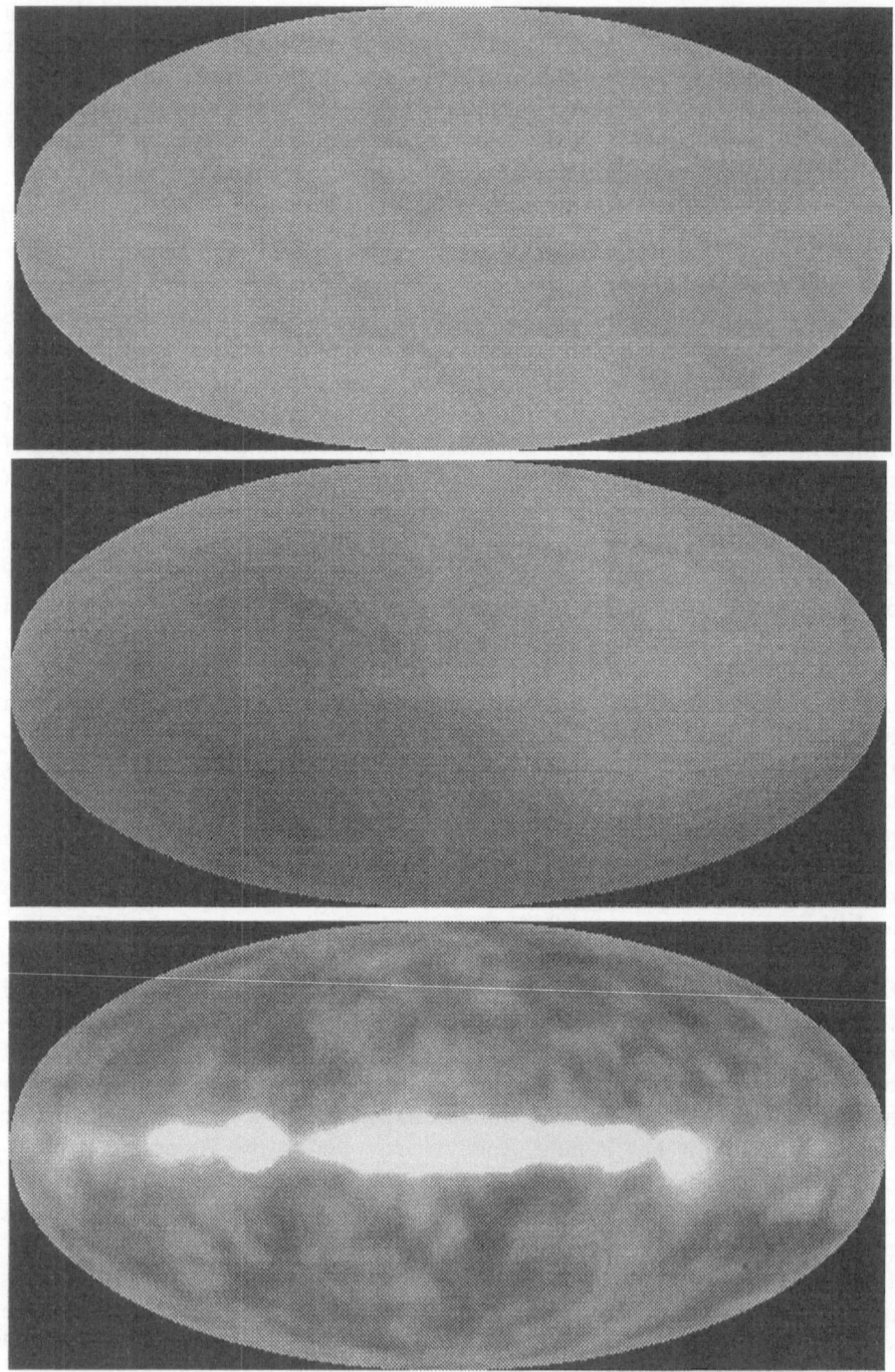

Fig. 4. 53 GHz COBE/DMR sky maps of two years of observations. a) After construction. b) Monopole subtracted. c) Dipole subtracted.

$RMS^2_{sky} = (RMS^2_{(A+B)/2} - RMS^2_{(A-B)/2})$ for the 53 GHz first plus second year map is $44 \pm 7.5 \mu K$ (Bennett et al. 1994).

4.3 Comparison with Cosmological Models

Cosmological models predict the average values of the multipole coefficients $C_l = < |a_{lm}|^2 >$ (average performed over the entire statistical ensemble) by solving the corresponding Boltzman equation (see Hu's constribution in these proceedings). However, experimental data sets have to deal with three main problems. The observed temperature fluctuations are always contaminated by instrumental noise. In the best case, our data would cover the whole celestial sphere, being such a sample only one realization of the whole statistical ensemble (over which the theoretical predictions are given). The error produced by this fact on any measurement is called cosmic variance. Moreover, not even in the case of the COBE/DMR experiment, is the whole sky covered (the Galactic plane is removed from the maps) and an additional error would appear in the data (sampling variance). In order to take into account these experimental constraints as well as all the experimental setup details, Monte Carlo realizations are required.

Considering that $a_{lm} = (1/\sqrt{2})(b_{l,m} - ib_{l,-m})$, the temperature fluctuations observed at each pixel (fixed direction $\boldsymbol{n}$) would be given by

$$\Delta T(\boldsymbol{n}) = [\sum_{l=2}^{l_{max}} K(b_{l,m} cos(m\phi) +$$

$$b_{l,-m} sin(m\phi)) N_l^m W_l P_l^m(cos\theta)] + \Delta_{noise}(\boldsymbol{n}) , \qquad (16)$$

where the constant $K = \sqrt{2}$ if $m \neq 0$ and $K = 1$ if $m = 0$, $N_l^m = ((2l+1)(l-m)!/4\pi(l+m)!)^{1/2}$, P_l^m are Legendre polinomials and Δ_{noise} represents the noise per pixel. In the case of COBE/DMR the window function is not exactly a Gaussian although it is often approximated by one with $FWHM = 7°$ (Wright et al. 1994b). l_{max} is approximately 40 for this experiment. Monte Carlo simulations are done by assuming a certain distribution for the multipole coefficients $b_{l,m}, b_{l,-m}$ and a Gaussian noise per pixel with dispersion $(RMS_{(A-B)/2}/N_{obs}(\boldsymbol{n}))^{1/2}$. As Inflation predicts a Gaussian distribution of the a_{lm}, the real and imaginary parts are usually drawn from a normal distribution with mean value zero and variance given by the theoretical predicted value C_l. Several realizations of sky maps are performed for a certain cosmological model and the values of different quantities on the simulated maps are statistically compared with the theoretical predictions. Many functions have been used to make predictions from the COBE/DMR data: correlation functions, RMS value, topological descriptors (number, area of spots, total curvature), etc.

From the Sachs-Wolfe effect and assuming a power law for the power spectrum of matter density fluctuations, the C_l are given by (10) and (11) as a function of Q_{rms-PS} and the spectral index n. The predictions of these two parameters have been worked out by different groups using different techniques.

There is a correlation between n and Q_{rms-PS} principally introduced by the noise and the variance (cosmic plus sampling variance) that affect the observational data. Some fits given in the literature are presented in Table 1.

Table 1. Q_{rms-PS}, n fits to the COBE/DMR data. Second column presents the results obtained including the quadrupole in the analysis. Third column gives the estimations based on analysis of data after subtracting the quadrupole (* Estimations based on the RMS)

References	$(Q_{rms-PS}\mu K, n)_{inc.quad.}$	$(Q_{rms-PS}\mu K, n)_{excl.quad.}$
Smoot et al. 1992	$(16 \pm 4, 1.1 \pm 0.5)$	
Seljack and Bertschinger 1993	$Q_{rms-PS} = (15.7 \pm 2.6)exp(0.46[1 - n])$	
Bennett et al. 1994	$(12.4^{+5.2}_{-3.3}, 1.59^{+0.49}_{-0.55})$	$(16.0+7.5-5.2, 1.21^{+0.6}_{-0.55})$
Banday et al. 1994	$(17^{+2.5}_{-2.1}, 1)$	$(19.4^{+2.3}_{-2.1}, 1)$
Smoot et al. 1994*	$(13.2 \pm 2.5, 1.7^{+0.3}_{-0.6})$	
Smoot et al. 1994	$Q_{rms-PS} = (15.7 \pm 2.2) - (6.6 \pm 0.3)(n - 1)$	
Torres et al. 1995	$Q_{rms-PS} = (22.2 \pm 1.7) - (4.7 \pm 1.3)n$	
Gorski et al. 1994	$(17^{+7.6}_{-4.8}, 1.22^{+0.46}_{-0.52})$	$(20.0^{+10.5}_{-6.5}, 1.02^{+0.53}-0.59)$
Wright et al. 1994c	$n = 1.46^{+0.41}_{-0.44}$	

Different cosmological models have been tested by COBE/DMR data analysis. These data predict a normalization for the Cold Dark Matter model of $Q_{rms-PS} \sim 20\mu K$. This normalization would produce matter density fluctuations at 8 Mpc that are larger than observed. In the case of Mixed Dark Matter models (incorporating a neutrino component) COBE/DMR would imply a RMS value of the matter density fluctuations at 8 Mpc in agreement with the observations. However, too small velocity dispersions are predicted (Stompor, Gorski and Banday 1995, Bunn, Scott and White 1995). As a possible solution to fit the large scale structure observations with the normalization given by the COBE/DMR data, a cosmological constant Λ has been considered in a flat model (Stompor, Gorski and Banday 1995, Bunn and Sugiyama 1995). The COBE/DMR data favour models with $\Lambda = 0$. Open models with different primordial power spectrum have been studied in comparison with COBE/DMR first and second year data (Gorski et al. 1995, Cayón et al. 1995). The conclusion is that this data can not rule out any $\Omega < 1$ models. The only theoretical models that seem to have very strong constraints in order to fit the COBE/DMR data are the Primordial Isocurvature Barionic models (Hu, Bunn and Sugiyama 1995).

4.4 Feature Identification

As all the CMB anisotropy experiments are constrained to observe the same sky, experiments working at comparable angular scales should detect the same features if these have a cosmological origin. The features appearing in the COBE/DMR sky maps are contaminated by noise and it is difficult to say which of them correspond to real cosmological signal.

The three COBE/DMR maps in Fig. 5 show all the connected pixels appearing above and below $\sigma_{(A-B)/2}$ in the 31, 53 and 90 GHz maps of two years of observations. In a first attempt to identify real spots, Cayón and Smoot 1995 considered all the spots that appear in coincidence at the three frequencies and compared their area (number of connected pixels) and signal-to-noise ratio (mean value of the ratio of the temperature over noise observed at each of the connected pixels) to predictions of noise Monte Carlo simulated maps. Two cold and one hot spots were pointed out as being produced by cosmological signal at the 95% confidence level. These are the spots located at $(l, b) = (-99°, 57°), (-21°, -45°)$ for the two cold spots and $(l, b) = (-81°, -33°)$ for the hot one.

Special methods to remove the noise from the data have been used to predict the spots that would be observed in the COBE/DMR maps assuming a certain cosmological model. Bunn et al. 1994 presented a Wiener filtering analysis (Press et al. 1992) of the first year data. The predictions of three models are presented considering Sachs-Wolfe effect and a power law for the matter density fluctuations with $(Q_{rms-PS}(\mu K, n) = (12.5, 1.5), (15.7, 1.0), (19.8, 0.5)$. Two of the more significant spots coincide with the two cold spots predicted to have a cosmological origin in Cayón and Smoot 1995.

4.5 Comparison with other Experiments

CMB anisotropy level observed by the COBE/DMR experiment require observations at large angular scales to be confirmed. The first confirmation appeared through a comparison with the FIRS (Far Infra-Red Survey) observations. This is a balloon experiment characterized by an antenna of $FWHM = 3°.8$. The observations were performed in October 6, 1989 at four different frequencies 170, 273, 477 and 682 GHz. An upper limit of the amplitude of the CMB temperature fluctuations assuming a Gaussian correlation function with coherence angle of 13° was set to $\sim 43\mu K$ (Meyer, Cheng and Page 1991). A posterior analysis comparing with theoretical models was presented in Ganga et al. 1994 setting $(Q_{rms-PS}, n) = (19\mu K, 1)$ using a maximum likelihood analysis. Cross-correlations with the COBE/DMR data that cover the same region of the sky as the one observed by the FIRS experiment, compared with Monte Carlo simulations, indicate that both experiments should be observing the same cosmological signal (Ganga, Cheng, Meyer and Page 1993).

The COBE/DMR data have also been cross-correlated with the Tenerife data (Lineweaver et al. 1995). The same feature seems to observed in both data sets in the region scanned by the Tenerife experiment. This hot spot appears above $\sigma_{(A-B)/2}$ at 53 GHz but it is not observed at the other two frequencies.

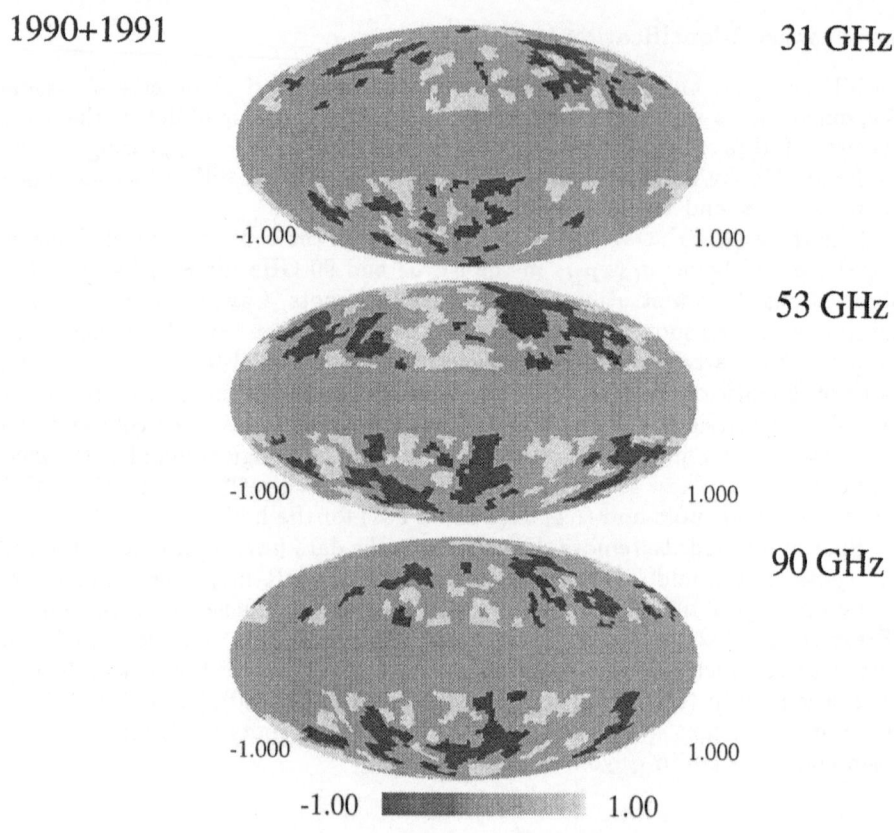

Fig. 5. Connected pixels above $1\sigma_{(A-B)/2}$ in the COBE/DMR sky maps of two years of data. a) 31 GHz. b) 53 GHz. c) 90 GHz.

Acknowledgements: The author is thankful for the opportunity given by the organizers to participate in this School as well as to the Universidad de Cantabria for organizing it. Many comments on experimental details were given by Joshua Gundersen (he also provided figure 1). George Smoot and Jon Aymon helped in the section dedicated to the COBE satellite.

References

Arendt, R.G. et al. (1994): ApJ Vol. 425, L85

Banday, A.J., Gorski, K.M., Tenorio, L., Wright, E.L., Smoot, G.F., Lineweaver, C.H., Kogut, A., Hishaw, G., Bennett, C.L. (1994): ApJ Vol. 436, L99

Bennett, C.L. et al. (1992): ApJ Vol. 396, L7

Bennett, C.L., Hinshaw, G., Banday, A.J., Kogut, A., Wright, E.L., Loewenstein, K., Cheng, E.S. (1993): ApJ Vol. 414, L77

Bennett, C.L. et al. (1994): ApJ Vol. 436, 423

Berriman, G.B., Boggess, N.W., Hauser, M.G., Kelsall, T., Lisse, C.M., Moseley, S.H., Reach, W.T., Silverberg, R.F. (1994): ApJ Vol. 431, L63

Bond, J.R. (1995): "Cosmology and Large Scale Structure", ed. by R. Schaeffer (Elsevier Science Publishers, Netherlands)

Bunn, E.F. (1994): ApJ Vol. 432, L75

Bunn, E.F., Scott, D., White, M. (1995): ApJ Vol. 441, L9

Bunn, E.F., Sugiyama, N. (1995): ApJ Vol. 446, 49

Cayón, L., Martínez-González, E., Sanz, J.L., Sugiyama, N., Torres, S. (1995): preprint

Cayón, L., Smoot, G.F. (1995): ApJ Vol. 452, 487

Clapp, A.C. et al. (1994): ApJ Vol. 433, L57

Cheng, E.S. et al. (1994): ApJ Vol. 422, L37

Cheng, E.S. et al. (1995): preprint

Christiansen, W.N., Hogbom, J.A. (1985): *Radio Telescopes*, (2nd Edition, Cambridge University Press)

De Bernardis, P. et al. (1994): ApJ Vol. 422, L33

Devlin, M.J. et al. (1994): ApJ Vol. 430, L1

Dragovan, M., Ruhi, J.E., Novak, G., Platt, S.R., Crone, B., Pernic, R., Peterson, J.B. (1994): ApJ Vol. 427, L67

Ganga, K., Cheng, E., Meyer, S., Page, L. (1993): ApJ Vol. 410, L57

Ganga, K., Page, L., Cheng, E., Meyer, S. (1994): ApJ Vol. 432, L15

Gorski, K.M., Hinshaw, G., Banday, A.J., Bennett, C.L., Wright, E.L., Kogut, A., Smoot, G.F., Lubin, P.M. (1994): ApJ Vol. 430, L89

Gorski, K.M., Ratra, B., Sugiyama, N., Banday, A.J. (1995): ApJ Vol. 444, L65

Gundersen, J.O. et al. (1993): ApJ Vol. 413, L1

Gundersen, J.O. et al. (1995): ApJ Vol. 443, L57

Hu, W., Bunn, E.F., Sugiyama, N. (1995): ApJ Vol. 447, L59

Kogut, A. et al. (1992): ApJ Vol. 401, 1

Kogut, A., Banday, A.J., Bennett, C.L., Hinshaw, G., Loewenstein, K., Lubin, P., Smoot, G.F., Wright, E.L. (1994): ApJ Vol. 433, 435

Lasenby, A.N., Davies R.D. (1988): "Large Scale Motions in the Universe", ed. by V.C. Rubin, G.V. Coyne (Vatican Press and Princeton Univ. Press), 277

Lineweaver, C.H. (1994): (PhD, U.C.Berkeley)

Lineweaver, C.H., Smoot, G.F., Bennett, C.L., Wright, E.L., Tenorio, L., Kogut, A., Keegstra, P.B., Hinshaw, G., Banday, A.J. (1994): ApJ Vol. 436, 452

Lineweaver, C.H., Hancock, S., Smoot, G.F., Lasenby, A.N., Davies, R.H., Banday, A.J., Gutiérrez de la Cruz, C.M., Watson, R.A., Rebolo, R. (1995): ApJ Vol. 448, 482

Meyer, S.S., Cheng, E.S., Page, L.A. (1991): ApJ Vol. 371, L7

Martin, B.R. (1971): *Statistics for Physicists*, (London: Academic)

Marton, L. (1976): *Methods of Experimental Physics: Astrophysics. Radio Observations*, (Academic Press, INC. London)

Netterfield, C.B., Jarosik, N., Page, L., Wilkinson, D., Wollack, E. (1995): ApJ Vol. 445, L69

Piccirillo, L., Calisse, P. (1993): ApJ Vol. 411, 529

Press, W.H., Teukolsky, S.A., Vetterling, W.T., Flannery, B.P. (1992): *Numerical Recipes*, (2nd Edition, Press Syndicate of the University of Cambridge)

Puget, J.L., Abergel, A., Bernard, J.P., Boulanger, F., Burton, W.B., Desert, F.X., Hartman, D. (1995): submitted to A.A.

Reach, W.T. et al. (1995): ApJ Vol. 451, 188

Ruhl, J.E., Dragovan, M., Platt, S.R., Kovac, J., Novak, G. (1995): preprint

Sachs, R.K., Wolfe, A.N. (1967): ApJ Vol. 147, 73

Seljak, U., Bertschinger, E. (1993): ApJ Vol. 417, L9

Smoot, G.F. et al. (1992): ApJ Vol. 396, L1

Smoot, G.F., Tenorio, L., Banday, A.J., Kogut, A., Wright, E.L., Hinshaw, G., Bennett, C.L. (1994): ApJ Vol. 437, 1

Stompor, R., Gorski, K.M., Banday, A.J. (1995): preprint

Torres, S., Cayón, L., Martínez-González, E., Sanz, J.L. (1995): MNRAS Vol. 274, 853

Weiland, J.L. et al. (1994): ApJ Vol. 425, L81

Wollack, E.J., Jarosik, N.C., Netterfield, C.B., Page, L.A., Wilkinson, D. (1993): ApJ Vol. 419, L49

Wright, E.L., Mather, J.C., Fixsen, D.J., Kogut, A., Shafer, R.A., Bennett, C.L., Boggess, N.W., Cheng, E.S., Silverberg, R.F., Smoot, G.F., Weiss, R. (1994a): ApJ Vol. 420, 450

Wright, E.L., Smoot, G.F., Kogut, A., Hinshaw, G., Tenorio, L., Lineweaver, C.H., Bennett, C.L., Lubin, P.M. (1994b): ApJ Vol. 420, 1

Wright, E.L., Smoot, G.F., Bennett, C.L., Lubin, P.M. (1994c): ApJ Vol. 436, 443

Observations of CMB Structure
with the Tenerife Experiments

Carlos M. Gutiérrez, Rafael Rebolo

Instituto de Astrofísica de Canarias, 38200 La Laguna. Spain

Abstract: The Tenerife Cosmic Microwave Background (CMB) experiment is mapping a region of 5000 square degrees on the sky. Several beam-switching radiometers at frequencies 10, 15 and 33 GHz have been operating in the Teide Observatory (Tenerife) for more than ten years. Our experiment discovered the first features in the CMB in a strip at Dec.=+40°. The region analyzed was free of strong radio sources and the spectral properties of the features ruled out the possibility of a Galactic origin. The most probably origin of the signal detected was cosmological, representing the first direct observation of intrinsic CMB structure. For the case of fluctuations described by a Gaussian auto-correlation function, a likelihood analysis of our combined results at 15 and 33 GHz implies an intrinsic rms fluctuation level of 48^{+21}_{-15} μK on a coherence scale of 4°; the equivalent analysis for a Harrison-Zel'dovich model gives a power spectrum normalisation of $Q_{RMS-PS} = 22^{+10}_{-6}$ μK. A comparison with the COBE DMR two-year data at Dec.=+40° shows the compatibility of the signal detected by both experiments and the presence of individual features common to both data sets. Several new instruments recently installed in Tenerife extending the angular scales and spectral range covered by our radiometers are briefly described.

1 Introduction

The anisotropy of the Cosmic Microwave Background (CMB) is one of the most powerful tools to investigate the physical conditions in the early Universe, providing direct access to the epoch $\sim 300,000$ years after the initial singularity. By observing different angular scales, we investigate different linear and mass scales; for example on the 5° scale of our Tenerife beam-switching experiments, we are probing the equivalent of $300-500$ Mpc structure in our contemporary Universe, which corresponds to the largest scales identified in the distribution of galaxies.

The standard picture envisages two components of the CMB structure on scales $\gtrsim 2°$. The first is scalar fluctuations arising from the Sachs-Wolfe effect, while the second is a tensor component due to gravitational wave radiation coming from the inflationary era. A detailed measurement of the spectrum of fluctuation amplitude on such scales will then establish the tensor contribution. A second feature of the angular spectrum of fluctuation amplitudes is the "Doppler peak" centred on scales of $1° - 2°$. These fluctuations are due to the motion of massive structures in the epoch of recombination, and are expected to have an amplitude of $2 - 3$ times that on large angular scales.

The Tenerife CMB experiments have the objective of detecting and mapping the primordial fluctuations in the CMB on angular scales of several degrees. In this contribution we present a description of the observing technique and data processing (Section 2), an analysis of the results obtained at Dec.=+40° (Section 3), some preliminary results at adjacent declinations (Section 4) and the future programme of observations (Section 5).

2 Observations and data reduction

The experiments are a collaboration between the University of Manchester, the IAC Tenerife and MRAO Cambridge, and consist of three independent two-channel receivers operating at frequencies of 10.4, 14.9 and 33 GHz, generically called the 10, 15 and 33 GHz experiments respectively. The instruments are located in the island of Tenerife at an elevation of 2400 m in a site which is usually above the inversion layer with three quarters of clear days and sunshine occuring during 2800 h of the year. The water-vapour content is about 2 mm for 30 % of the summer time, and our experience over the years shows that even at our higher frequency of 33 GHz it is possible to operate for a significant fraction of time. The three instruments operate following a double-difference technique: with a beam response of the form $-0.5, +1, -0.5$ with three beams (FWHM$\sim 5°$) the positive in the meridian and the other two displaced $8°.1$ in declination (Davies *et al.* (1992), 1996). The resultant configuration is sensitive to the linear part of the spectrum of CMB fluctuations (multipoles $l \lesssim 40$), having a peak response at $l \sim 20$ (Watson *et al.* 1992). The instruments use high electron mobility transistors (HEMT) amplifier, with bandwidths of ~ 10 % and have equivalent system brightness temperatures in the range $70 - 100$ K. The existence of two independent channels in each instrument gives further confidence in identifying atmospheric and systematic effects. The resulting theoretical sensitivities, including both channels, are 5.6, 3.4 and 2.2 mK$\times$Hz$^{-1/2}$ at 10, 15 and 33 GHz respectively. As primary calibrators we have used the Sun and the Moon, and an ambient temperature absorber. A continuous calibration is achieving by the injection of a signal of 4 K modulated at 16 Hz. Also the strong daily crossing of the Galactic plane acts as a secondary calibrator. The fraction of data lost due to atmospheric humidity effects throughout the year is about 30 % at 10 and 15 GHz, and a greater proportion at 33 GHz. In addition we have discarded any data taken when the Sun or Moon are less than 50° and 30° respectively from the beam. We have also removed data when the standard error is larger than three times the average value for the day. The rest of the time is lost due to maintenance, calibration or system failure.

3 Stacked data at Dec.=+40°

When the data are stacked at each frequency and declination we obtain very sensitive scans with a noise reduced by a factor $\sim 1/\sqrt{N}$ compared with individual observations. We will concentrate in this section in the analysis of the data at Dec.=+40°; at this declination we have analyzed the region at RA= $161° - 230°$ which is at Galactic latitudes $b \gtrsim 50°$. The search of CMB fluctuations must be done well away of the Galactic plane to avoid as much as possible the contribution of Galactic synchrotron and free-free emission. This region is also free of strong radio-sources (Kühr *et al.* (1981), VLA calibrators). Figure 1 represents the results at 15, 33 GHz and their weighted addition (15+33) respectively. The subsets A and B represent two independent splits of the data covering a similar number of observations. Cosmological features should appear with the same amplitude in each subset and frequency. A preliminary inspection of these plots indicates the presence of obvious common structure at RA $\sim$ 185° with an amplitude at the peak of about 80 μK; there is also some evidence of a second feature at RA $\sim$ 225°. A χ^2 test reveals the clear presence of signal: the probability that the results are a consequence of pure noise is less than 0.02, 1×10^{-4} and 3×10^{-6} for the 15, 33 GHz and their weighted addition respectively. It is possible to compute the mean level of the signal from the (A+B)/2 and (A−B)/2 results as $\sigma_s^2 = \sigma_{(A+B)/2}^2 - \sigma_{(A-B)/2}^2$; for data binned at 1° intervals in RA, we obtain 43 ± 12 μK in the (15+33) scans.

Our multifrequency approach allows us to estimate the Galactic contribution to the signal detected; this contribution is expected to be larger in our lower frequency, and therefore we can use the 10 GHz data to constrain the Galactic signal at the two higher frequencies by considering the case in which all the signals at 10 GHz were Galactic. Assuming a model with a single component we obtain a maximum contribution at 33 GHz of 4 and 2 μK for the free-free and synchrotron processes respectively which is a small fraction of the signal detected. We believe that our scans at 15 and 33 are dominated by CMB signal and that the features on these scans represent the first detection of individual primordial fluctuations in the CMB.

The amplitude of the signal in specific cosmological models can be determined using a likelihood approach (Watson *et al.* (1992), Gutiérrez *et al.* 1995). Most of the models assume a Gaussian field for CMB fluctuations and therefore it is possible to describe entirely their statistical properties using the ACF (Auto-correlation Function) of the field $C_{intr}(\theta) = < \delta T(\mathbf{n_1})\delta T(\mathbf{n_2}) >$, where δT are the fluctuations in temperature, and $\mathbf{n_1}$ and $\mathbf{n_2}$ two directions in the sky separated an angle θ. A useful model which allows a comparison of results obtained using different experimental configurations is given by the ACF $C_{intr}(\theta) = C_0 \exp\{-\theta^2/2\theta_c^2\}$, where C_0 and θ_c are the amplitude and coherence

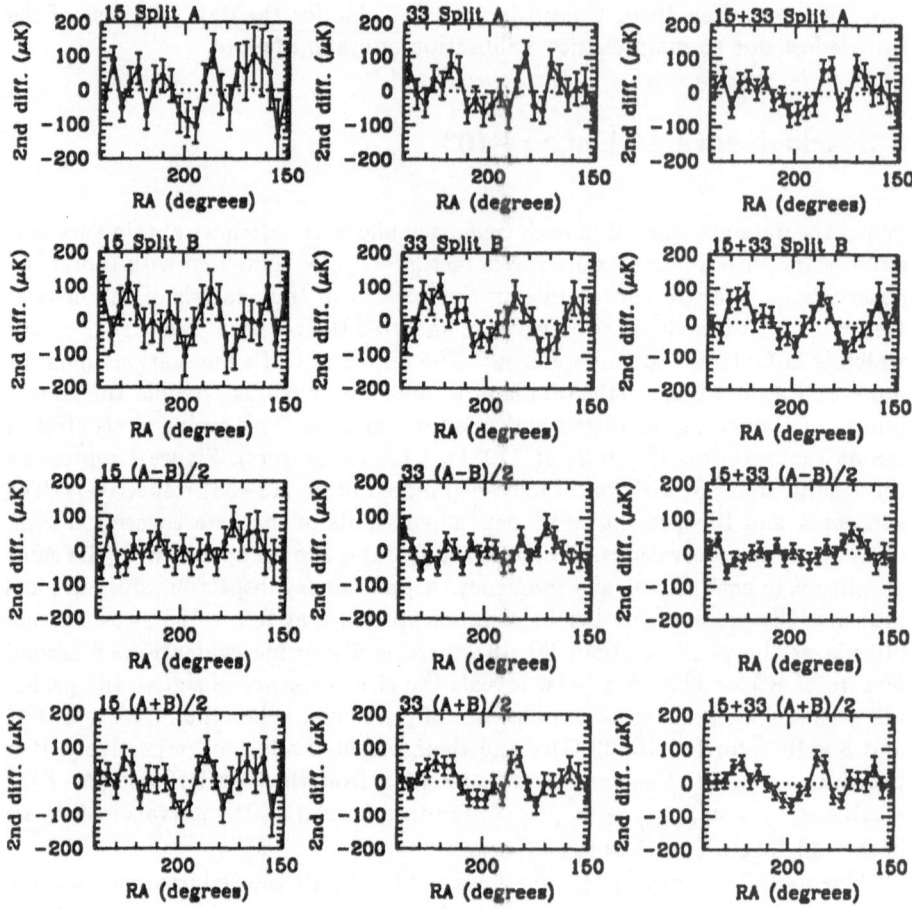

Fig. 1. Results (binned into 4° intervals in RA) at Dec.=+40° in the region of RA 150°-240°. We have split the data into the two subsets A and B, each covering a similar number of observations. A real feature should appear in each split and at higher significance in the addition (A+B)/2. A cosmological feature should appear with the same amplitude at 15 GHz and 33 GHz.

angle of the fluctuations respectively. The results for $\theta_c = 4°$ are presented in the third column of Table 1.

Another more insteresting model from a cosmological point of view, is the prediction in the case of a power law spectrum of primordial fluctuations $P = Ak^n$, where n is the spectral index. A model with $n = 1$ corresponds to the prediction of most of the inflationary scenarios. We have applied a likelihood analysis to this model, and we obtain the results shown in the second column of Table 1 which presents the amplitude of the quadrupole (see Hancock *et al.* (1994) for the notation used) deduced from each data set analyzed. The results look consistent, showing similar amplitudes for the A, B and (A+B)/2 data set, and a result compatible with null signal for the (A−B)/2 data set.

Table 1. Results of the likelihood analysis for a Harrison-Zel'dovich spectrum of fluctuations (second column) and for a Gaussian ACF (third column).

ν (GHz)	Q_{RMS-PS} (μK)	$\sqrt{C_0}$ (μK)
15A	27^{+16}_{-16}	54^{+37}_{-30}
15B	17^{+9}_{-17}	24^{+21}_{-24}
15	21^{+12}_{-9}	44^{+26}_{-19}
33A	22^{+14}_{-10}	45^{+32}_{-24}
33B	28^{+12}_{-9}	57^{+28}_{-25}
33	24^{+11}_{-8}	49^{+27}_{-17}
15+33	22^{+10}_{-6}	48^{+21}_{-15}

As has been shown in Lineweaver *et al.* (1995), the feature at RA∼ 185° is evident also in the COBE DMR two-year data at 53 and 90 GHz and has the Planckian spectrum expected for CMB anisotropy. Futhermore, the correlation function between the Tenerife and COBE DMR scans is also indicative of common structure across the range RA=160°−230°. The combination of the spatial and spectral information from the two data sets is consistent with the statistical level of fluctuations claimed by each experiment and strongly supports the cosmological origin of this structure.

4 Adjacent declinations

We have observed strips of the sky separated 2°.5 covering declinations 30° to 45° which are close to the zenith of Tenerife. The separation between adjacent scans is similar to the beam-width of the instruments and therefore the combinations of these strips allow us to construct a two-dimensional map. Table 2 presents the sensitivity per beam area in the stacked second differences scans at each declination and frequency in the section RA=161°−250°.

Table 2. The rms sensitivity (μK) in a 5° beam in the section RA=161°−250°

ν (GHz)	30°.0	32°.5	35°.0	37°.5	40°.0	42°.5	45°.0
10	-	69	74	39	43	75	76
15	20	24	20	20	18	21	24
33	-	38	30	29	21	47	-

A full analysis is not possible until we have observed these declinations with appropriate sensitivity at 33 GHz, but meanwhile some interesting preliminary results can be obtained. When comparing the data taken at 10 and 15 GHz in the region at high Galactic latitude, we see a common structure at Dec.=+45°, RA=225° with a reduced amplitude at 15 GHz; this structure also extends to the adjacent declinations, the relative amplitude at 10 and 15 GHz seems to be consequence of its Galactic origin, and is responsible for the detection of structure claimed by Davies *et al.* (1987) using an early version of the 10 GHz radiometer with a beam-width of 8°. At lower declinations we have observed a clear feature at Dec.=+35°, RA~215° which will require a detailed analysis when the data at 33 GHz will be obtained. At Dec.=+30° in the data at 15 GHz there is a feature at RA~ 200° which corresponds to the amplitude expected from the radio source 1328+30 (Gutiérrez *et al.* 1996). In Figure 2 we show our data at 15 GHz in this region and the modeled radio sources; the agreement between both in shape and amplitude confirms the consistency of our data and the fact that we are able to detect structures with amplitudes of ~ 50 μK.

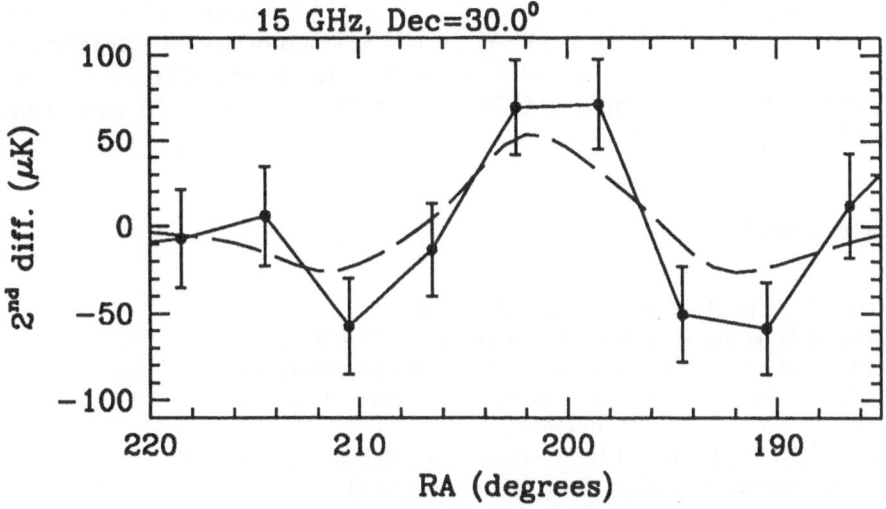

Fig. 2. A comparison between the predicted contribution of the radio source 1328+30 at 15 GHz (dashed line) and our measurements at Dec=+30° (solid lines).

5 The future programme

Data collection with the radiometers will continue in order to give a coverage of Dec.=+30° to 45° at the full sampling separation of 2°5 (half the FWHM). We plan to reach *rms* sensitivities in a 5° beam of 20 μK at 15 and 33 GHz, and 50 μK at 10 GHz. This combination of sensitivities will enable us to detect CMB fluctuations, and at the same time to determine the Galactic contribution to better than 5 μK at the highest frequency.

A 33 GHz two-element interferometer has been constructed at Jodrell Bank and installed at the Teide Observatory in collaboration with IAC. This interferometer has a resolution of 2°5 with full sine and cosine correlation in a 3 GHz bandwidth. The low noise amplifiers used are cryogenically cooled HEMTs, and the anticipated sensitivity is 0.7 mK×Hz$^{-1/2}$.

Measuring the Galactic emission at several frequencies ranging from 408 MHz to 15 GHz is the objective of The Galactic Emission Measurement (GEM) telescope, an experiment in collaboration with the Lawrence Laboratory which has been recently installed in the Teide Observatory. A collaboration between IAC and the Bartol Research Institute with CMB observations in the range 100-300 GHz at angular resolution ~ 2° started in the year 1993. The instrument have observed a strip on the sky centered at Dec.=+40°. The data taken in 1993 and 1994 are under analysis.

The collaboration between MRAO, NRAL and IAC continues with a proposal to build the Very Small Array (VSA) for CMB structure studies. Operating at frequencies between 28 and 34 GHz on the Tenerife site, the VSA will have the capability of imaging primordial CMB structure to a sensitivity of 5 μK over the angular range 10′ to 2°5.

References

Davies, R. D. *et al.*: *Nature* **326** 462 (1987)

Davies, R.D. *et al.*: *Mon. Not. Roy. Astron. Soc.* **258** 605 (1992)

Davies, R.D. *et al.*: *Mon. Not. Roy. Astron. Soc. (in press)* (1996)

Gutiérrez, C.M., Davies, R.D., Rebolo, R., Watson, R.A., Hancock, S., & Lasenby, A.N.: *Astrophys. J.* **442** 10 (1995)

Gutiérrez, C.M., Davies, R.D., Watson, R.A., Rebolo, R., Hancock, S., & Lasenby, A.N.: *Astron. Astrophys. Trans. (in press)* (1996)

Kühr, H., Witzel, A., Pauliny-Toth, I. I. K. and Nauber, U.: *Astron. Astrophys. Suppl. Ser.* **45** 367 (1981)

Lineweaver, C.H. *et al.*: *Astrophys. J.* **448** 482 (1995)

Hancock, S. *et al.*: *Nature* **367** 333 (1994)

Watson, R.A. *et al.*: *Nature* **357** 660 (1992)

Concepts in CMB Anisotropy Formation

Wayne Hu

Institute for Advanced Study, Princeton, NJ 08540

"To divide is to leave something undivided. To discriminate between alternatives is to leave something which is neither alternative."

Chuang-tzu

Abstract:

These lecture notes form a primer on the theory of cosmic microwave background (CMB) anisotropy formation. With emphasis on conceptual aspects rather than technical issues, we examine the physical foundations of anisotropy evolution in relativistic kinetic and perturbation theory as well as the manifestation of these principles in primary and secondary anisotropies. We discuss gauge choice and gauge invariance and their use in understanding the CMB. Acoustic, gravitational redshift and ionization effects have robust signatures in the CMB spectrum and may allow determination of classical cosmological parameters as well as reveal general distinctions between models for structure formation. We develop the tight and weak coupling approximations as analytic tools to help understand these effects and the robustness of their signatures.

1 Introduction

More than three years has passed since cosmic microwave background (CMB) anisotropies were first detected at large angular scales in the *COBE* DMR sky maps (Smoot *et al.* 1992). Since then much progress has been made in probing the anisotropy spectrum at smaller scales (see *e.g.* the compilations by Bond 1995; Scott, Silk & White 1995; Steinhardt 1995). Satellite missions now being proposed would be able to map the anisotropy definitively down to a fraction of a degree. To exploit these recent and potential experimental advances, we need an accurate and accessible theory of anisotropy formation. The first part of this task was largely accomplished in the 1980's with the ground-breaking work of Wilson & Silk (1981), Vittorio & Silk (1984), and Bond & Efstathiou (1984). Detailed numerical calculations in perturbation theory allow accurate predictions of the anisotropy in most models for structure formation. Summaries of the current state of the art in numerical codes can be found in Bond (1995), Hu *et al.* (1995) and Ma & Bertschinger (1995).

Numerical work however is notoriously inaccessible to the uninitiated. Moreover, it obscures the true potential of CMB anisotropy spectrum for cosmology by limiting itself to specific models. Indeed, CMB anisotropies are sensitive to classical cosmology parameters such as the matter and radiation content, baryon fraction, expansion rate, curvature, and cosmological constant (Bond *et al.* 1994; Seljak 1994; Hu & Sugiyama 1995a) as well as the model for structure formation (Hu & Sugiyama 1995b; Albrecht *et al.* 1995, Crittenden & Turok 1995; Durrer *et al.* 1995; Magueijo *et al.* 1995). In these lecture notes, we will review the basic concepts of anisotropy formation and their consequence for cosmology.

We begin in §2 with relativistic kinetic and perturbation theory which give the basic physical laws that govern anisotropy formation. Pure reductionism of this sort however would miss the true use of CMB anisotropies. In §3 and §4, we map out the manifestations of the basic principles before and after recombination, *i.e.* the primary and secondary anisotropies. These anisotropies possess information on both the cosmological model and structure formation. From this decomposition and reconstruction, the resulting complex structure of anisotropies can be understood in a general model-independent manner.

2 Physical Foundation

Two ingredients are necessary to describe the evolution of cosmic microwave background anisotropies: relativistic kinetic and perturbation theory. Kinetic theory describes the radiation transport properties of the CMB photons in the metric perturbed by gravitationally unstable density fluctuations. These fluctuations are evolved through relativistic perturbation theory. Combined, the two considerations yield a complete system that is responsible for anisotropy formation in any model where structure formation proceeds by gravitational instability.

2.1 Relativistic Kinetic Theory

Conceptually, relativistic kinetic theory is identical to the familiar non-relativistic case: the phase space distribution is conserved along geodesics save for a collision term due to scattering,

$$\frac{df}{dt} \equiv \frac{\partial f}{\partial t} + \frac{\partial f}{\partial x^i}\frac{dx^i}{dt} + \frac{\partial f}{\partial p}\frac{dp}{dt} + \frac{\partial f}{\partial \gamma^i}\frac{d\gamma^i}{dt} = C[f], \tag{1}$$

Here γ_i are the direction cosines of the photon momentum p. Before a redshift of $z_* \approx 1000$, CMB photons were hot enough to ionize hydrogen. Consequently, the dominant interaction process for the CMB at early times was Compton scattering off free electrons. Due to the higher electron density in the early universe, the Compton mean free path was quite short, much smaller than the particle horizon at that time. The horizon at z_* subtends a degree or less for $\Omega_0 + \Omega_\Lambda \lesssim 1$. Thus on scales relevant for observable anisotropies, the photons were tightly coupled to the electrons, which in turn were tightly coupled to the protons by Coulomb interactions. At z_*, neutral hydrogen formed through "recombination" and the photons last scattered. Unless the universe suffered reionization at high redshift, fluctuations in the CMB at recombination were frozen in at z_* and await observation today.

There are several features of Compton scattering worth noting

1. Scattering couples the photons to the baryons and forces perturbations in their number and hence energy density to evolve together.
2. Scattering isotropizes the photons in the electron rest frame thus coupling the local CMB dipole to the electron velocity.
3. In the Thomson limit, there is no energy transfer in scattering. Energy exchange only occurs to $\mathcal{O}(\langle v_e^2 \rangle)$, i.e. $\mathcal{O}(T_e/m_e)$.
4. There is no change in photon number through Compton scattering.

Scattering thus governs the intrinsic temperature perturbations at last scattering and the dipole or bulk velocity perturbation. Since it does not change the net energy or photon number in the CMB to lowest order, spectral distortions to the blackbody do not arise in linear theory. Unless the electrons have been heated significantly above the temperature of the CMB or photons and/or energy has been dumped into the CMB from an external source, the spectral information can be ignored.

The remaining subtlety is that the photons propagate in a space-time that is distorted by density fluctuations: we must employ the geodesic equation in the presence of perturbations. This leads to gravitational redshift effects from the dp/dt term in equation (1) which can also generate fluctuations in the CMB. The $d\gamma^i/dt$ term represents gravitational lensing and gives a first order contribution only in a curved universe (see §3.4). By integrating equation (1) over frequencies, one obtains the *Boltzmann equation* for the evolution of temperature perturbations $\Theta(\eta, \mathbf{x}, \gamma) \equiv \Delta T/T$,

$$\left[\frac{\partial}{\partial \eta} + \dot{x}^i \frac{\partial}{\partial x^i} + \dot{\gamma}^i \frac{\partial}{\partial \gamma^i} \right] \Theta = S_G + S_C \tag{2}$$

where S_G and S_C are the sources from the gravitational redshift and Compton scattering and overdots are derivatives with respect to conformal time $\eta = \int dt/a$. For a derivation from first principles of the Compton collision term S_C, see Hu, Scott & Silk (1994), Dodelson & Jubas (1995), and Kosowsky (1995). If the angular dependence and polarization of Compton scattering is ignored it reduces to $\dot{\tau}[\Theta_0 - \Theta - \gamma_i v_e^i]$. Here Θ_0 is the isotropic temperature fluctuation and the differential Compton optical depth $\dot{\tau} = x_e n_e \sigma_T a$ with x_e as the ionization fraction, n_e as the electron density and σ_T as the Thomson cross section. Implicit in S_G and S_C are the three fundamental sources of anisotropies in the CMB:

1. Gravitational redshifts from the presence *and* evolution of metric fluctuations.
2. Hot and cold spots from the intrinsic temperature at last scattering.
3. The Doppler effect due to the velocity of the last scatterers.

Of course, this high level description is not very practical. Gravitational instability controls the evolution of density, velocity, and metric perturbations in the universe. All of these effects are thus related by relativistic perturbation theory.

2.2 Relativistic Perturbation Theory

General relativity tells us that matter moves in a space-time perturbed by fluctuations in the matter density itself. Thus the Einstein equations reduce conceptually into two pieces. The stress-energy tensor of the total matter is covariantly conserved in the *perturbed* metric, $T^{\mu\nu}{}_{;\mu} = 0$, and matter fluctuations are the source of metric perturbations via a generalized Poisson equation. For the former, $\nu = 0$ gives number or energy density conservation, *i.e.* the continuity equation. The spatial components give momentum conservation, *i.e.* the Euler equation.

2.2.1 Gauge Choice

To explicitly implement these principles, we need to treat a subtlety due to *gauge freedom* in general relativistic perturbation theory. In order to define a perturbation, we must specify the relation between the physical spacetime and the hypothetical unperturbed background. The difference between quantities at the *same coordinate values* is deemed a perturbation. Since this choice is not unique, one must generally fix a gauge before making any calculations or interpretations regarding perturbation evolution.

Practically speaking, a gauge choice involves fixing the constant-time hypersurfaces and the spatial grid on these surfaces. Suppose we warp the time

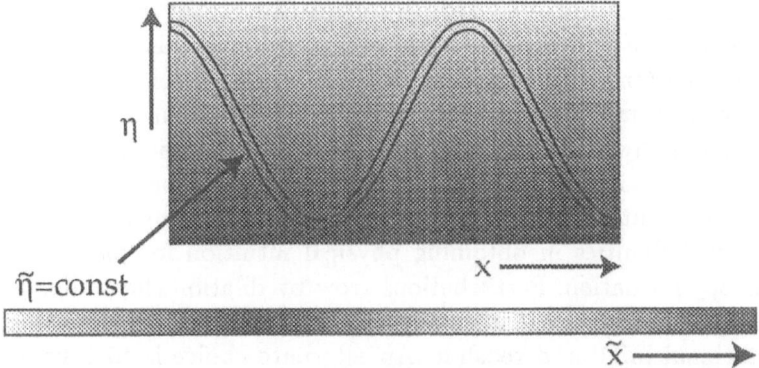

Fig. 1. Gauge ambiguity. Due to the change in density with the expansion, even a homogeneous FRW universe can appear to have a density perturbation if the time slicing is warped. The synchronous and total matter gauges choose the time slicing to correspond to the rest frame of the freely falling and total matter respectively. The Newtonian gauge chooses slicing such that the expansion rate is shear free.

slicing from some fiducial choice by $\tilde{\eta} = \eta + Te^{i\mathbf{k}\cdot\mathbf{x}}$ (see Fig. 1). The density of a fluid of particle X on this slicing becomes $\tilde{\rho}_X = \rho_X - \dot{\rho}_X Te^{i\mathbf{k}\cdot\mathbf{x}}$ for the same coordinate value. Because the expansion makes densities decrease in time as

$$\dot{\rho}_X = -3(\rho_X + p_X)\frac{\dot{a}}{a},$$

in the new frame the density fluctuation is

$$\tilde{\delta}_X = \delta_X + 3(1 + p_X/\rho_X)\frac{\dot{a}}{a}Te^{i\mathbf{k}\cdot\mathbf{x}}. \tag{3}$$

As a simple example of gauge ambiguity, consider a Friedmann-Robertson-Walker universe with *no* perturbations $\delta_X = 0$. Even in this case, a density fluctuation appears in the warped time slicing. One might argue that this involves a silly choice of time slicing and that observers may set up a sensible coordinate system that removes ambiguities by exchanging light signals. This is true. All "sensible" choices of coordinate systems agree on the nature and value of the density perturbation within the horizon by requiring that the warping satisfy $\delta\eta/\eta = T/\eta \ll \delta$ if $k\eta \gg 1$. However outside the horizon, there is no "sensible" way to set up the coordinate system. In particular, different choices of time slicing can produce seemingly different results for perturbations and their evolution in this regime.

There are two basic approaches to fixing a gauge or coordinate system. We can choose a set of preferred observers and set the coordinate frame to be their rest frame, *i.e.* employ "Lagrangian" coordinates. The popular and useful *synchronous* gauge is one such representation. Here we choose freely falling observers to define the frame, *e.g.* the collisionless cold dark matter. This rest frame analysis is convenient in that it simplifies the evolution equations and their solution. An extension of this idea is employed in the *total matter* gauge. Here the coordinate system coincides with the rest frame of the combined relativistic and non-relativistic matter. This choice is computationally useful in the early universe where radiation dominates and does not necessarily follow the perturbations in the non-relativistic matter.

The computational convenience of these "Lagrangian" techniques is offset by difficulties in obtaining physical intuition for processes such as anisotropy formation. Perturbations grow by dilation effects due to distortions of the coordinate grid rather than simple causal mechanisms such as gravitational infall and redshift. An alternate choice is to take an "Eulerian" coordinate system. More specifically, instead of employing a set of preferred observers, fix the coordinate system by a geometric property. One useful choice is the Newtonian gauge. The coordinate frame here corresponds to zero-shear hypersurfaces where the expansion appears isotropic. Inside the horizon, this analysis reduces to the standard non-relativistic Newtonian treatment. Thus, our physical intuition from the non-relativistic theory most easily carries over to a complete treatment in this gauge.

Since there is a one-to-one mapping between gauges whose coordinates are entirely fixed, these treatments are entirely equivalent and give the same result for any *observable* quantity, *e.g.* CMB anisotropies. Ambiguities only arise if the coordinate system is not entirely fixed. In this case, the one-to-many mapping produces unphysical *gauge modes* which historically has produced much confusion (see Press & Vishniac 1980). Gauge modes can be avoided by completely fixing the coordinates. The definitions of the total matter and Newtonian gauges already do so. On the other hand, a synchronous coordinate system, defined by a set of freely falling observers, requires us to specify *which* set through the initial conditions. In particular, if the initial velocity of the observers is fixed then there are *no* gauge modes. In practice, one usually takes the rest frame of the CDM.

2.2.2 Gauge Invariance

Due to the computational convenience of the "Lagrangian" gauges and the intuitive nature of the "Eulerian" ones, it is often useful to switch between gauges in the midst of a calculation or even employ different gauge representations in a single set of evolution equations. A useful technique for this purpose is provided by the *gauge invariant* method (Bardeen 1980; Kodama & Sasaki 1984; Mukhanov, Feldman & Brandenberger 1992). It is called gauge invariant *not* because a density perturbation, say, remains a density perturbation under a gauge transformation. Rather, a density perturbation in a *specific* gauge is represented in a coordinate free manner, *i.e.* in a manner valid in an arbitrary gauge. In practical terms, this just means that whatever gauge is chosen, one has a systematic way to mix in quantities that represent perturbations in a different gauge.

Let us make these considerations more concrete. The most general form of a metric perturbed by scalar fluctuations a given Fourier mode k is (Bardeen 1980)

$$g_{00} = -a^2[1 + 2A^G e^{i\mathbf{k}\cdot\mathbf{x}}],$$

$$g_{0j} = a^2 B^G i\hat{k}_j e^{i\mathbf{k}\cdot\mathbf{x}},$$

$$g_{ij} = a^2\{\delta_{ij} + [2H_L^G \delta_{ij} + 2H_T^G(-\hat{k}_i\hat{k}_j + \delta_{ij}/3)]e^{i\mathbf{k}\cdot\mathbf{x}}\},$$

(4)

where for illustration purposes, we have taken a flat geometry and superscript G is meant to remind the reader that these quantities are gauge dependent. By fixing the gauge, we can eliminate two out of the four metric terms. In the Newtonian gauge, $B^N = H_T^N = 0$, and it is traditional to label the Newtonian potential $A^N = \Psi$ and the space curvature perturbation $H_L^N = \Phi$. Synchronous gauge requires $A^S = B^S = 0$, whereas the total matter gauge takes $B^T = V_T^T$ and $H_T^T = 0$ where V_T^T is the velocity of the total matter. Ordinarily, one now writes down the Einstein equations $G_{\mu\nu} = 8\pi G T_{\mu\nu}$ which contain the conservation $T^{\mu\nu}{}_{;\mu} = 0$ and "Poisson" equations, with a specific choice of gauge. For a fluid of particle X, the stress energy momentum tensor is

$$T^0{}_0 = -(1 + \delta_X^G e^{i\mathbf{k}\cdot\mathbf{x}}),$$

$$T^0{}_i = (\rho_X + p_X)V_X^G(-i\hat{k}_i)e^{i\mathbf{k}\cdot\mathbf{x}},$$

(5)

$$T^i{}_j = (p_X + \delta p_X^G e^{i\mathbf{k}\cdot\mathbf{x}})\delta^i{}_j + p_X \Pi_X(-\hat{k}^i\hat{k}_j + \delta^i{}_j)e^{i\mathbf{k}\cdot\mathbf{x}},$$

where Π_X is the anisotropic stress of the fluid.

As an example of the gauge invariant program, let us write the total matter gauge density perturbation in a gauge-independent way. The total matter gauge condition is satisfied by shifting the time slicing from an arbitrary gauge by $T = (V_T^G - B^G)/k$ [Kodama & Sasaki 1984, eq. (3.3a); Hu 1995, eq. (4.88)]. Equation (3) tells us that the density perturbation transforms as

$$\delta_X^T = \delta_X^G + 3(1 + p_X/\rho_X)\frac{\dot{a}}{a}(V_T^G - B^G)/k.$$

(6)

In an arbitrary gauge G, this represents what the density perturbation would be in the total matter rest frame. This "gauge invariant" definition can be used to employ total matter gauge variables in a Newtonian or synchronous gauge treatment:

$$\delta_X^T = \delta_X^N + 3\frac{\dot{a}}{a}(1 + p_X/\rho_X)V_T^N/k$$

$$= \delta_X^S + 3\frac{\dot{a}}{a}(1 + p_X/\rho_X)V_T^S/k.$$

(7)

If fluctuations are *adiabatic*, the number density fluctuations of the matter and radiation evolve together, *i.e.* their bulk velocities are equal. In this case, V_T^S is equal to the freely-falling cold dark matter velocity, commonly defined to be zero in synchronous gauge. Equation (7) then tells us that the total matter gauge and synchronous gauge density perturbation are numerically equivalent. In a matter dominated (pressureless) universe, there is no fundamental scale in these rest frame evolution equations and density fluctuations evolve in a scale free manner. This is not true for the Newtonian gauge. Here fluctuations grow by the gravitational infall of the matter into potential wells. Due to causality, this introduces a fundamental scale, the *horizon* scale, into the evolution equations. For example in the case of adiabatic fluctuations, density perturbations are constant outside the horizon and only grow (or decay) after the horizon has grown larger than the wavelength.

Now let us see how gauge modes creep in. The mapping of synchronous perturbations onto any gauge with fixed coordinates is unambiguous as we have seen. However, since it is a many-to-one operation, its inverse leaves additional gauge freedom. Notice from equation (7), knowledge of the total matter or Newtonian gauge perturbations only fixes a certain combination of density and velocity perturbations in synchronous gauge. To remove the ambiguity, one must fix the initial synchronous velocity. Since an initial velocity decays with the expansion as a^{-1}, equation (7) tells us that a different choice will alter the behavior of densities by an additional term proportional to $\dot{a}/a^2$ or a^{-2} during radiation domination and $a^{-3/2}$ during matter domination. This is a *gauge mode*.

In summary, the "gauge invariant" approach does nothing to solve the gauge ambiguity; no solution is necessary since gauge choice poses no fundamental problems. Yet even though there is nothing particularly deep about the "gauge invariant" program, it is often useful. It allows us to borrow Newtonian and rest frame concepts for use in any gauge. In addition, by providing a systematic way of mapping perturbations in a specific gauge onto an arbitrary gauge, one automatically determines whether the coordinates have been completely fixed, *i.e.* whether gauge modes have been entirely eliminated.

2.3 Newtonian Gauge Equations

Now we are ready to state the evolution equations in an explicit form. To avoid unnecessary confusion, we will stay in the pure Newtonian gauge throughout this treatment. We will hereafter drop the superscript N with the understanding that all perturbation variables are in the Newtonian gauge unless otherwise specified. The continuity and Euler equations for the photon temperature in flat space are

$$\dot{\Theta}_0 = -\frac{k}{3}\Theta_1 - \dot{\Phi},$$

$$\dot{\Theta}_1 = k[\Theta_0 + \Psi - \frac{1}{6}\Pi_\gamma] - \dot{\tau}(\Theta_1 - V_b),$$

$$(8)$$

where the term proportional to the Compton differential optical depth $\dot{\tau}$ comes from momentum conservation in the scattering. Recall that $\Theta_0 = \frac{1}{4}\delta_\gamma$ is the isotropic temperature fluctuation, and $\Theta_1 = V_\gamma$ is the amplitude of the photon dipole or bulk velocity. One can see that when the optical depth to scattering is high, the photons become isotropic in the electron-baryon rest frame, i.e. the dipole moment $\Theta_1 = V_b$. The anisotropic stress of the photons Π_γ is directly proportional to its quadrupole moment. Since scattering makes the photons isotropic in the baryon rest frame, the photon anisotropic stress is negligible before recombination. These equations can also be directly obtained from kinetic theory. In fact, they are contained in the Legendre decomposition of the Boltzmann equation (2) in Fourier space which also gives the evolution of Π_γ through a hierarchy of higher angular moments (Wilson & Silk 1981).

Aside from the usual velocity divergence source in the continuity equation, there is a term dependent on the metric. This is due to the gravitational redshift effects of time dilation. As the form of the metric $g_{ij} = -a^2\delta_{ij}(1 + 2\Phi e^{i\mathbf{k}\cdot\mathbf{x}})$ implies, it is entirely analogous to the cosmological redshift. Heuristically, the presence of matter curves or stretches space taking the wavelength of the photon with it. In the Euler equation, the Newtonian potential Ψ acts as a source of the dipole. Gradients in the potential also induce gravitational blue and red shifts as the photons fall into and climb out of potential wells ($\Psi < 0$). This is countered by photon pressure from Θ_0. As the temperature rises so does the pressure which opposes the fall of a photon into the potential well. As the photons free stream, power in the dipole is converted into the higher multipole moments through the quadrupole via the Π_γ term.

The baryon continuity and Euler equations take on a similar form

$$\dot{\delta}_b = -kV_b - 3\dot{\Phi},$$
$$\dot{V}_b = -\frac{\dot{a}}{a}V_b + k\Psi + \dot{\tau}(\Theta_1 - V_b)/R. \tag{9}$$

Momentum conservation in Compton scattering gives the form of the coupling. From equation (5), the effective momentum density of a general fluid X is $(\rho_X + p_X)V_X$. Since $p_\gamma = \frac{1}{3}\rho_\gamma$ and $p_b \ll \rho_b$, conservation implies $(4\rho_\gamma/3)\delta\Theta_1 = \rho_b\delta V_b$. Thus the Compton coupling for the baryons takes on a similar form to the protons but is of opposite sign and is altered by a factor of $R = 3\rho_b/4\rho_\gamma$.

Again, the continuity equation is modified by gravitational effects from the stretching of space associated with Φ. Since the density decreases as the length scale cubed, the total differential effect becomes $3\dot{\Phi}$. In the Euler equation, the velocity is damped by the expansion and enhanced by infall into potential wells. Particle momenta scale as a^{-1} due to the expansion. For non-relativistic particles, this causes the peculiar velocity to scale similarly in the absence of sources. In the fully-relativistic case, it causes the temperature to decrease as a^{-1}.

Decoupled components such as the massless neutrinos and cold dark matter follow identical evolution equations save for the absence of Compton coupling. The metric perturbations on the other hand feel the influence of the total matter perturbations, $\rho_T \delta_T = \sum_i \rho_i \delta_i$, $(\rho_T + p_T)V_T = \sum_i (\rho_i + p_i)V_i$, and $p_T \Pi_T = \sum_i p_i \Pi_i$, where i runs through all the particle species, through the generalized Poisson equations,

$$k^2 \Phi = 4\pi G a^2 \rho_T [\delta_T + 3\frac{\dot{a}}{a}(1 + p_T/\rho_T)V_T/k],$$

$$k^2 (\Psi + \Phi) = -8\pi G a^2 p_T \Pi_T,$$

$$(10)$$

which arise from the time-time + time-space and traceless space-space components of the Einstein equations respectively. The presence of a^2 in the first equation represents the conversion from physical to comoving coordinates. Notice that if we rewrite the equation in terms of the density fluctuation on the total matter rest frame, the first equation would take on the familiar non-relativistic form of the Poisson equation and simplify perturbation calculations. The additional term represents a relativistic effect that is important outside the horizon. When anisotropic stress $p_T \Pi_T$ may be ignored, the second of equations (10) reduces to $\Psi = -\Phi$ as one would expect for the Newtonian potential.

2.4 Gauge Tricks: Sachs-Wolfe Example

Examining the photon conservation equations (8) and the Poisson equations (10), we see that the former are simpler in the Newtonian gauge, whereas the latter are simpler in the total matter gauge. Let us see how gauge tricks developed

in §2.2 can help us understand their joint evolution. In the matter dominated epoch, the Poisson and continuity equations reduce to

$$\Phi = \frac{6}{(k\eta)^2}\delta_T^T, \qquad \dot{\delta}_T^T = -kV_T,$$

as one expects from a non-relativistic analysis. Since δ_T^T is the density perturbation on the matter rest frame, its evolution in the growing mode also satisfies the non-relativistic relation $\delta_T^T \propto \eta^2$. Hence $V_T = -2\delta_T^T/k\eta$. Recalling its relation to the Newtonian density perturbation, we obtain

$$\delta_T = \delta_T^T - \frac{6}{k\eta}V_T = \left[1 + \frac{12}{(k\eta)^2}\right]\delta_T^T$$

$$\approx \frac{12}{(k\eta)^2}\delta_T^T = 2\Phi, \qquad k\eta \ll 1.$$

For adiabatic fluctuations, the intrinsic temperature fluctuation, $\Theta_0 \equiv \frac{1}{4}\delta_\gamma = \frac{1}{3}\delta_T = \frac{2}{3}\Phi$. We shall see that after a gravitational redshift of Ψ, this implies that the effective temperature $\Theta_0 + \Psi = \frac{2}{3}\Phi + \Psi = \frac{1}{3}\Psi$ which is the famous Sachs-Wolfe (1968) result.

3 Primary Anisotropies and the Tight Coupling Approximation

Before recombination, the Compton scattering time was so short that the photons and baryons behaved as a single fluid. This allows us to greatly simplify the treatment of anisotropy formation. Specifically, since the mean free path is much shorter than a wavelength of the fluctuation, the optical depth through a wavelength $\sim \dot{\tau}/k$ is large. Thus the evolution equations may be expanded in $k/\dot{\tau}$. Employing the baryon Euler equation (9), we may eliminate the baryon velocity from the photon evolution equation to obtain the first order equation (Peebles & Yu 1970; Hu & Sugiyama 1995a)

$$\frac{d}{d\eta}(1 + R)\dot{\Theta}_0 + \frac{k^2}{3}\Theta_0 = -\frac{k^2}{3}(1 + R)\Psi - \frac{d}{d\eta}(1 + R)\dot{\Phi}. \tag{11}$$

where we have dropped the higher order correction $\Pi_\gamma = \mathcal{O}(k/\dot{\tau})$ (see §3.2). Conceptually, this equation reads: the change in momentum of the photon-baryon fluid is determined by a competition between the pressure restoring and the gravitational driving forces. If we ignore the time dependence of the baryon-photon momentum ratio R from the expansion, this equation describes a forced harmonic oscillator. Since scattering requires the bulk velocities of the photons and baryons to be equal, the effective dimensionless mass of the fluid is given by $m_{\text{eff}} = 1 + R$ to account for the inertia of the baryons. Baryons also contribute gravitational mass to the system as is evident in the appearance of m_{eff} in the infall and dilation terms on the right hand side. They do not however contribute significantly to the pressure or restoring force of the system.

3.1 Acoustic Oscillations

As an instructive first approximation, let us ignore time variations in the potentials Φ and Ψ and also the baryon-photon momentum ratio R compared with changes at the oscillation frequency $\omega = kc_s$, where the sound speed (Doroshkevich, Zel'dovich & Sunyaev 1978)

$$c_s = \frac{1}{\sqrt{3(1 + R)}}.$$

Equation (11) then reduces to the familiar form

$$\ddot{\Theta}_0 + k^2 c_s^2 \Theta_0 = -\frac{1}{3} k^2 \Psi.$$

This is a simple harmonic oscillator under the constant acceleration provided by gravitational infall and can immediately be solved as

$$\Theta_0(\eta) = [\Theta_0(0) + (1+R)\Psi]\cos(kr_s) + \frac{1}{kc_s}\dot{\Theta}_0(0)\sin(kr_s) - (1+R)\Psi, \quad (12)$$

where the sound horizon $r_s = \int c_s d\eta \approx c_s \eta$. The two initial conditions $\Theta_0(0)$ and $\dot{\Theta}_0(0)$ govern the form of the acoustic oscillation. We shall see below that they represent the *adiabatic* and *isocurvature* modes respectively. Equation (12) also implies through the photon continuity equation (8) that

$$\Theta_1(\eta) = 3[\Theta_0(0) + (1+R)\Psi]c_s\sin(kr_s) + 3k^{-1}\dot{\Theta}_0(0)\cos(kr_s). \quad (13)$$

In equations (12) and (13), lie the main acoustic and redshift effects which dominate primary anisotropy formation.

3.1.1 Gravitational Infall and Redshift

In the early universe, photons dominate the fluid and $R \to 0$. In this limit, the oscillation takes on an even simpler form. For the adiabatic mode, $\dot{\Theta}_0(0) = 0$ and $\Theta_0(\eta) = [\Theta_0(0) + \Psi]\cos(kr_s) - \Psi$. This represents an oscillator with a zero point which has been displaced by gravity. The zero point represents the state at which gravity and pressure are balanced. The displacement $-\Psi > 0$ yields hotter photons in the potential well since gravitational infall not only increases the number density of photons but also their energy through gravitational blueshifts.

However, photons also suffer a gravitational redshift from climbing out of the potential well after last scattering. This precisely cancels the $-\Psi$ blueshift. Thus, the effective temperature perturbation is $\Theta_0(\eta) + \Psi = [\Theta_0(0) + \Psi]\cos(kr_s)$ (see Fig. 2). The phase of the oscillation at last scattering determines the effective fluctuation. Since the oscillation frequency $\omega = kc_s$, the critical wavenumber $k = \pi/r_s(\eta_*) \approx \pi/c_s\eta_*$ is essentially at the scale of the *sound horizon*. If there is a spectrum of k-modes, there will be a harmonic series of temperature *fluctuation* peaks with $k_m = m\pi/r_s(\eta_*)$ for the mth peak. Odd peaks thus represent the compression phase (temperature crests), whereas even peaks represent the rarefaction phase (temperature troughs), inside the potential wells.

As first calculated by Sachs & Wolfe (1967), the effective temperature in the matter dominated limit goes to $\Theta_0 + \Psi = \frac{1}{3}\Psi$. In the Newtonian frame, this is a direct consequence of the adiabatic initial conditions. In general, if $k\eta \ll 1$ equation (8) implies

$$[\Theta_0 + \Psi](\eta_*) = [\Theta_0 + \Psi](0) + [\Psi - \Phi]\Big|_0^{\eta_*}.$$

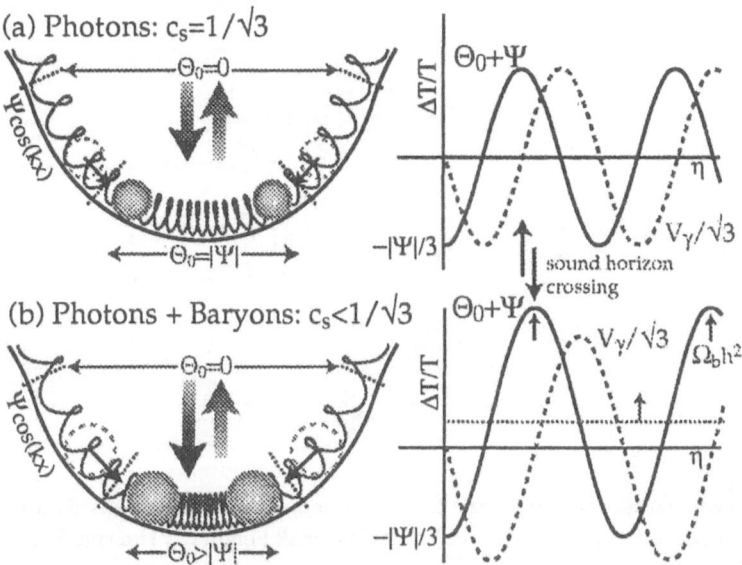

Fig. 2. Acoustic oscillations. Fluid compression through gravitational infall is resisted by photon pressure setting up acoustic oscillations. Without baryons the photon blueshift on infall is equal and opposite to the redshift upon leaving the potential leading to symmetric oscillations around zero with a Doppler effect of equal magnitude but 90 degrees out of phase. Baryons increase the mass of the fluid causing more infall and a net zero point displacement. Temperature crests (compression) are enhanced over troughs (rarefaction) and velocity contributions.

In a full calculation, the small variation in the potential at equality due to the change in the equation of state brings the effective temperature from $\frac{1}{2}\Psi$ to $\frac{1}{3}\Psi$. As this does not affect the qualitative picture, we will encorporate this effect as $[\Theta_0 + \Psi](0) = \frac{1}{3}\Psi$ and $\Delta(\Psi - \Phi) = 0$. Notice that if $\Theta_0(0) = \Psi(0) = -\Phi(0) = 0$, then $[\Theta_0 + \Psi](\eta_*) = \Psi(\eta_*) - \Phi(\eta_*) \approx 2\Psi(\eta_*)$ which yields a factor of 6 enhancement for isocurvature models. An alternate way of deriving these results is to employ gauge tricks as shown in §2.4. The Sachs-Wolfe effect is a combination of an intrinsic temperature and a gravitational redshift. Since the photon density and thus the intrinsic temperature is a gauge dependent concept, this breakdown, but not the observable result, will also depend on gauge.

3.1.2 Baryon Drag

Though effectively pressureless, the baryons still contribute to the inertial and gravitational mass of the fluid $m_{\text{eff}} = 1 + R$. This decreases the sound speed and changes the balance of pressure and gravity. Gravitational infall now leads to greater compression of the fluid in a potential well, *i.e.* a fur-

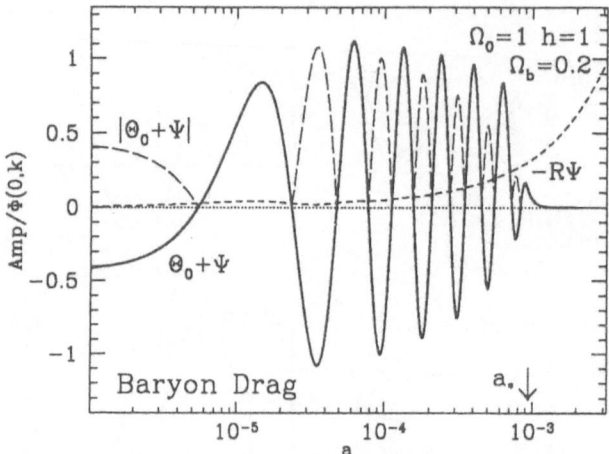

Fig. 3. Baryon Drag. Baryons drag the photons into potential wells leading to a zero point displacement of $|R\Psi|$ and alternating peak heights in the rms fluctuation which provide a measure of $\Omega_b h^2$. The fall off near a_* of the amplitude is due to diffusion damping.

ther displacement of the oscillation zero point (see Fig. 2). Since the redshift is not affected by the baryon content, this relative shift remains after last scattering to enhance all peaks from compression over those from rarefaction. If the baryon-photon ratio R were constant, the effective temperature perturbation would become

$$\Theta(\eta) + \Psi = \frac{1}{3}\Psi(1 + 3R)\cos(kr_s) - R\Psi,$$

with compressional peaks a factor of $(1+6R)$ over the $R = 0$ case. In reality, the effect is reduced since $R \to 0$ at early times (see Fig. 3).

The *evolution* of the effective mass has another effect on its own. In classical mechanics, the ratio of energy $\frac{1}{2}m_{\text{eff}}\omega^2 A^2$ to frequency ω of an oscillator is an adiabatic invariant. Thus for the slow changes in $m_{\text{eff}} \propto c_s^{-2}$, the amplitude of the oscillation varies as $A \propto c_s^{1/2} \propto (1 + R)^{-1/4}$ since $\omega \propto c_s$. The fundamental solutions of the oscillator equation are modified to be $(1 + R)^{-1/4}\cos(kr_s)$ and $(1 + R)^{-1/4}\sin(kr_s)$.

3.1.3 Doppler Effect

Since the turning points are at the extrema, the fluid velocity oscillates 90 degrees out of phase with the density (see Fig. 2). Its motion relative to the observer causes a Doppler shift. Whereas the observer velocity creates a pure dipole anisotropy on the sky, the fluid velocity causes a spatial temperature variation $\Theta_1/\sqrt{3}$ on the last scattering surface from its line of

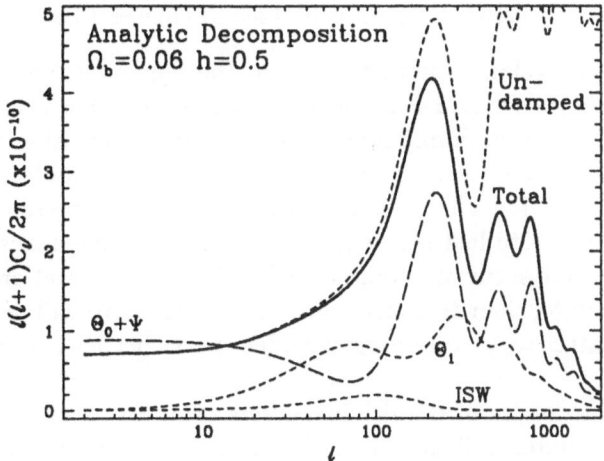

Fig. 4. Analytic decomposition ($\Omega_0 = 1$ scale invariant adiabatic model). The effective temperature after gravitational redshift dominates the primary anisotropy. Peak heights are enhanced and modulated by baryon-photon ratio R as well as experience a boost crossing the equality scale at $\ell \approx \sqrt{3}(2\Omega_0 H_0^2/a_{eq})^{1/2}\eta_0 \approx 400$. The Doppler effect is smaller and out of phase with the temperature. The ISW effect due to potential decay after recombination is small here but can be significant for low matter content universes. Diffusion damping cuts off the acoustic spectrum at small scales.

sight component. For a photon-dominated $c_s \approx 1/\sqrt{3}$ fluid, equation (13) tells us the velocity contribution is equal in amplitude to the density effect. This photon-intrinsic Doppler shift should be distinguished from the electron-induced Doppler shift of reionized scenarios.

The addition of baryons significantly changes the relative velocity contribution. As the effective mass increases, conservation of energy requires that the velocity decreases for the same initial temperature displacement. Thus the *relative* amplitude of the velocity scales as c_s. In the toy model of a constant baryon-photon density ratio R, the oscillation becomes $\Theta_1/\sqrt{3} = \frac{1}{3}\Psi(1 + 3R)(1 + R)^{-1/2}\sin(kr_s)$. Notice that velocity oscillations are symmetric around zero unlike the temperature ones. Thus compressional peaks will rise clearly above the velocity oscillations if R is large. Even in a universe with $\Omega_b h^2$ given by nucleosynthesis, R is sufficiently large to make velocity contributions subdominant (see Fig. 4).

3.1.4 Acoustic Driving Effects and Isocurvature Models

Whenever the non-relativistic matter is not the dominant dynamical component, the potentials Φ and Ψ become time-dependent. For example, when the universe is radiation dominated, pressure and entropy alter the behavior of the gravitational potential. External sources from topological defects can also play a role. The effects of potential evolution can be separated into those that occur before last scattering and thus affect the acoustic oscillations and those that occur afterwards which affect the gravitational redshift of the photons along their free-streaming trajectories. We will defer consideration of the latter to §4 where we consider effects between recombination and the present.

As a simple example of the driving effects of potential evolution, let us again consider adiabatic fluctuations where the initial temperature $\Theta_0(0)$ and curvature fluctuations $\Phi(0)$ are finite constants related by the Poisson equation and $\dot{\Theta}_0(0) = 0$. Inside the sound horizon, pressure prevents gravitational infall in the photon-baryon and neutrino systems. Unless CDM dominates, energy density perturbations $\delta\rho_T$ decay with the expansion and consequently can no longer maintain a constant gravitational potential. Counterintuitively, this decaying potential can actually enhance temperature fluctuations through its near resonant driving force. Since the potential decays after sound horizon crossing, it drives the first compression without a counterbalancing effect on the subsequent rarefaction stage (see Fig. 5).

Furthermore, the dilation effect from a decaying space curvature Φ also enhances the acoustic fluctuations. Heuristically, the overdensities which establish the potential well "stretch" the space-time fabric. As the potential well decays, it re-contracts. Photons which are caught in this contraction find their wavelength similarly contracted, *i.e.* blueshifted. Thus a differential change in Φ leads to a dilation effect, $\dot{\Theta}_0 = -\dot{\Phi}$. Combined, the driving effects of infall and dilation yield an effect of order $\frac{1}{3}\Psi - \Psi + \Phi \approx \frac{5}{3}\Psi$ or 5 times the Sachs-Wolfe effect when the perturbation crosses the sound horizon (see Fig. 4,5 and Hu & Sugiyama 1995c). Since this effect only occurs for modes which cross before CDM domination, the amplitude of this boost is sensitive to the matter-radiation ratio of the universe. In particular, lowering $\Omega_0 h^2$, raising the massless neutrino number or energy density, or making some of the cold dark matter relativistic at early times through eV mass neutrinos (see Fig. 6b; Ma & Bertschinger 1995; Dodelson, Gates & Stebbins 1995), can move the boost to larger scales and thus enhance the first few acoustic peaks.

Similar resonant effects can occur in other situations. For example, many alternate scenarios follow a typical *isocurvature* behavior. The hallmark of isocurvature type models is that curvature fluctuations Φ are zero or at least suppressed outside the horizon. The curvature fluctuation grows by causal processes as the fluctuation crosses the horizon. This isocurvature behavior occurs for entropy fluctuations where matter and radiation perturbations

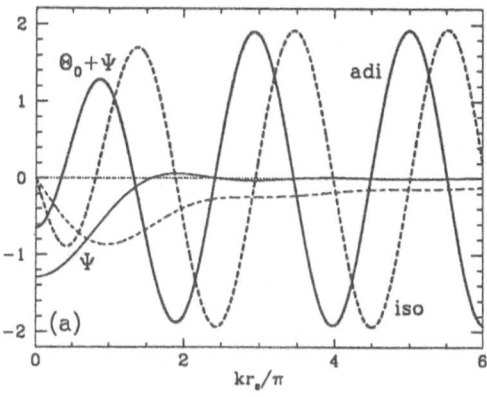

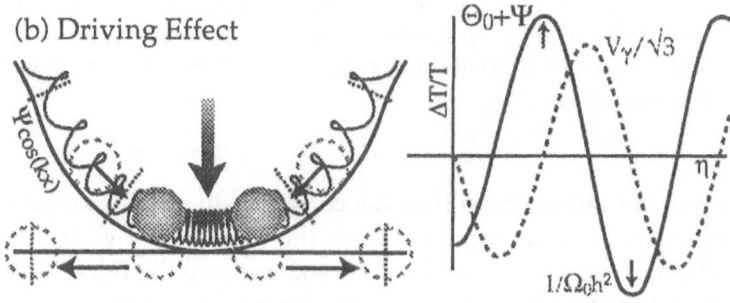

Fig. 5. Driving effects. The time evolution of the potential can enhance the amplitude of the acoustic oscillation by driving effects. In the adiabatic case [(a) solid lines, numerical results and (b) heuristic picture], the potential enhances the first compression through infall and decays leaving the oscillator strongly displaced from the zero point. In the (baryon) isocurvature case [(a) dashed line], the potential grows from zero and stimulates a sine mode. The first extrema here is suppressed due to the fact that the gravitation driving begins near sound horizon crossing. Notice that for the first cycle of the adiabatic and isocurvature oscillations, the gravitational force mimics a driving force of approximately twice the natural period.

are counterbalanced in the initial state, whether through the baryons (Hu & Sugiyama 1995b), hot dark matter (de Laix & Scherrer 1995) or axions (Sugiyama & Gouda 1992), texture models (Crittenden & Turok 1995, Durrer, Gangui & Sakellariadou 1995), and is at least part of the story for string models (Hindmarsh, private communication, but see also Albrecht, Coulson & Ferreira 1995). The qualitative effect on isocurvature conditions is easy to see. Since the potential grows from zero to a maximum near sound horizon crossing and then decays due to radiation pressure, the force drives the sine harmonic of oscillations (see Fig. 5a). The result is that isocurvature acoustic peaks are 90 degrees out of phase with their adiabatic counterparts. The

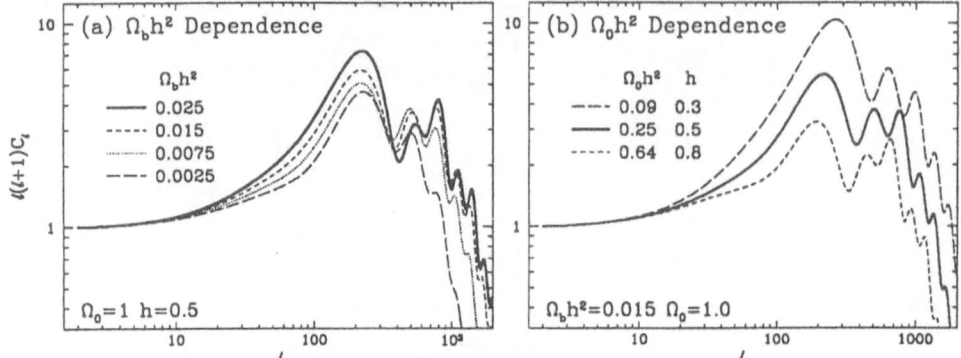

Fig. 6. Heights of the peaks in a scale invariant adiabatic model. The heights of the peaks is determined by the baryon drag and acoustic driving effects and so are sensitive to $\Omega_b h^2$ and $\Omega_0 h^2$. Baryons increase compression in potential wells causing the peak heights to alternate. Lower matter content causes more potential decay driving the oscillation.

first peak occurs at larger scales that the corresponding adiabatic peak but generally has a suppressed amplitude since the corresponding mode is just at the sound horizon $kr_s = \pi/2$ and has not experienced the full effect of the driving term. These features may serve to distinguish the two basic types of scenarios in a manner independent of the details of the given model.

Notice that in these examples, oscillations give a coherent spectrum of peaks in integral or half integral multiples of $\int \omega d\eta = kr_s(\eta_*)$. Since each k mode evolves independently, this coherence in the temporal phase is the result of a special timing in the gravitational impulse. In all the examples considered above, the gravitational force kicks in between horizon crossing and sound horizon crossing. This explains the coherence in k and the fundamental scale in the problem $r_s(\eta_*)$. More exotic scenarios may not preserve this coherence. If the source of the gravitational potential has a more random temporal behavior, the sine and cosine acoustic modes will be stimulated incoherently with equal likelihood. Combined, the two modes would lead to a smoothed out rms temperature fluctuation below the sound horizon. Preliminary calculations indicate that this may be case for cosmic string models due to complicated behavior of the string network inside the horizon (Albrecht *et al.* 1995; Magueijo *et al.* 1995).

These considerations can be made quantitative by solving the oscillator equation under the influence of an arbitrary but known gravitational forcing function either analytically (Hu & Sugiyama 1995a) or numerically by modeling the gravitational source (e.g. Seljak 1994; Crittenden & Turok 1995). Simple closed form solutions for standard adiabatic and isocurvature models are presented in Hu & Sugiyama (1995c).

3.2 Diffusion Damping

In reality, the photons and baryons are not perfectly coupled since the photons possess a mean free path in the baryons $\lambda_C \approx \dot{\tau}^{-1}$ due to Compton scattering. As the photons random walk through the baryons, hot spots and cold spots are mixed. Fluctuations thereafter remain only in the unscattered fraction causing a near exponential decrease in amplitude as the diffusion length $\lambda_D \sim \sqrt{N}\lambda_C = \sqrt{\eta\lambda_C}$ or $k_D \sim \sqrt{\dot{\tau}/\eta}$ overtakes the wavelength.

To be more specific, diffusion causes heat conduction and generates viscosity in the fluid (Weinberg 1972). As photons from regions of different temperature meet, anisotropies form leading to a quadrupole moment or anisotropic stress in the fluid. If the diffusion length is well under the horizon, it overtakes the wavelength of the fluctuation when $\dot{\tau}/k = k\eta \gg 1$. Thus the optical depth through a wavelength is still high and the perturbative expansion of §3.1 still holds. This fact allows us to extend the tight coupling approximation for acoustic modes to last scattering. For small scale modes in which the optical depth through a wavelength $\dot{\tau}/k$ is *not* high at recombination, all acoustic oscillations will already have damped away (Hu & Sugiyama 1995c).

Photon diffusion is a second order effect and damps acoustic oscillations as $\exp[-(k/k_D)^2]$ with the damping wavenumber

$$k_D^{-2} = \frac{1}{6} \int d\eta \frac{1}{\dot{\tau}} \frac{R^2 + 4f_2^{-1}(1+R)/5}{(1+R)^2},$$

where f_2 accounts for the subtle effects in the generation of anisotropic stress (see Fig. 4). Kaiser (1983) showed that $f_2 = 9/10$ due to the angular dependence and 3/4 if the additional effects of polarization are included. The two processes aid the generation of the quadrupole moment and hence increase the viscous damping of the acoustic oscillations.

The baryons are dynamically coupled to the photons by momentum exchange in Compton scattering from $\dot{\tau}_d = \dot{\tau}/R$ in equation (9). If $\dot{\tau}_d/k \gg 1$, the baryons will be dragged in and out of potential wells with the photons. This process destroys the baryonic acoustic oscillations as well and is known as Silk (1968) damping. Since coupling requires coevolution in the number density $\delta_b = \frac{3}{4}\delta_\gamma = 3\Theta_0$, only entropy fluctuations $S = \delta_b - \frac{3}{4}\delta_\gamma$ survive diffusion damping.

Notice also that the ionization history factors in the diffusion length through the mean free path $\dot{\tau}^{-1}$. As we shall now see, at recombination the mean free path and hence the diffusion length increases substantially but does not approach the horizon scale.

3.3 Decoupling

The CMB anisotropy today is simply the acoustic fluctuation at last scattering, modified by diffusion damping and free streamed to the present:

$$[\Theta + \Psi](\eta_0, k, \mu) \approx [\Theta_0 + \Psi - i\mu\Theta_1](\eta_*, k)\mathcal{D}_\gamma(k)e^{ik\mu(\eta_* - \eta_0)}, \qquad (14)$$

for flat space, where $\mathbf{k} \cdot \gamma = k\mu$,

$$\mathcal{D}_\gamma(k) = \int_0^{\eta_0} d\eta \, \mathcal{V}_\gamma e^{-[k/k_D(\eta)]^2}, \qquad (15)$$

is the diffusion damping factor and the visibility function $\mathcal{V}_\gamma = \dot{\tau}e^{-\tau}$ is the probability of last scattering within $d\eta$ of η. Its peak is near $\tau(z_*) = 1$. The damping factor is dependent only on the background cosmology and fits across the whole range of parameter space are presented in Hu & Sugiyama (1995c). We shall examine the meaning and implications of each of these terms below.

The baryons do not decouple from the photons precisely at last scattering. As we have seen in §3.2, the coupling strength is altered by a factor of R, $\dot{\tau}_d = \dot{\tau}/R$. By analogy to the photon case, we can define a drag optical depth τ_d and the end of the drag epoch as $\tau_d(z_d) = 1$ (Hu & Sugiyama 1995c). After this point, Compton drag on the baryons can no longer prevent the gravitational infall of the baryons. Because $R(z_*) \sim 0.3$ for standard recombination and a big bang nucleosynthesis value for $\Omega_b h^2$, z_* and z_d approximately coincide for the standard case. However for reionized scenarios $R(z_*) \gg 1$ and the baryons decouple from the photons long before last scattering. It is thus no longer appropriate to consider the photons and baryons as tightly coupled at last scattering. We will develop new techniques to handle this situation in §4. From the drag depth τ_d, a drag visibility function $\mathcal{V}_b \approx \dot{\tau}_d e^{-\tau_d}$ can be constructed (see Hu & Sugiyama 1995c for the small correction due to expansion damping). The final scale for Silk damping is obtained from the visibility function and the diffusion length as in equation (15) for the photons.

3.4 Projection Effects and the Anisotropy Spectrum

The presence of the $\exp(ik\mu\Delta\eta)$ term in equation (14) represents the free streaming of the photons in flat space. Photons travel at the speed of light so that the number of wavelengths traveled between last scattering and today is $k\Delta\eta$. However, the phase of the wavefront changes only in the perpendicular direction so that the observed phase change will depend on the line of sight $\phi = k\Delta\eta\mu$. Since inhomogeneities at the last scattering surface appear as anisotropies on the sky today, this can equivalently be viewed as a simple projection of the plane wave on the sphere. The anisotropy is expressed through the decomposition of the plane wave into multipole moments $\ell \sim \theta^{-1}$.

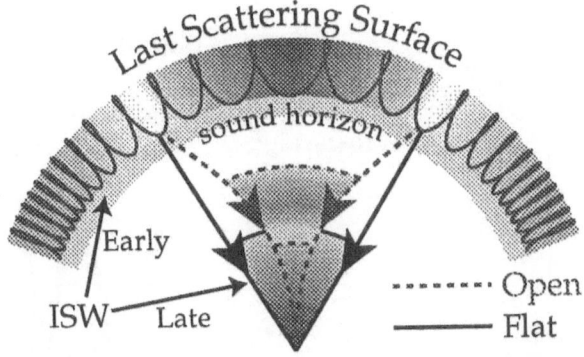

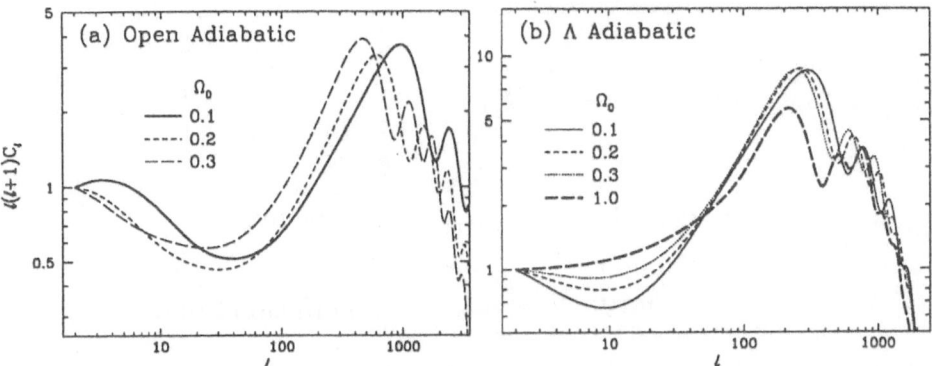

Fig. 7. Location of the peaks. The location of the peaks is determined by the angle the sound horizon subtends at last scattering. In an open universe, geodesic deviation makes this angle much smaller than in a flat universe. As this angle involves a ratio of the sound horizon to the (angular diameter) distance to last scattering, distance scale changes through Λ and h have little effect on its angle. The drop in the lowest multipoles for the open $\Omega_0 = 0.1$ is due to the lack of supercurvature scale power in the otherwise scale invariant initial conditions assumed here. Models here have $h = 0.5$ and $\Omega_b = 0.05$.

$$\frac{\Theta_\ell(\eta_0, k)}{2\ell + 1} = \left\{ [\Theta_0 + \Psi](\eta_*, k) + \Theta_1(\eta_*, k)\frac{1}{k}\frac{d}{d\eta} \right\} \mathcal{D}_\gamma(k) j_\ell(k\Delta\eta). \quad (16)$$

Since j_ℓ peaks at $\ell \sim k\Delta\eta$, it is clear that free streaming just projects the physical scale $k \sim \lambda^{-1}$ onto an angular scale as $\theta \sim \lambda/\Delta\eta$. The total power in the ℓth multipole is obtained by integration over k modes, $C_\ell = (2/\pi) \int k^3 |\Theta_\ell(\eta_0, k)|^2/(2\ell + 1)^2 d\ln k$.

The physical content of the generalization to open universes is now obvious. Since free streaming merely represents a projection, one replaces the comoving distance $\Delta\eta$ with the comoving angular diameter distance $r_\theta = |K|^{-1/2} \sinh[|K|^{1/2}\Delta\eta]$, where the curvature $K = -H_0^2(1 - \Omega_0 - \Omega_\Lambda)$.

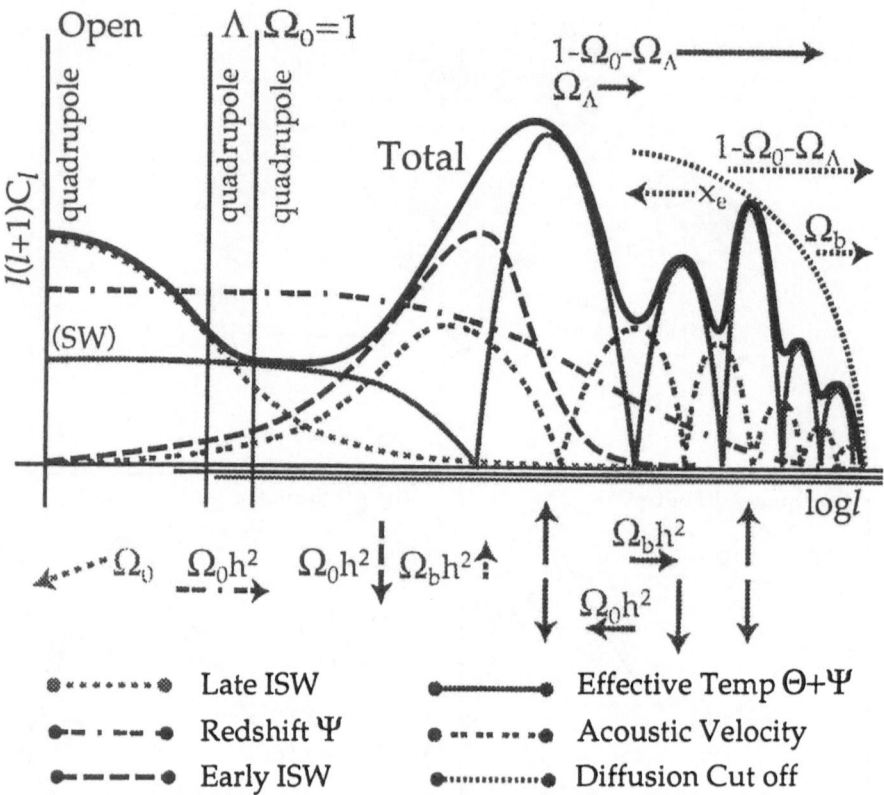

Fig. 8. Parameter sensitivity. A schematic representation based on a scale invariant adiabatic scalar model. Features in an open model are shifted to significantly smaller angles compared with Λ and $\Omega_0 = 1$ models, represented here as a shift in the ℓ axis. Isocurvature models behave similarly with respect to the acoustic oscillations save that the peaks are 90 degrees out of phase. The spectrum may also be tilted by changing the initial power spectrum from the scale invariant assumption.

For a $K > 0$ closed geometry, merely replace the sinh with sin. Notice that in the closed case, the location of the first peak has oscillates with the curvature (White & Scott 1995). Putting this together, the acoustic peaks occur at

$$\ell_j = k_j |r_\theta(\eta_*)| = \left|\frac{r_\theta}{r_s}\right|_{\eta_*} \times \begin{cases} j\pi, & \text{adiabatic} \\ (j - 1/2)\pi. & \text{isocurvature} \end{cases}$$

Formally, this arises by replacing j_ℓ with the radial eigenfunction of an open universe (Wilson 1983). Notice that the peak locations only depend on the background cosmology once the adiabatic or isocurvature nature of the fluctuations has been established. Indeed, the ratio of the peak locations themselves can be employed to separate the two (Hu & Sugiyama 1995b).

The dominant factor in the peak locations is the curvature of the universe which can make the same physical scale subtend a much smaller angle on the sky (see Fig. 7). Here we have a classical cosmological test of the curvature: knowing the physical scale (sound horizon at last scattering) and the redshift ($z_* \approx 1000$) of the acoustic peak, we measure its angular extent. The inferred angular diameter distance tells us the curvature. If the peak location is known to sufficient accuracy, we can also measure $\Omega_b h^2$, $\Omega_0 h^2$ and Λ through their effects on the sound horizon and angular diameter distance (see Figs. 6 & 7). Analogous information can be obtained from the location of the diffusion damping scale even if oscillations are not apparent due to random driving effects.

The heights of the peaks also contain important, if less model independent information. As we have seen, every other peak is boosted by the baryon content due to baryon drag and scales inside the horizon at equality probe $\Omega_0 h^2$ or more generally the matter-radiation ratio which includes information on the neutrino content and mass. Since these two effects have relatively robust features, alternating peak heights and a strong height boost at matter-radiation equality, there is hope of obtaining information out of the heights of the peaks as well as their location for a wide class of models. As an example, we present a schematic picture for the anisotropy spectrum in Fig. 8 based on the popular scale invariant adiabatic model. Note that changing the overall dynamics from $\Omega_0 = 1$ through flat Λ models to open models is similar to shifting the spectrum in angular space toward smaller angles.

4 Secondary Anisotropies and the Weak Coupling Approximation

We have been assuming up to this point that CMB fluctuations are frozen in at a redshift $z_* \approx 1000$. Several processes in the foreground of recombination could alter the anisotropy spectrum. These are generally known as *secondary* anisotropies. Just as the tight coupling approximation assisted in the calculation and interpretation of primary anisotropies, the weak coupling approximation helps in understanding secondary anisotropies (Hu & White 1995).

In abstract form, anisotropy generation is governed by internal sources from photon fluctuations at last scattering S_{prim} and sources external to the photon system in the foreground of recombination S_{sec} by

$$\frac{\Theta_\ell(\eta_0, k)}{2\ell + 1} = \int_0^{\eta_0} [S_{prim} + S_{sec}] j_\ell(k\Delta\eta) d\eta, \tag{17}$$

in flat space with an appropriate generalization of the radial eigenfunction for an open geometry (Wilson 1983). Equation (17) just states that we observe the projection of the source at η on a shell at a distance $\Delta\eta = \eta_0 - \eta$.

Whether the sources or the phase of the wave varies more rapidly distinguishes the tight from the weak coupling regime. For primary anisotropies, the source is localized over a short range of time around recombination. Since $\dot{\tau}/k \gg 1$, j_ℓ can be taken out of the integral (17) leading to the simple projection of equation (16).

Secondary anisotropies do not necessarily possess this property. Most cosmological effects aside from decoupling take on the order of an expansion time at the relevant epoch to be completed. In these cases, S_{sec} may be taken to be slowly varying at small scales and removed from the integral in (17). Since

$$\int_0^{\eta_0} j_\ell(k\Delta\eta)d\eta \approx \frac{\sqrt{\pi}}{2k}\frac{\Gamma[\frac{1}{2}(\ell+1)]}{\Gamma[\frac{1}{2}(\ell+2)]} \approx \sqrt{\frac{\pi}{2\ell}}\frac{1}{k}, \tag{18}$$

in the weak coupling limit where $\dot{S}_{sec}/(kS_{sec}) \ll 1$, anisotropies fall with ℓ more rapidly than a simple projection of the source would imply. This just reflects the fact that it contributes across many wavelengths of the fluctuation allowing contributions from crests and troughs to cancel. We will now discuss the main sources of secondary anisotropies S_{sec}.

4.1 Gravitational Effects

Even in the absence of reionization, secondary anisotropies can be generated by gravitational redshift effects between recombination and today. The differential redshift from $\dot{\Psi}$ and dilation from $\dot{\Phi}$ discussed above must be integrated along the trajectory of the photons $S_{sec} = \dot{\Psi} - \dot{\Phi}$. Tensor fluctuations can also give rise to anisotropies through this mechanism. We thus call the combination the *integrated* Sachs-Wolfe (ISW) effect. Notice that these effects only occur if the metric fluctuation is time varying. We can separate this general mechanism for anisotropy formation into categories based on the reason for the time evolution. There are four possibilities to consider: time evolution due to the radiation content (early ISW effect), due to the expansion (late ISW effect), due to non-linear evolution (Rees-Sciama effect) and any more exotic sources such as gravity waves or defects (sourced ISW). Additionally, beyond linear theory gravitational lensing may affect the CMB. Since this merely shuffles power in anisotropies around scales at the arcminute level for a CDM type model, we will not further consider the effect and refer the interested reader to Seljak (1995a) and references therein.

4.1.1 Early ISW Effect

The early ISW effect is the direct analogue of the acoustic driving effect of §3.1.4 except that the photons are in the free streaming rather than the tight coupling regime. For adiabatic conditions, the potential decays after horizon crossing in the radiation dominated limit. For isocurvature conditions, it grows outside the horizon and then decays as in the adiabatic case. In general, this effect will smoothly match onto the acoustic peaks due to similarities in the cause of their generation. Due to the later time of generation, this effect influences larger scales than the acoustic peak but is cut off above the equality scale. Furthermore, since it arises from a distance closer to ourselves, the same physical scale subtends a larger angle on the sky. Together these considerations imply that the early ISW effect fills in the anisotropy on scales just larger than the first acoustic peak (see Fig. 8). It serves to broaden the rise and shifts the location of the first peak in the spectrum to larger scales. Unfortunately, this effect satisfies neither the tight nor the weak coupling approximation since the decay time $\sim \eta$ and the wavelength $\sim k^{-1}$ are by definition comparable for this horizon crossing effect.

4.1.2 Late ISW Effect

If the universe has a non-vanishing curvature or cosmological constant, eventually these will dominate the expansion rate. Under the rapid expansion, density fluctuation $\delta \rho_T$ decays taking the potential with it. The decay takes on the order of an expansion time at the end of matter domination independent of the wavelength. Since the photons free stream, they travel across many wavelengths of the perturbation on scales smaller than the horizon. If the potential decays while the photon is in an underdense region, it will suffer an effective redshift rather than a blueshift. Contributions from overdense and underdense regions will cancel and damp the ISW effect on small scales. This is the hallmark of the weak coupling regime and is mathematically expressed through cancellation in the integral (18).

For a fixed Ω_0, the decay epoch occurs much later in flat $\Omega_\Lambda + \Omega_0 = 1$ models than open ones. Thus Λ models will suffer cancellation of late ISW contributions at a much larger scale than open models. In fact, for reasonable $\Omega_0 \gtrsim 0.1$ decay the decay has already started at the quadrupole (see Fig. 7,8; Kofman & Starobinskii 1985; Hu & White 1995). In summary, the epoch that the universe exits the matter dominated phase is imprinted on the CMB by the late ISW effect.

4.1.3 Rees-Sciama Effect

Even for an $\Omega_0 = 1$ CDM dominated universe, potentials only remain constant in linear perturbation theory. The second order contribution has been shown to be negligibly small (Martinez-Gonzalez, Sanz & Silk 1992). The effect in the non-linear regime has been estimated with N-body simulations through power spectrum techniques (Seljak 1995) and ray tracing techniques (Tuluie, Laguna & Anninos 1995). The general conclusion is that this effect does not become comparable to the primary signal until well into the diffusion damping tail. In the absence of reionization, it should therefore present no problem for the extraction of main features such as the curvature and the baryon content from the acoustic peaks. It may however complicate the extraction of cosmological parameters from more subtle effects, for example the neutrino mass.

4.1.4 Sourced ISW Effect

This catch-all category contains all other gravitational redshift effects on the CMB. Tensor perturbations in the metric, *i.e.* gravity waves, give rise to dilation effects as the photons free stream. Due to the nature of the metric distortion, they leave a quadrupole signature in the CMB which thereafter is projected onto higher multipoles. Combined with cancellation effects, this implies that the tensor spectrum thus typically exhibits a sharp drop in power from the quadrupole. Since contributions only arise between last scattering and the present, there is also a cut off at the angle the horizon subtends at last scattering. Below this scale, the gravity waves have already redshifted away by last scattering. Detailed calculations of the spectrum were first carried out by Crittenden *et al.* (1994) and the spectrum shown in Fig. 9 is from Hu *et al.* (1995). For inflationary models, there exists a consistency relationship between the amplitudes of the scalar and tensor modes and the slope of the power spectrum (see *e.g.* Steinhardt 1995). There is hope that detailed measurements of the CMB spectrum can thus provide a strong test of the inflationary scenario (see Lindsey *et al.* 1995 and references therein).

Cosmological defects may also act as an external source of gravitational potential perturbations. Pen, Spergel & Turok (1993) find that in such models a scale invariant form for the anisotropy results from this effect and can mimic the inflationary prediction. As in the case of the late ISW effect, the magnitude drops from 2Ψ to zero as the characteristic time scale for the change in Ψ becomes longer than the light travel time across a wavelength. For defect models Pen, Spergel & Turok (1993) estimate the amplitude as approximately $\frac{4}{3}\Psi$. Since this effect occurs around horizon crossing for a given mode due to the isocurvature conditions, the largest spatial modes have not been enhanced and can lead to a drop in power at the lowest multipoles.

4.2 Scattering Effects

An early round of structure formation may be able to reionize all or part of the universe at high redshift. If this occurs at

$$z \gtrsim z_* \approx 10^2 (\Omega_0 h^2/0.25)^{1/3} (x_e \Omega_b h^2/0.0125)^{-2/3},$$

where $\tau(z_*) = 1$ then CMB anisotropies will be drastically influenced by scattering. The model-independent information stored in the primary

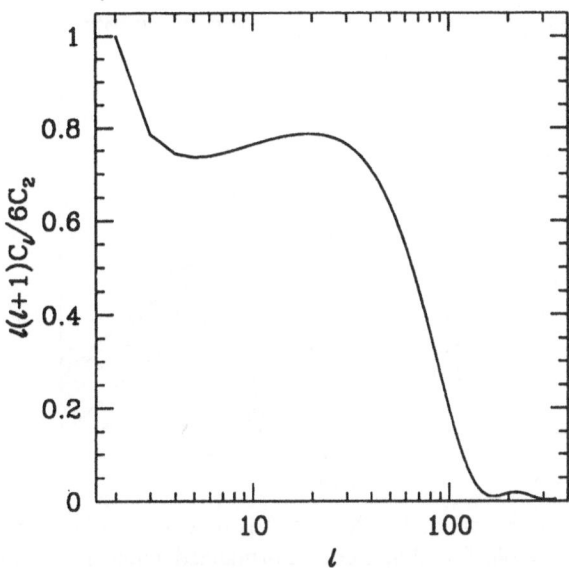

Fig. 9. Gravitational Wave ISW Effect. The metric distortion caused by a gravitational wave induces a redshift/dilation effect on CMB photons. The spectrum is cut off below the angle subtended by the horizon at last scattering since gravity waves only affect free streaming photons significantly. Due to the signature of the metric distortion, it originates as a source of the quadrupole anisotropy.

anisotropy spectrum will be greatly reduced. It is replaced by more detailed information and clues about the evolution of structure in the universe.

4.2.1 Reionization Damping

The most dramatic effect of reionization is the increase in the photon diffusion length. Since last scattering is delayed until the Compton mean free path approaches the horizon (or equivalently when the scattering rate $\dot{\tau}$ drops below the expansion rate $\dot{a}/a$), the diffusion length at last scattering is the horizon scale $\lambda_D \approx \sqrt{\eta/\dot{\tau}} \approx \sqrt{\eta a/\dot{a}} \approx \eta$. Intrinsic photon fluctuations such as the acoustic oscillations will be damped by diffusion below the horizon scale. Moreover, since acoustic oscillations appear only below the sound horizon no oscillations will be apparent in the reionized spectrum.

If reionization is not sufficiently early, some trace of the acoustic oscillations may remain in the spectrum. Recall that diffusion damping works since rescattering of photons arriving from directions with different intrinsic temperatures varies destroys the anisotropy. Fluctuations are only retained in the unscattered fraction $e^{-\tau}$. If the total optical depth between recombination and the present is $\tau \lesssim 1$, primary and secondary scattering anisotropies may be present in the spectrum (see Fig. 10). In this case, the information contained in the CMB would truly be enormous but also difficult to extract.

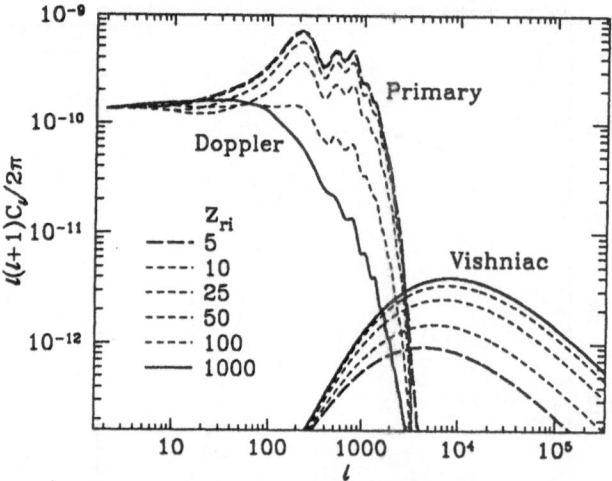

Fig. 10. Reionization damping, Doppler, and Vishniac effects (scale invariant adiabatic $\Omega_0 = 1$, $\Omega_b = 0.05$, $h = 0.5$, COBE normalized model). As the ionization level between recombination and the present increases, primary anisotropies are reduced by diffusion and rescattering. For sufficiently high ionization, here $z_i \gtrsim 100$, the Doppler effect on the new last scattering surface can regenerate some fluctuations at intermediate scales. The Vishniac effect dominates at very small angular scales. Due to its second order nature, it contributes even if the ionization redshift is low. The $z_i = 5, 10$ curves are indistinguishable for the primary anisotropies as are the $z_i = 100, 1000$ for the Vishniac effect.

4.2.2 Cancelled Doppler Effect

After the Compton drag epoch $\frac{3}{5} z_d \approx 165 (\Omega_0 h^2)^{1/5} x_e^{-2/5}$ where $\tau_d(z_d) = 1$ (Hu & Sugiyama 1995c), baryonic gravitational instability can no longer be prevented by the photons. Collapse of the baryon density fluctuations implies that baryon peculiar velocities will create a Doppler effect in the CMB. However the weak coupling approximation tells us that these too will be damped as the photon traverses many crests and troughs of the perturbation at last scattering. In fact, cancellation for the Doppler effect is particularly severe. If the perturbation wavevector is perpendicular to the line of sight $\mathbf{k} \perp \boldsymbol{\gamma}$, cancellation is avoided. However, flows are irrotational in linear theory so $\mathbf{v}_b \parallel \mathbf{k}$. Since the velocity is then perpendicular to the line of sight, no Doppler effect arises $\boldsymbol{\gamma} \cdot \mathbf{v}_b = v_b \boldsymbol{\gamma} \cdot \hat{\mathbf{k}} = 0$ (see Fig. 11). Fluctuations only survive if there are variations in the velocity or optical depth through last scattering. Thus the Doppler source is related to the derivative of these quantities as $S_{sec} = e^{-\tau} [V_b \dot{\tau} + V_b \ddot{\tau}]/k$ at small scales where recall $\mathbf{v}_b(\mathbf{x}) = -i V_b e^{i\mathbf{k}\cdot\mathbf{x}} \hat{\mathbf{k}}$. This leads to a suppression by a factor of approximately $(k\eta_*)^{-1}$ (Kaiser 1984). Cancellation under the weak coupling approximation yields a further suppression via equation (18). In general, later last scattering implies a greater $k\eta_*$, a larger cancellation scale, and hence smaller Doppler fluctuations. Nevertheless, in Fig. 10 the slight upturn at intermediate scales for the high ionization case is due to this effect.

4.2.3 Vishniac Effect

Since the first order Doppler fluctuation is severely suppressed due to "coincidental" geometrical reasons other effects may dominate the anisotropy at small scales. Variations in the optical depth due to density fluctuations across the last scattering surface may alow the Doppler effect to escape severe cancellation. The increased probability of scattering causes a preferential generation of Doppler fluctuations in overdense regions. The ensuing anisotropy is not as severely damped as the first order contribution. This is because the velocity field can be parallel to the line of sight while the density variation is perpendicular to the line of sight (see Fig. 11). This effect can also be calculated under the weak coupling approximation. It is slightly more complicated than linear effects since it entails mode coupling between the density and velocity fields. An explicit expression is given in Hu & White (1995). For a CDM model, we find that Vishniac contributions do indeed exceed linear contributions at sufficiently small scales. Furthermore, since second order effects are strongly peaked to late times, even minimally ionized scenarios carry a significant fraction of the maximum signal (see Fig. 10).

4.2.4 Cluster Sunyaev-Zeldovich Effect

Clusters provide a non-linear analogue of the Vishniac effect. Here the hot cluster provides the variation in the optical depth which causes preferential scattering. The Doppler effect due to the peculiar velocity of the cluster yields an anisotropy known as the *kinematic* Sunyaev-Zel'dovich (SZ) effect (Sunyaev & Zel'dovich 1972). For an individual cluster the temperature fluctuation is of order $\tau_c v_c$, where the optical depth through a cluster is of order $\tau_c \approx 0.1 - 0.01$ and the peculiar velocity $v_c \approx$ few $\times 10^{-3}$. This provides an interesting way of measuring the peculiar velocity of a cluster without introducing the problems associated with determining the distance scale (see *e.g.* Sunyaev & Zel'dovich 1980, Haehnelt & Tegmark 1995). The average effect is much smaller however and probably does never dominates the anisotropy spectrum. Persi *et al.* (1995) estimate the effect in a CDM model from hydrodynamic simulations and find that this non-linear Doppler effect is small in comparison to the second order Doppler (Vishniac) effect

Compton scattering off hot electrons also produces spectral distortions which look like anisotropies to any experiment confined to low frequency measurements. As discussed in §2.1, Compton scattering exchanges energy between the electrons and photons to order T_e/m_e. This upscatters photons from the Rayleigh-Jeans to the Wien regime and leads to a temperature distortion of $(\Delta T/T)_{RJ} = -2y \approx -2\tau_c T_e/m_e \approx 10^{-5} - 10^{-3}$ for an individual

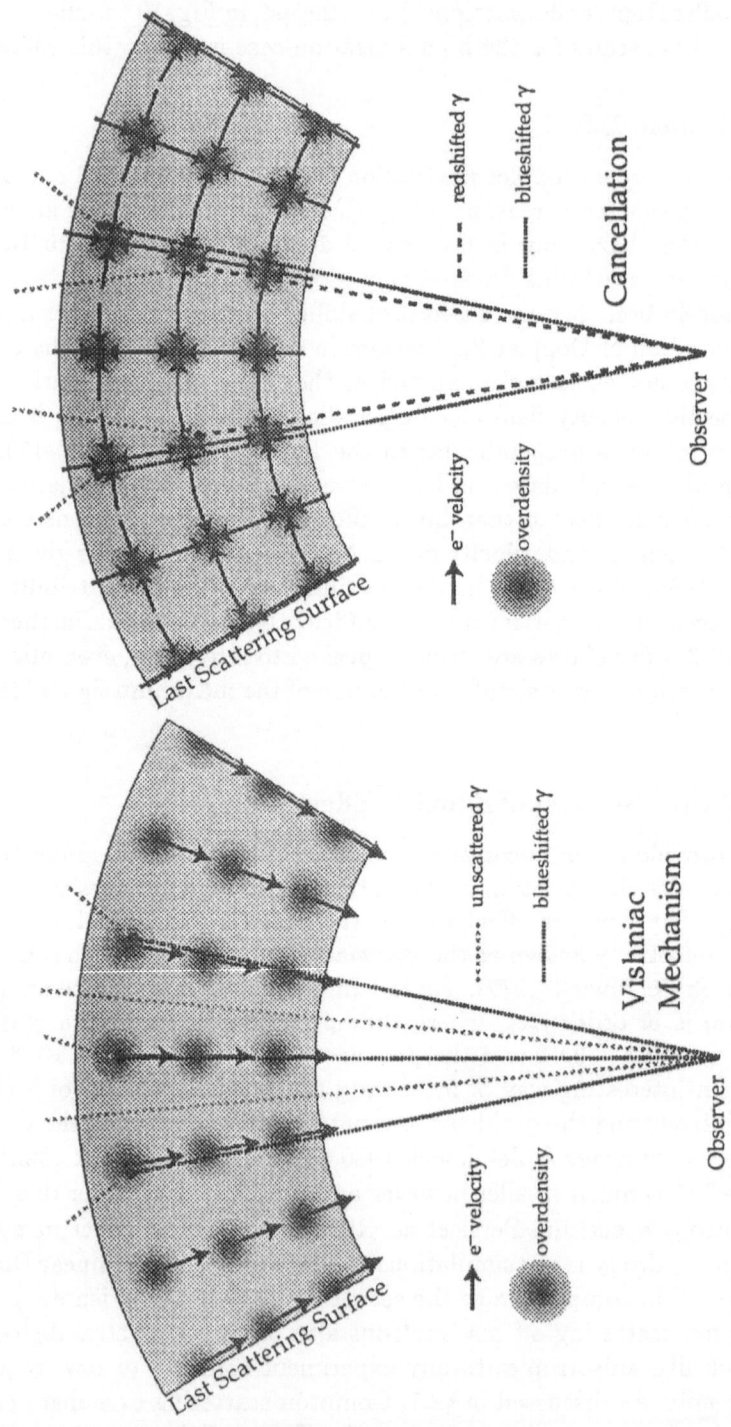

Fig. 11. Cancellation vs. Vishniac mechanisms

cluster. Again, the rms fluctuations would be much smaller. Much effort has been expended to estimate the fluctuations caused by this *thermal* SZ effect with varying results (Barbosa *et al.* 1995; Makino & Suto 1993; Cole & Kaiser 1988). Recently, empirical modeling of clusters has predicted that the anisotropy at arcminutes should be on the order of $(\Delta T/T)_{RJ} \lesssim 10^{-7}$ (Ceballos & Barcons 1994). Moreover the signal is in large part due to bright and easily identifiable clusters. If such known clusters are removed from the sample, the anisotropy drops to an entirely negligible level.

4.2.5 Inhomogeneous Reionization

Finally, let us mention another variant of the Vishniac-Sunyaev-Zel'dovich mechanism. The optical depth could vary due to inhomogeneities in the ionization fraction x_e near last scattering. Note that inhomogeneities well before optical depth unity have little effect on the CMB due to the cut off from the visibility function. Details of this effect will of course depend on the exact model for reionization, *i.e.* the extent of the inhomogeneities. Small inhomogeneities may be expected to behave as the Vishniac effect; large inhomogeneities like the kinetic SZ effect.

5 Discussion

It should now be clear that a great wealth of information about cosmology and the model for structure formation lies waiting to be observed in the CMB anisotropy spectrum. The location of the first acoustic peak provides a robust classical test for curvature in the universe if the fluctuations are known to be either adiabatic or isocurvature. The only caveat here is that if the universe suffered early reionization the acoustic effect may be so suppressed that it becomes unobservable. In this case, the CMB will place a lower limit on the epoch of reionization. It is very likely that the curvature and/or thermal history of the universe will be measured in the near future.

Barring early reionization, once we obtain precise measurements of the first peak and beyond, we should be able to extract much more information. The relative locations of the higher peaks can cleanly separate adiabatic and isocurvature models independently of the curvature and thermal history. A precise measurement of the location of the peaks can supply information on Λ and $\Omega_0 h^2$. The heights of the peaks yield even more information if some minimal assumptions are made for the theory of structure formation. Indeed, requiring consistency with large scale structure measurements should eventually fix the model quite precisely. The relative heights of the peaks give a robust probe of the baryon content $\Omega_b h^2$. In any given model, the absolute heights yield a constraint on the matter content of the universe

$\Omega_0 h^2$ and perhaps the number, temperature and mass of the neutrinos. Combining these pieces of information we can infer the Hubble constant.

It would seem that all the fundamental cosmological parameters are encoded in the CMB anisotropy spectrum. Yet, even discounting the possibility of early reionization, how likely is it that we will precisely measure them with the next generation of experiments? The results of Jungman *et al.* (1995) suggest that at least the curvature can be measured to better than 5% accuracy with a full sky map to half a degree resolution. Precisely how much information can be extracted will depend in the end on how severe foreground contamination from synchotron radiation, free-free emission, interstellar cold dust, and radio point sources will be. Many of the effects described here will require $\Delta T/T$ to 10^{-6} accuracy to measure definitively. With sufficient frequency coverage, there is hope of distinguishing the background from the foreground signal to employ at least the clean patches of the sky for cosmology.

Acknowledgments: I would like to thank E. Bunn, J. Silk, D. Scott, N. Sugiyama, M. White as the ideas and results presented here arise from our many collaborations. In particular, numerical results presented here are from the code of N. Sugiyama unless otherwise stated. M. Tegmark provided useful comments on a draft of these notes. I would also like to thank the organizers of this school E. Martinez-Gonzalez and J.L Sanz for an enjoyable and productive meeting.

References

Albrecht, A., Coulson, D., Ferreira, P. & Magueijo, J. (1995): astroph-9505030

Barbosa, D., Bartlett, J.G., Blanchard, A. & Oukbir, J. (1995): astroph-9511084

Bardeen, J.M. (1980): Phys. Rev. D **22** 1882

Bond, J.R. (1995): "Theory and Observations of the CBR", in Cosmology and Large Scale Structure, ed. by Schaeffer (Elsevier, Netherlands)

Bond, J.R. & Efstathiou, G. (1984): ApJ Lett. **285** L45

Bond, J.R. *et al.* (1994): Phys. Rev. Lett. **72** 13

Cole, S. & Kaiser, N. (1988): MNRAS **233** 637

Crittenden, R. *et al.* (1994): Phys. Rev. Lett. **71** 324

Crittenden, R. & Turok, N. (1995): astroph-9505120

de Laix, A.A. & Scherrer, R.J. (1995): astroph-9509075

Dodelson, S., Gates, E. & Stebbins, A. (1995): astroph-9509147

Dodelson, S. & Jubas, J. (1995): ApJ **439** 503

Doroshkevich, A.G., Zel'dovich, Ya.B. & Sunyaev, R.A. (1978): Sov. Astron. **22** 523

Durrer, R., Gangui, A. & Sakellariadou (1995): astroph-9507035

Haehnelt, M.G. & Tegmark, M. (1995): astroph-9507077

Hu, W. (1995): astroph-9508126

Hu, W., Scott, D. & Silk, J. (1994): Phys. Rev. D. **49** 648

Hu, W., Scott, D., Sugiyama, N. & White, M. (1995): Phys. Rev. D (in press, astroph-9505043)

Hu, W. & Sugiyama, N. (1995a): ApJ **436** 456

Hu, W. & Sugiyama, N. (1995b): Phys. Rev. D **51** 2599

Hu, W. & Sugiyama, N. (1995c): astroph-9510117

Hu, W. & White, M. (1995): A&A (in press, astroph-9507060)

Jungman, G., Kamionkowski, M., Kosowsky, A. & Spergel, D.N. (1995): astroph-9507080

Kaiser, N. (1983): MNRAS **202** 1169

Kaiser, N. (1984): ApJ **282** 374

Kodama, H. & Sasaki, M. (1984): Prog. Theor. Phys. Supp. **78** 1

Kofman, L.A. & Starobinskii, A.A. (1985): Sov. Astron. Lett. **9** 643 (1985)

Kosowsky, A. (1995): astroph-9501045

Lidsey, J.E. *et al.* (1995): astroph-9508078

Ma, C.-P. & Bertschinger, E. (1995): astroph-9506072

Magueijo, J., Albrecht, A., Coulson, D. & Ferreira, P. (1995): astroph-9511042

Makino, N. & Suto, Y. (1993): ApJ **405** 1

Martinez-Gonzalez, E., Sanz, J.L. & Silk, J. (1992): Phys. Rev. D **46** 4193

Mukhanov, V.F., Feldman, H.A., & Brandenberger, R.H. (1992): Phys. Rep. **215** 203

Ostriker, J.P. & Vishniac, E.T. (1986): ApJ **306** 51

Peebles, P.J.E. & Yu, J.T. (1970): ApJ **162** 815

Pen, U.-L., Spergel, D.N. & Turok, N. (1994): Phys. Rev. D **49** 692

Persi, F.M., Spergel, D.N., Cen, R. & Ostriker, J.P. (1995): ApJ **442** 1

Press, W. & Vishniac, E.T. (1980): ApJ **239** 1

Rees, M.J. & Sciama, D.N. (1968): Nature **519** 611

Sachs, R.K. & Wolfe, A.M. (1967): ApJ **147** 73

Scott, D., Silk, J. & White, M. (1995): Science **268** 829

Seljak, U. (1994): ApJ Lett. **419** L47

Seljak, U. (1995a): astroph-9506048

Seljak, U. (1995b): astroph-9505109

Silk, J. (1968): ApJ Lett. **151** 459

Smoot, G. *et al.* (1992): ApJ Lett. **396** L1

Steinhardt, P.J. (1995): astroph-9502024

Sugiyama, N. & Gouda, N. (1992): Prog. Theor. Phys. **88** 803

Sunyaev, R.A. & Zel'dovich, Ya. B. (1972): Comm. Astrophys. Space Phys. 4 73

Sunyaev, R.A. & Zel'dovich, Ya. B. (1980): MNRAS **190** 413

Tuluie, R. & Laguna, P. (1995): ApJ Lett. **445** L73

Vishniac, E.T. (1987): ApJ **322** 597

Vittorio, N. & Silk, J. (1984): ApJ Lett. **285** L39

Weinberg, S. (1972): Gravitation and Cosmology (Wiley, New York)

White, M., & Scott, D. (1995): astroph-9508157

White, M., Scott, D. & Silk, J. (1994): ARA&A **32** 319 (1994)

Wilson, M.L. (1983): ApJ **273** 2

Wilson, M.L. & Silk, J. (1981): ApJ **243** 14

$\frac{\Delta T}{T}$ Beyond Linear Theory

Enrique Martínez-González

Instituto de Física de Cantabria, Consejo Superior de Investigaciones
Científicas–Universidad de Cantabria, Facultad de Ciencias, Avda. Los
Castros s/n, 39005 Santander, Spain

Abstract: The major contribution to the anisotropy of the temperature of the Cosmic Microwave Background (CMB) radiation is believed to come from the interaction of linear density perturbations with the radiation previous to the decoupling time. Assuming a standard thermal history for the gas after recombination, only the gravitational field produced by the linear density perturbations present on a $\Omega \neq 1$ universe can generate anisotropies at low z (these anisotropies would manifest on large angular scales). However, secondary anisotropies are inevitably produced during the nonlinear evolution of matter at late times even in a universe with a standard thermal history. Two effects associated to this nonlinear phase can give rise to new anisotropies: the time-varying gravitational potential of nonlinear structures (Rees-Sciama RS effect) and the inverse Compton scattering of the microwave photons with hot electrons in clusters of galaxies (Sunyaev-Zeldovich SZ effect). These two effects can produce distinct imprints on the CMB temperature anisotropy. We discuss the amplitude of the anisotropies expected and the relevant angular scales in different cosmological scenarios. Future sensitive experiments will be able to probe the CMB anisotropies beyong the first order primary contribution.

1 Introduction

Cosmic Microwave Background (CMB) temperature anisotropies carry information on the large scale distribution of matter via the gravitational imprint of linear density fluctuations on large angular scales ($\gtrsim (2\Omega^{1/2})°$), the so called Sachs-Wolfe effect for a $\Omega = 1$ universe (Sachs and Wolfe 1967) or the generalized Sachs-Wolfe effect for a $\Omega \neq 1$ universe or in the presence of a cosmological constant Λ (Anile and Motta 1967, Wilson 1983, Gouda et al. 1991). In the first case the gravitational potential is static and the temperature fluctuations are produced by the potential fluctuations at recombination whereas in the other cases they are also produced by an integrated effect of the time-varying potential along the photon path from recombination to the observer.

On smaller angular scales the major contribution to the CMB anisotropy comes from photon fluctuations and the Doppler effect due to the velocity of the last scatterers at recombination. These primary anisotropies can be erased if the gas in the universe is reionized at a sufficiently early time ($z \gtrsim 100$). However, in this case new secondary anisotropies are generated by the gas motion which produces a Doppler effect and by the coupling of velocity and density fluctuations (Vishniac 1987). All the mentioned effects characterize the shape of the radiation power spectrum which contains information on the cosmological parameters Ω, Λ, the baryonic content Ω_b, the Hubble parameter, the initial matter density distribution and the thermal history of the universe (see Hu in this volume for a general review on the CMB anisotropies).

CMB temperature anisotropies, detected for the first time by Smoot et al. (1992) with the COBE-DMR experiment, are believed to be dominated by the interaction of matter density perturbations and radiation to first order in perturbation theory. In the standard scenario these primary anisotropies are produced before neutral hydrogen formed at recombination ($z \approx 1000$) and they are preserved until today. However, secondary anisotropies are inevitably produced during the nonlinear evolution of matter at late times. Even in a flat universe with a standard thermal history (i.e. with no reionization of the gas content after recombination), the time-varying gravitational potential due to the evolving nonlinear structures generates new temperature fluctuations (Rees-Sciama RS effect, Rees and Sciama 1968). Their amplitude is estimated to be $\sim 10^{-6}$ and the signatures imprinted in the radiation power spectrum may distinguish them from the primary anisotropies. We will discuss the possible detectability of these secondary fluctuations with future very sensitive experiments.

The Compton scattering of microwave photons by hot electrons in the intra-cluster gas also produces new anisotropies at very late times after the formation of rich clusters of galaxies (Sunyaev and Zeldovich 1972). This Sunyaev-Zeldovich effect (SZ) produces spectral distortions which translate into anisotropies when the angular distribution of temperatures in the sky is measured by an experiment at a given frequency band. The peculiar velocity of clusters also generates anisotropy via Doppler effect but the temperature amplitude of this kinematic Sunyaev-Zeldovich effect (KSZ) is expected to be at least an order of magnitude below the thermal SZ one in the Raileygh-Jeans region of the spectrum. The SZ effect is several orders of magnitude greater than the RS effect for a rich cluster. However, since the SZ effect is restricted to the central regions of rich clusters with a high concentration of hot gas whereas the RS one is a much more general effect the statistical average of many structures in the sky may give comparable amplitudes of the two effects or may even be dominant for the later effect.

2 The Rees-Sciama Effect

Rees and Sciama (1968) first noticed that the time-varying gravitational potential of evolving nonlinear structures will produce secondary anisotropies in the temperature of the CMB. That seminal paper, where an incorrect Newtonian approach was used to study the anisotropy generated by a compensated cluster, motivated many works which used a General Relativistic treatment to model the nonlinear evolution of spherical over-densities and under-densities in a uniform background. A simple model used was that of the "Swiss-cheese" where it is assumed a central spherical over-density surrounded by a compensated under-density (Notale 1984, Martínez-González and Sanz 1990). Also a similar model consisting of a central spherical under-dense region compensated by a thin shell was used for cosmic voids (Thompson and Vishniac 1987, Dyer and Ip 1988, Martínez-González and Sanz 1990). A more sophisticated spherical model based on the Tolman solution of the Einstein equations allows a general density profile (Panek 1992, Arnau et al. 1993, Fang and Wu 1993, Fullana et al. 1996).

A useful expression to calculate the RS effect produced by non-spherical structures and general realistic matter distributions can be derived within the potential approximation (Martínez-González et al. 1990, see below). Chodorowski (1992, 1994) and Atrio-Barandela and Kashlinski (1992) used that expression to calculate the effect of pancake-like collapse of ellipsoids. The results of all these works for the most prominent structures seen in our local universe can be summarized as follows. For the large Bootes void of size $\approx 60\,h^{-1}$ Mpc located at a distance of $\approx 150\,h^{-1}$ Mpc and with a density of galaxies $\approx 25\%$ of the background density (Kirshner et al. 1981, de Lapparent et al. 1986), it is estimated $\Delta T/T \approx a\,few \times 10^{-7}$ with small variation for different models (Martínez-González and Sanz 1990, Panek 1992, Arnau et al. 1993). For the Great Attractor located at a distance of $\approx 60\,h^{-1}$ Mpc and a density contrast of ≈ 1 inside a radius of $\approx 40\,h^{-1}$ Mpc (Lynden-Bell et al. 1987) the estimated RS effect is $\approx a\,few \times 10^{-6}$ and again the results are fairly insensitive to the model used (Martínez-González et al. 1990, Panek 1992, Arnau et al. 1994). In the case of the large concentration of galaxies known as the Great Wall which forms a flattened structure of dimensions $\approx 5 \times 60 \times 170\,h^{-1}$ Mpc at a distance of some $75\,h^{-1}$ Mpc (Geller and Huchra 1989) the effect is also $a\,few \times 10^{-6}$ (Atrio-Barandela and Kashlinski 1992, Chodorowski 1994).

If similar structures are assumed to be at different distances from the observer in low density universes then it is possible to increase the RS effect in almost an order of magnitude. For distances $z \sim 1 - 10$ and universes with $\Omega \approx 0.2$ great attractor-like structures can produce an effect of the order of 10^{-5} (Arnau et al. 1994) and Bootes-like voids of $a\,few \times 10^{-6}$ (Fullana et al. 1996), with angular scales of a few degrees. Thus, only in open universes the presence of such large inhomogeneities might leave an observable imprint in the CMB anisotropy maps.

Below we will describe a useful expression for the RS effect in realistic situations and use it to calculate the statistical effect produced by a distribution of matter.

2.1 The Potential Approximation

Martínez-González et al. (1990), based on the potential approximation to general relativity, derived a useful expression for the temperature fluctuation generated when microwave photons cross a nonlinear density perturbation $\Delta(t, \boldsymbol{x})$. If $\varphi(t, \boldsymbol{x})$ is the corresponding gravitational potential then the RS effect can be simply written as

$$(\Delta T/T)_{secondary} = 2 \int_{ls}^{o} dt \frac{\partial \varphi}{\partial t}(t, \boldsymbol{x}) \quad , \quad \nabla^2 \varphi = 6\Omega a^{-1} \Delta(t, \boldsymbol{x}) \quad , \qquad (1)$$

where $a(t)$ is the scale factor normalized to the present time ($a_o = 1$) and our units are $c = 8\pi G = 1$ and the Hubble length at present is $d_o = 2/H_o^{-1}$. Thus, the Rees-Sciama effect is basically the work performed by a photon travelling through the time-dependent gravitational potential $\varphi(t, \boldsymbol{x})$ from recombination to the observer. The previous equation was initially derived for a flat universe, however, since the scales relevant for nonlinear structures are always much smaller than the curvature scale it also applies to open universes. The integral must be performed along the geodesic associated to the corresponding background

$$\boldsymbol{x}(t, \boldsymbol{n}) = \lambda(a)\boldsymbol{n} \quad , \quad \lambda(a) = \frac{1 - (\Omega/a + 1 - \Omega)^{-1/2}}{1 - (1 - \Omega)(\Omega/a + 1 - \Omega)^{-1/2}} \qquad (2)$$

An order of magnitude estimate of the effect can easily be obtained from equation (1). For a photon crossing a cluster the anisotropy is

$$\frac{\Delta T}{T} \sim 2\frac{\varphi}{c^2}\frac{t_c}{t_d} \sim 2\frac{\varphi}{c^2}\frac{v}{c} \sim 2\left(\frac{v}{c}\right)^3 \qquad (3)$$

where φ is the potential of the cluster, t_c and t_d are the crossing time and the dynamical time and v the velocity of collapse. For a cluster with $v = 1000$ km/s the effect is $\Delta T/T \sim 10^{-7}$. A statistical distribution of lumps can therefore increase the effect to values of cosmological interest.

2.2 Second Order Perturbation Theory

We will now estimate the RS effect in an open or flat background for a cold dark matter (CDM) cosmogony. This effect has been recently calculated by Sanz et al. (1996) within the second order perturbation theory. We will follow that paper below.

For vanishing pressure, by perturbing the Einstein field equations and considering perturbations in density up to the second order we can obtain the following expression for the density fluctuations

$$\Delta(t, \textbf{\textit{x}}) = D\delta + D^2\left[\frac{5}{7}\delta^2 + \nabla\delta\cdot\nabla\xi + \frac{2}{7}\xi_{,ij}\xi^{,ij}\right] \quad , \quad \nabla^2\xi = \delta \qquad (4)$$

where $\delta(\textbf{\textit{x}}) \equiv \delta_r(1+z_r)$. Equation (4) is exact within the 2nd order perturbation theory for a flat background (see Peebles 1980) and it is a good approximation for open universes with $\Omega \gtrsim 0.1$ as has been shown by Bouchet et al. (1993) and Catelan et al. (1995). By using equations (2,3) in the expression for the second order effect (eq. 1), we obtain the following result

$$\left(\frac{\Delta T}{T}(\textbf{\textit{n}})\right)_{secondary} = \int_{\lambda_{l_s}}^{o} d\lambda\, W(\lambda)\phi(\textbf{\textit{x}} = \lambda\textbf{\textit{n}}) \quad ,$$

$$W(\lambda) = 24\Omega D^2(2f - 1)\frac{1 - (1 - \Omega)\lambda}{(1 - \lambda)^3} \qquad (5)$$

where the new potential related to the second order desity perturbation δ_2 satisfies the following Poisson equation

$$\nabla^2\phi(\textbf{\textit{x}}) = \delta_2 \equiv \frac{5}{7}\delta^2 + \nabla\delta\cdot\nabla\xi + \frac{2}{7}\xi_{,ij}\xi^{,ij} \quad . \qquad (6)$$

In the equation above D is the growing mode of the perturbations normalized to 1 at the present time. For a flat universe: $D = a$, whereas for an open universe (Peebles, 1980):

$$D = \frac{g(x)}{g(x_o)} \quad , x = \left(\frac{1}{\Omega} - 1\right)a \quad , f(x) = \frac{d\ln D}{d\ln a} \quad ,$$

$$g(x) = 1 + \left(\frac{3}{x}\right)\left[1 + \left(1 + \frac{1}{x}\right)^{1/2}\ln\left((1 + x)^{1/2} - x^{1/2}\right)\right] \quad . \qquad (7)$$

It is common to expand the temperature fluctuations in terms of spherical harmonics Y_{lm}, $\Delta T(\textbf{\textit{n}}) = \sum_l \sum_{m=-l}^l a_{lm} Y_{lm}(\textbf{\textit{n}})$. The temperature correlation function is given in terms of the radiation power spectrum as $C(\Theta) = 1/(4\pi)\sum(2l+1)C_l P_l(\cos(\alpha))$, with $C_l = <a_{lm}^2>$. In the second order perturbation theory the multipole component C_l is (we shall not consider the monopole and dipole in the calculations of the temperature anisotropy below):

$$C_l = \frac{2}{\pi}\int dk k^{-2} P_{(2)}(k) R_l^2(k) \quad , R_l(k) \equiv \int_0^{\lambda_{l_s}} d\lambda W(\lambda) j_l(kp) \quad . \qquad (8)$$

Here, $p \equiv \lambda/[1 - (1 - \Omega)\lambda^2]$, j_l is the Bessel function of fractional order and λ_{l_s} is the distance from the observer to the last scattering surface. This equation generalizes the result obtained by Martínez-González et al. (1992) to open universes. The function $P_{(2)}(k)$ is the power spectrum associated with the 2nd order density perturbation δ_2 and is related to the power spectrum P_ξ of the time derivative of the potential ξ by $P_\xi = \frac{1}{k^4}P_{(2)}(k)$. The second order perturbation power spectrum $P_{(2)}(k)$ is given in terms of the first order power spectrum $P(k)$ by the equation (Goroff et al. 1986, Suto and Sasaki, 1991)

$$P_{(2)}(k) = \frac{k^3}{98(2\pi)^2} \int_0^\infty dr P(kr) \int_{-1}^1 dx P\big(k(r^2 + 1 - 2rx)^{1/2}\big)$$

$$\left(\frac{3r + 7x - 10rx^2}{r^2 + 1 - 2rx}\right)^2 . \tag{9}$$

In the limit of small k, there is a cancellation of the three terms contributing to δ_2, implying that $P_{(2)}(k)$ has very little power on large scales and in this regime goes like $P_{(2)}(k) \propto k^4$ independently of the primordial power spectrum.

Equation (8) for the multipoles C_l can be further simplified by realizing that the integral to calculate $R_l(k)$ is the product of a slowly changing function of λ, the window $W(\lambda)$, and a spherical Bessel function (Seljak 1996). Thus, fixing the value of the argument of W and applying the large l approximation $\int_0^x j_l(x') = \sqrt{(\pi/2l)}S(x - l)$, S being the step function, we obtain

$$C_l = \frac{1}{l^4} \int_0^{\lambda_{l*}} d\lambda W^2(\lambda) P_2[l(1 - (1 - \Omega)\lambda^2)/\lambda] . \tag{10}$$

This approximation works very well and in the worst situation for the quadrupole moment the error is $\lesssim 30\%$ for all Ω values.

In relation to a possible secondary contribution coming from 3rd order density perturbations for open models (in the case of flat models this does not exist because the 1st order gravitational potential is static), we have the following comment: the coupling of the 1st order gravitational potential $\varphi^{(1)} \propto \frac{D}{a}$ with the 3rd order potential $\varphi^{(3)} \propto \frac{D^3}{a}$ gives a kernel for the integrated gravitational effect proportional to $\frac{D^3}{a^2}(f - 1)(3f - 1)$, whereas the coupling of the 2nd order gravitational potential $\varphi^{(2)} \propto \frac{D^2}{a}$ with itself gives a kernel for the integrated gravitational effect proportional to $\frac{D^3}{a^2}(2f - 1)^2$. This second function is always greater than the first one. Moreover, for quasi-flat models we expect a negligible contribution from the coupling of 1st-3rd order perturbations because $f \approx 1$, whereas for low-Ω models the integrated gravitational effect due to 2nd-2nd order perturbations is produced at high-z (75% of the final effect is produced in the interval $[10, 10^3]$ for $\Omega = 0.1$, see next section) where $f \approx 1$, and so there is practically no 1st-3rd order contribution. At smaller z some contribution due to the 1st-3rd coupling is produced but it is estimated to be always bounded by that due to the 2nd-2nd coupling at low-z, and this is a small fraction of the final contribution.

2.3 Results

The equations derived in the previous section have been applied to calculate the predicted amplitudes of the multipole components C_l in open universes. We assume an open or flat CDM model with a Harrison-Zeldovich primordial spectrum of matter density fluctuations $P(k) = Ak$ and $\Omega_b = 0.05$, $H = 50$km s^{-1} Mpc^{-1}. In figure 1 we display the multipoles C_l for $\Omega = 1, 0.3, 0.1$ due to primary anisotropies (upper curves) and due to 2nd order anisotropies (lower curves). Primary anisotropies are normalized to the 2-year COBE-DMR maps as given by the analysis of Cayón et al. (1996) for the HZ spectrum. Since secondary anisotropies are generated by the relatively small scale structure, $\lesssim 100$Mpc, it is more appropriate to use the $\sigma_{16} = 1$ normalization, i.e. the rms density fluctuation is unity at 16 Mpc. The secondary radiation power spectrum peaks at $l \approx 250$ for all Ω values, contrary to the linear one where the multipole order, l, of the peak increases for low Ω values. For the secondary contribution, the shift of the maximum of the 2nd order power spectrum $P_{(2)}(k)$ towards large scales (small k) for low Ω values is balanced by the opposite shift of the maximum of the function $R_l(k)$ in equation (8).

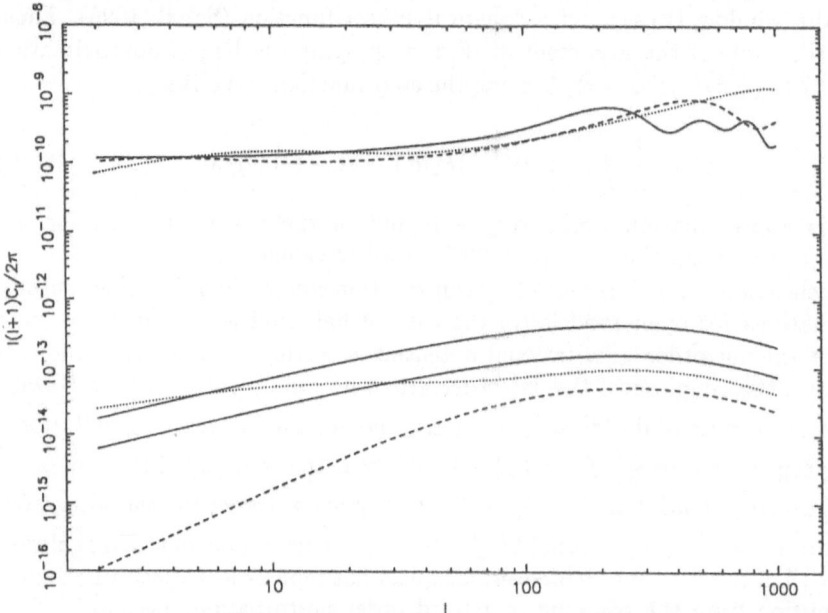

Fig. 1. Radiation power spectrum for the primary (upper curves) and secondary (lower curves) contributions. Solid, dashed and dotted lines correspond to $\Omega = 1, 0.3, 0.1$, respectively. The secondary contribution is normalized to $\sigma_{16} = 1$, except for the solid upper curve which represents the flat case with the COBE-DMR normalization.

The generation of the secondary radiation power spectrum with redshift is shown in figures (2a,b) for $\Omega = 1, 0.1$ respectively. For $\Omega = 1$ more than 90% of the anisotropy is produced at $z \lesssim 10$ (Martínez-González et al. 1992). In the case of low Ω models the anisotropy is produced at a relatively higher redshift, for $\Omega = 0.1$ about 80% of the effect is generated at a redshift $z < 30$. In any case, reionization of the matter in the universe cannot substantially erase the secondary anisotropy at such low redshifts.

Nonlinear scales do not enter in the second-order calculation (eq. 8) because for any Ω value the maximum of $R_l(k)$ is at $k \approx l\Omega$ and the scales that are contributing to the multipole l have $k \lesssim l$. Thus, for the multipoles C_l up to $l = 1000$ only scales ≥ 12Mpc contribute to the second-order effect. This result

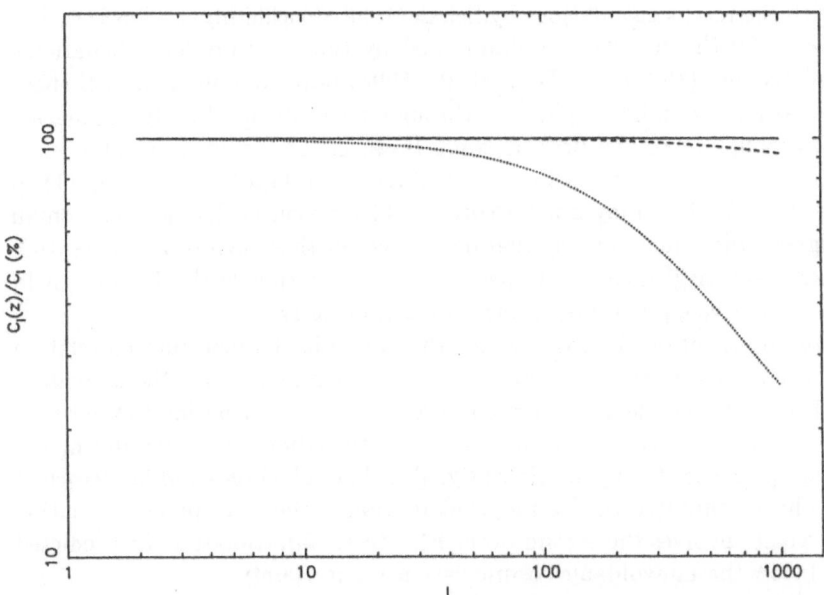

Fig. 2a. Generation of the secondary radiation power spectrum with redshift for a flat universe. Solid, dashed and dotted lines correspond to the anisotropy generated from redshifts $z = 100, 10, 1$ to the present, respectively.

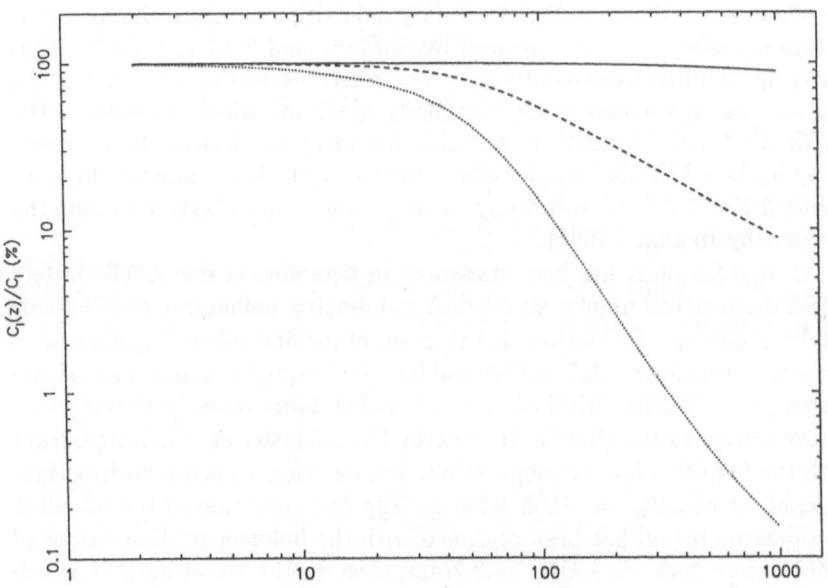

Fig. 2b. The same as figure (2a) but for an open universe with $\Omega = 0.1$.

agrees with the recent calculations by Seljak (1996) considering nonlinear scales, up to $l \sim 1000$ the RS effect is dominated by the second-order calculations. Tului and Laguna (1995) and Tului et al. (1996) have computed the RS effect by propagating a bundle of photons through an evolving N-body simulation from the decoupling to the present time. However, one must be careful when calculating the nonlinear effect because at scales $\lesssim 10$Mpc the use of eq. (1) to compute the RS effect may not be correct. The reason is that at those small scales higher orders including spacial derivatives of the gravitational potential, which have been neglected in the perturbative expansion of the Einstein field equations to derive eq. (1), may contribute significantly.

The results mentioned above are for the two-point temperature correlation function where clearly the primary contribution dominates over the secondary RS one. However, the linear primary effect gives a null contribution to the three-point correlation since the primordial matter perturbations generated during the inflationary phase are Gaussian. Recently, Munshi et al. (1995) and Mollerach et al. (1995) have computed the leading contribution to the three-point correlation function which involves the second-order RS effect. Unfortunately the predicted signal is below the unavoidable cosmic variance uncertainty.

3 The Sunyaev-Zeldovich Effect

Clusters of galaxies not only induce secondary anisotropies in the CMB by means of gravitational effects. As first noticed by Sunyaev and Zeldovich (1970,1972) the scattering of microwave photons by hot electrons in the intracluster gas produces spectral distortions in the blackbody spectrum, which is known as the thermal SZ effect. Additionally, the peculiar motion of the clusters also produces anisotropy in the CMB via Doppler effect, known as the kinematic SZ effect. In subsections 3.1 and 3.2 we will study in detail these two effects (see also the recent review by Rephaeli 1995).

The thermal SZ effect has been measured in a number of rich Abell clusters using three distinct techniques: single-dish radiometry, bolometric observations and interferometry (see Birkinshaw 1993). Some of the first convincing detections were those of the clusters A665, A2218 and 0016+16 using the single-dish OVRO radio telescope at 20GHz (Birkinshaw et al. 1984). More recently Herbig et al. (1995) have measured the effect in the nearby Coma cluster at a high significant level with the OVRO 5.5m telescope at 32GHz, deriving a maximum Rayleigh-Jeans decrement of $\Delta T_{RJ} = -505 \pm 92\mu K$. The first detection of the SZ effect in the millimetric region has been obtained with the bolometric observations of the SUZIE experiment on A2163 at 2.2mm, close to the wavelength at which the SZ decrement is maximum (Wilbanks et al. 1994). With these bolometric observations it is expected to measure the positive effect at wavelengths < 1.4mm (as we will see below the thermal SZ effect changes sign at that wavelength). The first image of the SZ effect was obtained for A2218 with the Ryle interferometer at 15Ghz (Jones et al. 1993). More recently, images of the rich clusters A773 and 0016+16 have been taken with the Ryle and OVRO arrays (Grainge et al. 1993, Saunders 1995, Carlstrom et al. 1996) and several other clusters are being imaged at present (very recently, an image of the SZ decrement towards A1413 has been taken by Grainge et al. 1996 with the Ryle Telescope). Interferometry makes possible to directly compare the SZ image with the X-ray image of a cluster and thus study the structure of the intracluster gas.

3.1 Thermal SZ Effect

The inverse Compton scattering of the isotropic CMB radiation ($h\nu \ll m_e c^2$) by a hot ($T_e \gg T$), nonrelativistic ($kT_e \ll m_e c^2$) Maxwellian electron gas is quantitatively described by a simplified version of the Kompaneets (1957) equation for the rate of change of the photon occupation number n (Sunyaev and Zeldovich 1980)

$$\frac{\partial n}{\partial y} = \frac{1}{x^2}\frac{\partial}{\partial x}x^4\frac{\partial n}{\partial x} \ ,$$

$$y = \int_0^\tau \frac{kT_e}{m_e c^2}d\tau \ , \ x = \frac{h\nu}{kT} \ , \tag{11}$$

where the nondimensional frequency x has been defined, the optical depth due to Thompson scattering is $d\tau = n_e \sigma_T c dt$ with σ_T being the Thomson cross section and y is the Comptonization parameter. Since for the richest clusters the radiation is only weakly scattered, $y \ll 1$, and so the deviations from the Planckian spectrum are small, then it is easy to solve equation (11) by inserting in the right-hand side of this equation the Planck function for the occupation number $n_P = 1/(e^x - 1)$ (Zeldovich and Sunyaev 1969). The solution for the change of spectral intensity $I = 2h\nu^3 n/c^2$ is

$$\Delta I = \frac{2(kT)^3}{(hc)^2}\frac{x^4 e^x}{(e^x - 1)^2}y\left[x\coth\frac{x}{2} - 4\right] \ . \tag{12}$$

The corresponding change in the thermodinamic temperature is

$$\Delta T = yT[x\coth\frac{x}{2} - 4] \ . \tag{13}$$

For the Raileygh-Jeans region of the spectrum, $x \ll 1$, equations (12,13) take the following simple form:

$$\Delta I_{RJ} \simeq -\frac{2(kT)^3}{(hc)^2}x^2 y \ , \Delta T_{RJ} \simeq -2yT \ . \tag{14}$$

Previous solutions are obtained from the Kompaneets equation which is based on a diffusion approach and thus is not in principle suitable for Compton scattering in clusters where the optical depth $\tau \ll 1$ and consequently most of the microwave photons are not scattered even once. Another limitation is the nonrelativistic treatment used which affects to the calculations for the richest clusters with temperatures $kT_e \sim 10keV$ and produces significant errors in the

Wien region (Fabbri 1981, Raphaeli 1995). Even so, except for the extreme cases in the distribution of clusters, solutions (12,13) provide an adequate description of the SZ effect in the whole spectral range of interest (for the richest clusters the solutions are still accurate in the Raileygh-Jeans region).

An interesting property of equations (12,13) is that the spectral change is zero at $\lambda \approx 1.4$mm ($\nu \approx 220$GHz) being negative and positive above and below that wavelength, respectively. This characteristic change in the sign of the SZ effect is still waiting to be observed (a tentative detection of the sign change has been recently claimed by Andreani et al. 1996 for the clusters A2744 and S1077 observed at 1.2 and 2 mm with the 15m SEST antenna in Chile).

We can now estimate the amplitude of the effect for a cluster of galaxies. Considering the nearby, well-studied Coma cluster, we know from X-ray data that the properties of the cluster atmosphere consist of a central density of $n_e \approx 2.9 \times 10^{-3}$cm^{-3}, a radius $r_c \approx 10'.5$ (≈ 0.43Mpc for $H = 50$Km s^{-1} Mpc^{-1}) and a temperature $T_e \approx 9.1$keV (Hughes et al. 1988, Briel et al. 1992). From these values we can easily derive an approximate optical depth of $\tau \approx n_e \sigma_T 2r_c \approx 5.1 \times 10^{-3}$ and a maximum decrement in the Raileygh-Jeans region of $\Delta T \approx 500\mu$K, very close to the value derived from the observations with the OVRO 5.5m telescope (Herbig et al. 1995).

Combining the SZ measurement with the parameters of the model atmosphere of the cluster derived from the X-ray data one can determine the angular diameter distance to a cluster and thus deduce the Hubble constant H_o without relying on any cosmic distance ladder. Estimates of H_o have already been made for several clusters of galaxies providing an accuracy for H_o similar to other methods (see for instance the recent estimate from the SZ measurement in Coma by Herbig et al. 1995). The major uncertainty of this method originates from systematic effects related to the X-ray model, departures from spherical symmetry and possible existence of clumpiness in the gas distribution (Birkinshaw et al. 1991).

3.2 Kinematic SZ Effect

The peculiar velocity of a cluster also modifies the intensity of the CMB via the Doppler effect. Because of its nature, this kinematic SZ effect produces fluctuations in the CMB temperature which are independent of frequency, in contrast to the thermal SZ. The change in temperature is therefore given by

$$\Delta T_k = -\tau T \frac{v_r}{c} \tag{15}$$

where v_r is the radial velocity of the cluster with positive sign for recession. The change in the intensity ΔI_k can be easily obtained from the temperature fluctuation using the Planck spectrum

$$\Delta I_k = -\frac{2(kT)^3}{(hc)^2} \frac{x^4 e^x}{(e^x - 1)^2} \tau \frac{v_r}{c} \ . \tag{16}$$

For a rich cluster like Coma and assuming a peculiar velocity of $500 \mathrm{km\,s^{-1}}$ the expected change in the temperature is only of $\Delta T_k \approx 20\mu K$, much smaller than the thermal effect. It is interesting to notice that the peculiar velocity of a cluster can be determined by measuring its thermal and kinematic effects and knowing the spatial distribution of the gas temperature from X-ray spectral data. Future very sensitive multifrequency ground-based, balloon and satellite experiments surveying large areas of the sky at a resolution of $\lesssim 10$ arcmin (see e.g. the proposed satellite COBRAS/SAMBA to ESA M3, Presentation of Assesment Study Results 1994) will be able to measure the thermal and kinematic SZ effects in ~ 1000 clusters. This information combined with X-ray spectral data of the same clusters will allow a reliable measurement of the bulk motion in a volume of ~ 1 Gpc (Haehnelt and Tegmark 1995).

3.3 Contribution to the CMB fluctuations

In the previous subsections we have studied the SZ effect expected when the microwave photons cross the intracluster gas of hot electrons in a rich cluster of galaxies. Below we will discuss the contribution of the emsemble of clusters to the CMB temperature maps.

Considerable work has been done to estimate the rms temperature fluctuations originated by the SZ effect. The usual approach is to consider an evolution for the cluster mass function that corresponds to a previously assumed theoretical model of density perturbations (Cole and Kaiser 1989; Markevitch et al. 1991, 1992; Makino and Suto 1993; Bartlett and Silk 1994; Colafrancesco et al. 1994; De Luca et al. 1995). The cluster mass function can be represented by the Press-Schechter formula (Press and Schechter 1974) which gives the comoving number density of collapsed objects per mass interval at a given redshift. A collapsed object is formed when its density contrast reaches the value $\delta_c = 1.68$ as given by the spherical collapse model. Assuming that the cluster is virialized to a given virial radius R_v the temperature of the gas can be related to the mass of the cluster. The electron density profile is usually parametrized with a β-model with the exponent $\beta \approx 0.75$ from observations. Typical rms values of the temperature fluctuations obtained with this formalism are of the order $\Delta T/T \sim 10^{-6}$.

A different approach has been considered by Ceballos and Barcons (1994) who used an empirically based model for the mass function of the intracluster mass. They use a parametrization of the X-ray luminosity function and its negative evolution (Edge et al. 1990) which supports the idea that the number of luminous clusters decreases with z (Giogia et al. 1990; Henry et al. 1992) and therefore the amount of hot gas available for the inverse Compton scattering of the CMB photons is consequently limitted. The result is a negligible SZ contribution to the rms temperature fluctuations of $\Delta T/T \lesssim 10^{-7}$. The contribution is dominated by the hottest clusters which are easily identified and thus its SZ signal can be removed from the temperature map.

The evidence for fewer X-ray luminous clusters at $z > 0.2$ than locally shown by the Einstein Extended Medium Sensitivity Survey (EMSS) should be confirmed using a complete, X-ray selected sample. The present ongoing cluster surveys based on ROSAT data will measure the X-ray luminosity function at redshifts > 0.2 and thus will determine the amount of hot gas at higher redshifts. Evidence of the marked decline in the volume density of luminous clusters with redshift has been recently presented with the X-ray RIXOS survey up to $z \approx 0.6$ by Castander et al. (1996) (this claim is, however, based on a sample of only 13 clusters).

4 Summary

The RS and SZ effects, associated with the late, nonlinear phase of structure formation, can give rise to secondary anisotropies. The RS effect is produced by the time-varying gravitational potential of the evolving matter density perturbations whereas the SZ one is due to the inverse Compton scattering of microwave photons by hot electrons in the intracluster gas. These two effects produce distinct signatures on the CMB temperature anisotropy: contrary to the RS effect the SZ one is frequency dependent.

The secondary anisotropy generated by the SZ effect is probably dominated by the more luminous clusters and is generated at low redshifts $z \lesssim 1$, if the negative evolution in the X-ray luminosity function of clusters with redshift shown by the EMSS is confirmed. The amplitude of the signal is very sensitive to the amount of hot gas available at moderate and high reshift, $z \gtrsim 0.5$. Thus, predictions based on theoretical models like the CDM one, which imply an increase in the amplitude of the X-ray luminosity function with redshift, give anisotropies $(\Delta T/T)_{SZ} \sim 10^{-6}$ whereas calculations which take into account the observed decline of luminous clusters with z suggest much smaller values.

The amplitude of the RS effect for realistic models of structure formation is $(\Delta T/T)_{RS} \lesssim 10^{-6}$, being higher for universes with high Ω. The signal is very sensitive to the normalization of the matter power spectrum being directly proportional to its amplitude. Early reionization of the matter in the universe, which can strongly affect the primary anisotropies, would not appreciably change the secondary ones. Future very sensitive satellite experiments might be able to test the gravitational instability theory beyond the linear theory.

Acknowledgements: The author would like to thank Jose Luis Sanz and Sergio Torres for their comments on the manuscript.

References

Andreani, P. et al. 1996, ApJ, in press

Anile, A.M. and Motta, S. 1967, ApJ, 207, 685

Arnau, J.V., Fullana, M.J., Monreal, L. and Sáez, D. 1993, ApJ, 402, 359

Arnau, J.V., Fullana, M.J. and Saez, D. 1994, MNRAS, 268, L17

Atrio-Barandela, F. and Kashlinski, A. 1992, ApJ, 390, 322

Bartlett, J. and Silk, J. 1994, ApJ, 407, L45

Birkinshaw, M., Gull, S.F., Hardebeck, H.E. 1984, Nature, 309, 34

Birkinshaw, M., Hughes, J.P. and Arnaud, K.A. 1991, ApJ, 379, 466

Birkinshaw, M. 1993, in Proc. Present and Future of the Cosmic Microwave Background, eds. J.L. Sanz, E., Martínez-González and L. Cayón (Springer-Verlag)

Blumenthal, G.R., Nicolaci Da Costa, L., Goldwirth, D.S., Lecar, M. and Piran, T. 1992, ApJ, 388, 234

Briel, U.G., Henry, J.P. and Böhringer, H. 1992, A&A, 259, L31

Bond, J.R., Carr, B. and Hogan, C.J. 1991, ApJ, 367, 420

Bouchet, F., Juszkievicz, R., Colombi, S. and Pellat, R. 1993, preprint

Carlstrom, J.E., Joy, M. and Grego, L. 1995, ApJ, 456, L75

Castander, F.J. et al. 1996, Nature, in press

Catelan, P., Lucchin, F., Matarrese, S. and Moscardini, L. 1995, preprint

Cayón, L. 1995, in this volume

Cayón, L., Martínez-González, E., Sanz, J. L., Sugiyama, N. and Torres, S. 1996, MNRAS, in press

Ceballos, M.T. and Barcons, X. 1994, MNRAS, 271, 817

Chodorowski, M. 1992, MNRAS, 259, 218

Chodorowski, M. 1994, MNRAS, 266, 897

Colafrancesco, S., Mazzotta, P., Rephaeli, Y. and Vittorio, N. 1994, ApJ, 433, 454

Cole, S. and Kaiser, N. 1988, MNRAS, 233, 637

Dyer, C.C. and Ip, P.S.S. 1988, MNRAS, 235, 895

Edge, A.C., Stwart, G.C., Fabian, A.C. and Arnaud, K.A. 1990, MNRAS, 245,559

Fabbri, R. 1981, Astrophys. Space Sci., 77, 529

Fang, L. and Wu, X. 1993, ApJ, , 408, 25

Fullana, M.J., Arnau, J.V. and Sáez, D. 1996, MNRAS, in press

Geller, M.J. and Huchra, J. 1989, Science, 246, 897

Giogia, I.M., Henry, J.P., Maccacaro, T., Morris, S.L., Stocke, J.T. and Wolter, A. 1990, ApJ, 356, L35

Goroff, H., Gristein, B., Rey, S.-J. and Wise, M.B. 1986, ApJ, 311, 6

Gorski, K. M., Ratra, B., Sugiyama, N. and Banday, A. J., 1995, ApJ, 444, L65.

Gouda, N., Sugiyama, N. and Sasaki, N. 1991, Prog. Theor. Phys., 85, 1023

Grainge, K., Jones, M., Pooley, G., Saunders, R. and Edge, A. 1993, MNRAS, 265, L57

Grainge, K., Jones, M., Pooley, G., Saunders, R., Baker, J., Haynes, T. and Edge, A. 1996, MNRAS, 278, L17

Haelhnelt, M.G. and Tegmark, M. 1995, preprint

Henry, J.P., Giogia, I.M., Maccacaro, T., Morris, S.L., Stocke, J.T. and Wolter, A. 1992, ApJ, 386, 408

Herbig, T., Lawrence, C.R. and Readhead, A.C.S. 1995, ApJ, 449, L5

Hu, W. 1995, in this volume

Hughes, J.P., Gorestein, P. and Fabricant, A. 1988, ApJ, 329, 82

Jones, M., Saunders, R., Alexander, P., Birkinshaw, M. and Dillon, N. 1993, Nature, 365, 320

Kirshner, R.P., Oemler, A., Schechter, P.L. and Schectman , S.A. 1981, ApJ, 248, L57

Kompaneets, A.S. 1957, Sov. Phys. JETP, 4,730

de Lapparent, V., Geller, M.J. and Huchra, J. 1986, ApJ, 302, L1

De Luca, A., Désert, F.X. and Puget, J.L. 1995, A&A, 300, 335

Lynden-Bell, D., Faber, S.M., Burstein, D., Davies, R.L. , Dressler, A., Terlevich, R. and Wegner, G. 1988, ApJ, 326, 19

Makino, N. and Suto, Y. 1993, ApJ, 405, 1

Markevitch, M., Blumenthal, G.R., Forman, W., Jones, C. and Sunyaev, R.A. 1991, ApJ, 378, L33

Markevitch, M., Blumenthal, G.R., Forman, W., Jones, C. and Sunyaev, R.A. 1992, ApJ, 395, 326

Martínez-González, E. and Sanz, J.L. 1990, MNRAS, 247, 473

Martínez-González, E. and Sanz, J.L. 1995, Astro. Lett. and Comm., 32, 89

Martínez-González, E., Sanz, J.L. and Silk, J. 1990, ApJ, , 355, L5

Martínez-González, E., Sanz, J.L. and Silk, J. 1992, Phys. Rev. D, 46, 4193

Martínez-González, E., Sanz, J.L. and Silk, J. 1994, ApJ, , 436, 1

Meszaros, A. 1994, ApJ, , 423, 19

Mollerach, S., Gangui, A., Lucchin, F. and Matarrese, S. 1995, ApJ, 453, 1

Munshi, D., Souradeep, T. and Starobinsky, A.A. 1995, ApJ, 454, 552

Nottale, L. 1984, MNRAS, 206, 713

Panek, M. 1992, ApJ, , 388, 225

Peebles, P. J. E. 1980, The Large Scale Structure of the Universe, (Princeton, Princeton University Press, 1980)

Press, W.H. and Schechter, P. 1974, ApJ, 187, 425

Raphaeli, Y. 1995, ApJ, 445, 33

Raphaeli, Y. 1995, Annu. Rev. Astron. Astrophys., 33, 541

Ratra, B. and Peebles, P. J. E., 1994, ApJ, , 432, L5.

Rees, M.J. and Sciama, D.W. 1968, Nature, 217, 355

Sachs, R.K. and Wolfe, A.N. 1967, ApJ, , 147, 73

Saez, D., Arnau, J.V. and Fullana, M.J. 1993, MNRAS, 263, 681

Sanz, J.L. and Cayón, L. 1996, Proc. Conf. Mapping, Measuring and Modelling the Universe, eds. P. Coles, V. Martínez and M.J. Ponz (ASP Conf. Series)

Sanz, J.L., Martínez-González, E., Cayón, L., Silk, J. and Sugiyama, N. 1995, ApJ, , in press.

Seljak, U. 1996, ApJ, , in press.

Smoot, G. et al. 1992, ApJ, 396, L1

Saunders, R. 1995, Astrophys. Lett. Commun., 32, 339

Sunyaev, R.A. and Zeldovich, Y.B. 1970, Astrophys. Space Sci., 7, 3

Sunyaev, R.A. and Zeldovich, Y.B. 1972, Comments Astrophys. Space Phys., 4, 173

Sunyaev, R.A. and Zeldovich, Y.B. 1980, Annu. Rev. Astron. Astrophys., 18, 537

Suto, Y. and Sasaki, M. 1991, Phys. Rev. D, 66, 265

Thompson, K.L. and Vishniac, E.T. 1987, ApJ, 313, 517

Tului, R. and Laguna, P. 1995, ApJ, 445, L73

Tului, R., Laguna, P. and Anninos, P. 1996, preprint

Vishniac, E.T. 1987, ApJ, , 322, 597

Wilbanks, T.M., Ade, P.A.R., Fisher, M.L., Holzapfel, W.L. and Lange, A.E. 1994, ApJ, 427, 75

Wilson, M.L. 1983, ApJ, 420, 1

Zeldovich, Y.B. and Sunyaev, R.A. 1969, Astrophys. Space Sci., 4, 301

Lecture Notes in Physics

For information about Vols. 1–439
please contact your bookseller or Springer-Verlag

New Series m: Monographs